COMPOUND SEMICONDUCTORS STRAINED LAYERS AND DEVICES

ELECTRONIC MATERIALS SERIES

This series is devoted to electronic materials subjects of active research interest and provides coverage of basic scientific concepts, as well as, relating the subjects to the electronic applications and providing details of the electronic systems, circuits or devices in which the materials are used.

The Electronic Materials Series is a useful reference source for senior undergraduate and graduate level students, as well as, for research workers in industrial laboratories who wish to broaden their knowledge into a new field.

Series Editors:

Professor A.F.W. Willoughby
Dept. of Engineering Materials
University of Southampton
UK

Professor R. Hull
Dept. of Material Science & Engineering
University of Virginia
USA

Series Advisor:

Dr. Peter Capper
GEC-Marconi Infra-Red Ltd.
Southampton
UK

Other Titles Available:

1. *Widegap II-VI Compounds for Opto-electronic Applications*
   Edited by E. Rúda

2. *High Temperature Electronics*
   Edited by M. Willander and H.L. Hartnagel

3. *Narrow-gap II-VI Compounds for Optoelectronic and Electromagnetic Applications*
   Edited by Peter Capper

4. *Theory of Transport Properties of Semiconductor Nanostructures*
   Edited by Eckehard Schöll

5. *Physical Models of Semiconductor Quantum Devices*
   Ying Fu; Magnus Willander

6. *Quantum Effects in Semiconductor Materials and Devices,*
   Edited by T. P. Pearsall

COMPOUND SEMICONDUCTORS STRAINED LAYERS AND DEVICES

by

S. Jain
IMEC vzw
Leuven, Belgium

M. Willander
Gothenburg University
Gothenburg, Sweden

R. Van Overstraeten
IMEC vzw
Leuven, Belgium

KLUWER ACADEMIC PUBLISHERS
Boston / Dordrecht / London

Distributors for North, Central and South America:
Kluwer Academic Publishers
101 Philip Drive
Assinippi Park
Norwell, Massachusetts 02061 USA
Telephone (781) 871-6600
Fax (781) 871-6528
E-Mail <kluwer@wkap.com>

Distributors for all other countries:
Kluwer Academic Publishers Group
Distribution Centre
Post Office Box 322
3300 AH Dordrecht, THE NETHERLANDS
Telephone 31 78 6392 392
Fax 31 78 6546 474
E-Mail <orderdept@wkap.nl>

 Electronic Services <http://www.wkap.nl>

Library of Congress Cataloging-in-Publication Data

Jain, S.C. (Suresh C.), 1926-
   Compound semiconductors strained layers and devices / by S. Jain, M. Willander, R. Van Overstraeten
      p. cm. -- (Electronic materials series)
   Includes bibliographical references and index.
   ISBN 0-7923-7769-9 (alk. paper)
   1.   Compound semiconductors. 2. Layer structure (Solids) I. Willander, M. II. Overstraeten, R. van. III Title. IV. Series.

TK7871.99.C65 J35 2000
621.3815'2--dc21

                                                              99-089333

Copyright © 2000 by Kluwer Academic Publishers.

All rights reserved. No part of this publication may be reproduced, stored in a retrieval system or transmitted in any form or by any means, mechanical, photocopying, recording, or otherwise, without the prior written permission of the publisher, Kluwer Academic Publishers, 101 Philip Drive, Assinippi Park, Norwell, Massachusetts 02061

Printed on acid-free paper.

Printed in the United States of America

Contents

Preface		**xi**
1	**Introduction**	**1**
	1.1 Evolution of strained layers	1
	1.2 Conventional III-V-based heterostructures	4
	1.2.1 Unstrained heterostructures	4
	1.2.2 Strained heterostructures	5
	1.3 III-Nitrides	6
	1.3.1 Properties	6
	1.3.2 Historical perspective	8
	1.4 Wide bandgap II-VI semiconductors	10
	1.4.1 Crystal structure, phase diagram and growth	10
	1.4.2 Historical perspective	13
	1.5 Material parameters	14
	1.5.1 Lattice constants and bandgaps	14
	1.5.2 Effective masses and mobilities	15
	1.5.3 Band alignments	15
	1.6 Scope and organization of this book	18
2	**Characterization and growth**	**19**
	2.1 Methods of characterization	19
	2.1.1 Electron Microscopy and X-ray diffraction	19
	2.1.2 Characterization by RHEED technique	21
	2.1.3 Optical and magnetic methods	22
	2.2 Epitaxial growth methods	24
	2.2.1 Substrates for the growth of II-VI and III-V semiconductors	24
	2.2.2 MBE	25
	2.2.3 MOVPE	27
	2.2.4 MOMBE and related methods	30
	2.2.5 ALE, MEMBE, and MMBE	32
	2.3 Growth of conventional III-V semiconductors	33
	2.3.1 Growth by MBE, MOVPE and other techniques	33
	2.3.2 Highly mismatched layers	35
	2.3.3 Growth in nitrogen carrier gas	35

	2.4	Growth of II-VI semiconductors	38
		2.4.1	Growth of ZnSe	38
		2.4.2	Growth of CdTe and ZnTe	40
		2.4.3	Growth of ZnS	41
		2.4.4	Photo-assisted growth of II-VI semiconductors	43
		2.4.5	Interfaces of II-VI heterostructures	44
	2.5	Growth of III-nitride epilayers	47
		2.5.1	Substrates for the growth of III-Nitrides	47
		2.5.2	Laterally Epitaxially Overgrown (LEO)- and Pendeo (PE)-epitaxial layers	48
		2.5.3	Growth of GaN and AlN by MOVPE	50
		2.5.4	Growth of GaN and AlN by MBE and related methods	51
		2.5.5	Comparison of III-Nitride layers grown on different substrates and by different methods	53
		2.5.6	Growth of InGaN and AlGaN alloys	54
3	Strain and critical thickness	59
	3.1	Strain and energies of epilayers	59
		3.1.1	Misfit strain	59
		3.1.2	Dislocation and strain energies: Periodic arrays of dislocations	60
		3.1.3	Non-periodic arrays of dislocations	63
	3.2	Processes involved in dislocation generation	65
		3.2.1	Propagation of dislocations	65
			3.2.1.1 Excess stress	66
			3.2.1.2 The kink model of dislocation propagation	67
			3.2.1.3 Measurements of dislocation velocity	69
		3.2.2	Nucleation, multiplication and blocking	72
			3.2.2.1 Nucleation	72
			3.2.2.2 Multiplication	74
			3.2.2.3 Blocking	75
	3.3	Critical thickness	77
		3.3.1	Theory of critical thickness	77
		3.3.2	Strain and critical thickness of superlattices	78
		3.3.3	Symmetrically strained superlattices	79
		3.3.4	Experimental values of critical thickness	79
		3.3.5	Critical thickness determined by islanding	84
4	Strain relaxation and defects	89
	4.1	Strain in GeSi layers	89
	4.2	Strain in III-V semiconductor layers	91
	4.3	Strain in II-VI layers	94
		4.3.1	Relaxation of strain in low mismatched layers	94
			4.3.1.1 Relaxation in CdZnSe and in the initial stages in ZnSe layers	94
			4.3.1.2 Relaxation with increase of layer thickness	95

            4.3.1.3   Strain relaxation by electron irradiation 97
      4.3.2   Relaxation of strain in highly mismatched layers 99
            4.3.2.1   Strain relaxation in the early stages of growth . 99
            4.3.2.2   Periodic distribution of dislocations 100
            4.3.2.3   Strain in magnetic superlattices 102
      4.3.3   Strain oscillations 104
  4.4 Strain and defects in III-Nitride layers 107
      4.4.1   Strain distribution and dislocations 107
      4.4.2   Structural defects 109
      4.4.3   Effect of defect clusters on the minority carriers in GaN . 109
      4.4.4   Interfaces in III-Nitride epilayers 110

5 Band structure and optical properties   115
  5.1 Band structure 115
      5.1.1   Zinc-blende semiconductors 115
      5.1.2   Wurtzite (WZ) III-Nitrides 118
  5.2 Band offsets 121
      5.2.1   General remarks 121
      5.2.2   Band offsets: III-V heterostructures 121
      5.2.3   Band offsets: II-VI heterostructures 122
      5.2.4   Band offsets: III-Nitride heterostructures 125
  5.3 Optical properties of III-V semiconductors 126
  5.4 Optical properties of II-VI semiconductors 127
      5.4.1   Reflectance, absorption, and luminescence 127
      5.4.2   Effect of strain on the optical properties of II-VI semiconductors 129
      5.4.3   II-VI quantum wells and superlattices 133
            5.4.3.1   Quantum confinement 133
            5.4.3.2 $ZnSe/ZnS_xSe_{1-x}$ superlattices 134
            5.4.3.3 $ZnSe/Zn_{1-x}Cd_xSe$ quantum wells 136
            5.4.3.4   CdTe/ZnTe superlattices 140
            5.4.3.5   Raman studies of strained CdTe/ZnTe superlattices 142
            5.4.3.6   Effect of hydrostatic pressure 144
  5.5 Optical properties of III-Nitrides 145
      5.5.1   Intrinsic luminescence 145
            5.5.1.1   PL and splitting of the valence band 145
            5.5.1.2   Binding energies and lifetimes of the excitons . . 147
      5.5.2   Yellow and DAP bands and persistent photoconductivity 147
      5.5.3   Effect of strain on III-Nitride layers 148
      5.5.4   Temperature dependence of the optical transitions 150
      5.5.5   Optical properties of InGaN alloys 151
            5.5.5.1   Bandgap and PL 151
            5.5.5.2   Bowing parameter of InGaN 152
            5.5.5.3   Nonuniform distribution of indium 153

- 5.5.6 AlGaN alloys: Optical absorption, bandgap and bowing parameter 154
- 5.5.7 III-Nitride quantum wells and superlattices 156

6 Electrical and magnetic properties — 159
- 6.1 Electrical properties of II-VI semiconductors 159
  - 6.1.1 n-type doping 159
  - 6.1.2 p-type doping 160
- 6.2 Electrical properties of n-type GaN 162
  - 6.2.1 Undoped GaN 162
  - 6.2.2 Si and Ge doped GaN: Carrier concentration 163
  - 6.2.3 Electron mobility in n-type GaN 165
    - 6.2.3.1 Measurements of electron mobility in n-doped GaN 165
    - 6.2.3.2 Monte Carlo simulations of electron mobility and velocity 169
- 6.3 Electrical properties of p-type III-Nitrides 170
  - 6.3.1 Resistivity, hole concentration and mobility 170
  - 6.3.2 Activation energy of Mg acceptors in GaN 172
- 6.4 Electrical properties AlN, InN and alloys 174
  - 6.4.1 AlN and AlGaN 174
  - 6.4.2 InN 176
- 6.5 Schottky barriers and ohmic contacts 176
  - 6.5.1 Contacts on II-VI heterostructures 176
  - 6.5.2 Contacts on III-Nitrides 179
    - 6.5.2.1 Schottky barriers 179
    - 6.5.2.2 Ohmic contacts 182
- 6.6 Effect of applied electric field 185
  - 6.6.1 Franz-Keldysh and Quantum Confined Stark Effects ... 185
  - 6.6.2 Experimental results: II-VI quantum wells 187
    - 6.6.2.1 CdTe/Cd$_{1-x}$Mn$_x$Te quantum wells 187
    - 6.6.2.2 Zn$_{1-x}$Cd$_x$Se/ZnSe quantum wells 188
    - 6.6.2.3 CdZnSSe multiple quantum wells 190
  - 6.6.3 Effect of electric field on GaN 192
- 6.7 Piezoelectric effect 192
  - 6.7.1 Piezoelectric effect in zinc blende and wurtzite semiconductors 192
  - 6.7.2 Experiments on piezoelectric effect in II-VI quantum wells and superlattices 194
    - 6.7.2.1 CdS/CdSe superlattices 194
    - 6.7.2.2 CdS/ZnSe superlattices 195
  - 6.7.3 Piezoelectric effects in III-Nitrides 196
- 6.8 Effect of magnetic field on semiconductors 199
  - 6.8.1 Magnetic polarons in diluted magnetic II-VI semiconductors 199
  - 6.8.2 Transport properties 200
  - 6.8.3 Magnetic and 0ptical properties 201

	6.8.3.1	Bulk crystals 201
	6.8.3.2	ZnTe and ZnSe epilayers 202
	6.8.3.3	Quantum wells and superlattices of diluted magnetic alloys 202
	6.8.3.4	Digital alloys 205
	6.8.3.5	Effect of magnetic field on the properties of Quantum dots 205
	6.8.3.6	EMP bifurcation and asymmetric quantum wells 206
	6.8.3.7	Cyclotron resonance measurements 206

7 Strained layer optoelectronic devices — 207

- 7.1 Conventional-III-V semiconductor lasers 207
  - 7.1.1 Suppression of Auger recombination by strain in semiconductor lasers 207
  - 7.1.2 Pump-probe and other methods for measurement of Auger lifetimes 210
  - 7.1.3 PL of superlattices containing Sb in the active layers . . . 211
  - 7.1.4 $InAs_{0.9}Sb_{0.1}/In_{0.85}Al_{0.15}As$ strained layer superlattice mid-IR lasers 212
  - 7.1.5 Summary of this section 213
- 7.2 ZnSe-based light emitters and other devices 214
  - 7.2.1 Light Emitting Diodes 214
  - 7.2.2 Photo-pumped lasers 216
  - 7.2.3 ZnSe-based electron-beam pumped lasers 219
  - 7.2.4 ZnSe-based laser diodes 220
    - 7.2.4.1 Structure and performance 220
    - 7.2.4.2 Degradation and reliability 223
- 7.3 Other II-VI semiconductor applications 228
  - 7.3.1 Modulators and switches 228
  - 7.3.2 Optical detectors 229
- 7.4 III-Nitride Light Emitting Diodes 231
  - 7.4.1 Fabrication and performance 231
  - 7.4.2 Comparison with other LEDs 235
- 7.5 GaN based Lasers 236
  - 7.5.1 General remarks 236
  - 7.5.2 Optically pumped III-Nitride lasers 237
  - 7.5.3 III-Nitride laser diodes 238
  - 7.5.4 Mechanism of laser emission 241
  - 7.5.5 Comparison of ZnSe- and GaN-based lasers 242
  - 7.5.6 Applications of LEDs and LDs 243

8 Transistors — 245

- 8.1 InGaAs transistors 245
  - 8.1.1 Early work 245
  - 8.1.2 Recent work 247
- 8.2 II-VI semiconductor transistors 251

	8.3	III-Nitride based transistors	253
		8.3.1 GaN/SiC HBT	253
		8.3.2 Field Effect Transistors on sapphire substrate	254
		8.3.3 FETs on SiC substrate	258
	8.4	Device Processing	260
		8.4.1 Etching	261
		8.4.1.1 Wet etching	261
		8.4.1.2 Dry etching	262
		8.4.2 Si ion implantation	262
		8.4.3 Implantation for isolation	264
9	**Summary and conclusions**		**265**
	9.1	Growth, defects and strain	265
	9.2	Band structure and electronic properties	267
		9.2.1 III-V semiconductors	267
		9.2.2 II-VI semiconductors	267
		9.2.3 III-Nitride semiconductors	269
	9.3	Applications and future work	271
		9.3.1 Applications	271
		9.3.2 Future work	272
Appendix A			**275**
Bibliography			**281**
Index			**333**

Preface

During the last 25 years (after the growth of the first pseudomorphic GeSi strained layers on Si by Erich Kasper in Germany) we have seen a steady accumulation of new materials and devices with enhanced performance made possible by strain. 1989-1999 have been very good years for the strained-layer-devices. Several breakthroughs were made in the growth and doping technology of strained layers. New devices were fabricated as a results of these breakthroughs. Before the advent of strain layer epitaxy short wavelength (violet to green) and mid-IR (2 to 5 μm) regions of the spectrum were not accessible to the photonic devices. Short wavelength Light Emitting Diodes (LEDs) and Laser Diodes (LDs) have now been developed using III-Nitride and II-VI strained layers. Auger recombination increases rapidly as the bandgap narrows and temperature increases. Therefore it was difficult to develop mid-IR (2 to 5 μm range) lasers. The effect of strain in modifying the band-structure and suppressing the Auger recombination has been most spectacular. It is due to the strain mediated band-structure engineering that mid-IR lasers with good performance have been fabricated in several laboratories around the world. Many devices based on strained layers have reached the market place. This book describes recent work on the growth, characterization and properties of compound semiconductors strained layers and devices fabricated using them.

The work that has been done on strain, dislocations, and mechanical properties of strained layers is very extensive. It is not possible to describe all this work in a monograph of this size. However work is treated in sufficient details to cover essential aspects of recent developments. References to reviews and books published on the subject are quoted. The discussion of this work should be useful to engineers and material scientists concerned with effects of strain on the mechanical properties of crystalline layers of any material. Though the book is devoted to compound semiconductors, work on GeSi strained layers is included either for comparison with the results of compound semiconductors or to clarify physics and prove results of theoretical calculations using experiments performed on GeSi layers. The effects of strain on band structure, transport, and optical properties of both the zinc blende and the wurtzite compound semiconductors are discussed. Piezoelectric Effects and Quantum Confined Stark Effects are included. Magnetic polarons in diluted II-VI magnetic polarons are also discussed. Among the applications, blue and green LEDs and LDs and mid-IR LDs are discussed. One whole chapter is devoted to these devices. Another chapter is devoted to Transistors based on conventional III-V, II-VI and III-Nitride semiconductors. The subject matter is treated at a level appropriate for students and senior researchers interested in material science, and in designing and modelling semiconductor devices. Several thousand papers of high quality have appeared on strained layers and devices in the last ten years. We have quoted some six hundred papers in the bibliography that are most relevant for a coherent discussion of the subject. To make the bibliography more useful, titles of the papers have been included.

Unfortunately Professor R. Van Overstraeten, one of the co-authors of the book, died on April 29, 1999. He was ill during the year and a half before this book was completed. He contributed significantly to the parts of the book written before he became ill. He did not see the manuscript in the final form.

We have benefited from interaction and collaboration with a large number of students and colleagues. It is impossible to mention them individually. Discussion with Dr. Uma Jain on strain and strain relaxation in lattice mismatched epilayers were very useful. Dr. K. van der Zanden and Dr. H. Hardtdegen read critically some parts of the book and made constructive criticism. We would like to thank sincerely Professor R. Mertens for useful discussions and for supporting this project. Madhulika Jain helped considerably in mechanical preparation of the manuscript. We are grateful to IMEC library, particularly to Greet Vanhoof and her colleague Griet Op de Beeck for the excellent work they did in providing bibliography and obtaining papers, books and Conference Proceedings from libraries abroad. Without their help it would have been difficult to write a monograph on a modern topic where the literature is scattered in numerous publications world wide.

Finally, we must thank sincerely our wives for their unfailing support and help during the preparation of this book.

S. C. Jain
IMEC, Kapeldreef 75
3001 Leuven, Belgium
June 26, 1999

M. Willander
Göthenberg University/Chalmers
University of Technology
Department of Physics
S-41296 Göteborg, Sweden
June 26, 1999

Chapter 1

Introduction

1.1 Evolution of strained layers

Frank and Van der Merwe [1] predicted in 1949 that a coherent (or pseudomorphic) epilayer of a crystal can be grown on a substrate of slightly different lattice constant. The epilayer is biaxially strained so that the lattice planes of the epilayer and the substrate are contiguous. If the lattice constant of the epilayer is bigger, the epilayer is compressed. On the other hand, if the lattice constant of the substrate is bigger, the epilayer is under biaxial tensile stress. The epilayer remains pseudomorphic up to a certain thickness known as the critical thickness h_c. The experimental critical thickness varies approximately as $(f_m)^{-n}$ where f_m is the lattice mismatch between the epilayer and the substrate and n varies between 1.5 and 2 (exact relation between f_m and h_c is given in chapter 3). Matthews and Blakeslee formulated the theory of critical layer thickness using an alternative approach [2]. It has been shown recently that their theory is equivalent to the theory of Frank and Van der Merwe and the two theories yield identical results [3]. Use of semiconductors of different bandgaps in a device was suggested by Shockley in a patent [4] in 1951. The two ideas, one of electrical heterostructure contained in Shockley's patent [4] and the other of mechanical heterostructure proposed by Frank and Van der Merwe were combined to fabricate strained layer heterostructures in mid 1970s [2, 5, 6]. Matthews and Blakeslee were produced $GaAs_xP_{1-x}/GaAs$ pseudomorphic strained layer heterostructures in 1974 [7]. Erich Kasper and his collaborators in Germany fabricated strained Ge_xSi_{1-x}/Si layer heterostructures about the same time [5, 6]. Extensive investigations of growth and properties of strained layers were made in the 1980s. Osbourn and collaborators [8, 9, 10, and references to earlier papers given in these paper] and others at Sandia Laboratories concentrated on InGaAs and other III-V strained layers. Bean and collaborators [11] at Bell Laboratories investigated extensively the growth and properties of Ge_xSi_{1-x} strained layers. Several other groups made important contributions to the development of strained layer heterostructures (see references given in [3, 12, 13].

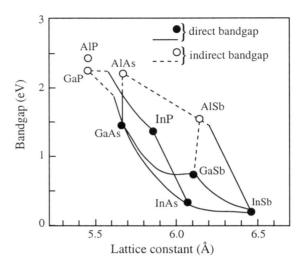

Figure 1.1: Bandgaps versus lattice constants of III-V semiconductors

Critical thickness, structure and stability of the strained layers, effect of strain on band structure, mobility, and optical properties were calculated theoretically and determined experimentally. These developments were very important. They showed that high quality defect free strained epilayers can be grown and the choice of substrate need not be limited by the condition that the lattice constants of the epilayer and the substrate should be lattice matched. They also showed that in many cases strain improves the material parameters and provides another degree of freedom for bandgap and band offset engineering.

In 1957 after the publication of Shockley's patent Kroemer [14] wrote a paper on the theory and advantages of Heterostructure Bipolar Transistor (HBT) with wide-gap emitters. In 1963 Kroemer [15] wrote a similar paper on the importance of heterostructure technology for improving the performance of an LD. Several experimental papers on wide-gap HBTs and heterostructure LDs appeared in the late 1960s and 1970s (see [16, 17] for references). By late 1970s heterostructure technology was firmly established. Both high performance HBTs and LDs were designed and fabricated. In 1970s, the emphasis was on finding a pair of semiconductors with differing bandgaps but with nearly the same lattice constants so that a good epilayer of one semiconductor could be grown on the other. Lattice constants of III-V semiconductors are shown in Fig. 1.1. AlAs and GaAs are closely lattice matched but have different bandgaps. These devices were based on AlGaAs/GaAs system. Use of lattice matched heterostructures severely limited the choice to conventional III-V semiconductors. It was not possible to fabricate HBTs using Si compatible technology. The bandgap of pseudomorphic GeSi/Si layers is reduced substantially by the strain produced by addition of $\sim 10\%$ Ge. An unstrained GeSi alloy of the same composition has negligible bandgap reduction and is useless for fabricating the

1.1. EVOLUTION OF STRAINED LAYERS

devices. Si HBTs with strained GeSi base layers have become important devices in recent years [3]. They have enhanced considerably the performance of Si bipolar transistors. They can be easily integrated on a Si chip along with other components.

Before the advent of strained layer-devices the Light Emitting Diodes (LEDs) and LDs could emit light only in a narrow region of the spectrum. Short wavelength (violet to green) and mid-IR (2 to 5 μm) regions of the spectrum were not accessible to the photonic devices. Short wavelength devices are required for full color display, laser printers, for high-density information storage and for under-sea optical communication (see section 7.5.6 for details). If the 650 nm light presently used in DVD and CD technology is replaced by 400 nm light, the data storage capacity will increase by a factor 2.64 [18]. III-Nitride- and ZnSe-based strained layers have enabled the development of LEDs and LDs in the blue and green regions of the spectrum. III-Nitride strained layers are also useful for high temperature and high power device applications (see section 1.3). High temperature and high power transistors are needed for automobile engines, future advanced power distribution systems, all electric vehicles and avionics.

Mid-IR lasers are needed for ambient air quality monitoring, sensing trace gases, toxic gas leakage detection, for research and for medical diagnostics. They are also needed for remote sensing, short-link high bandwidth optical communication, night vision devices and IR counter measures for defense applications. It has been difficult to achieve lasing in the mid-IR range of the spectrum because Auger recombination limits the performance of the narrow bandgap III-V semiconductors. Strained layer Mid-IR lasers with improved performance have been fabricated by several groups in 1996 and 1997 [19]–[22].

There has been intense activity in the field of strained layer-heterostructure devices, particularly III-Nitride devices, during the recent years (see recent papers [23]-[40] and references given in these papers.. Conventional III-V strained layers [12, 42], II-VI strained layers [43] and III-Nitride strained layers and devices [44]–[47] have been studied extensively. By way of introduction salient features and historical development of III-V-, III-Nitride- and II-VI-semiconductors are discussed in the next three sections. GeSi strained layers have also been investigated extensively. Because of the degenerate conduction band, even a small strain has a large effect on the band structure. Bandgap decreases rapidly with strain. Strain, dislocation velocities and strain relaxation have been investigated more thoroughly in GeSi strained layers than in other systems. GeSi strained layers have made possible the development of the HBTs compatible with Si technology. Heterostructure photodetectors for the IR region of the spectrum have also been developed. A good account of the properties of GeSi strained layers and heterostructure devices is given in the book by Jain [3]. GeSi strained layers is not the topic to be covered in this book. However physics of the strained layers is common to all systems. We discuss the work on GeSi strained layers in this book whenever it helps to elucidate the physics of the strained layers.

Steel played a dominant role in the technological revolution in the early 19th century. Si semiconductor technology has played a similar role in the 20th century [17]. Strained layer heterostructures have extended the areas of device applications and have made the semiconductor technology more useful.

The two concepts, one of mechanical heterostructure contained in the paper of Frank and Van der Merwe [1] and the other of electrical heterostructure contained in the patent of Shockley [4] in 1951 were many years ahead of their times, for their full impact on the device technology has been realized only in recent years.

1.2 Conventional III-V-based heterostructures

1.2.1 Unstrained heterostructures

As mentioned earlier AlGaAs/GaAs system has been a major player in the field of heterostructure devices. Though these devices do not involve strain, we discuss briefly their developments in this section. We also discuss later in this section InGaAs/GaAs strained layer heterostructures. A high quality interface without defects can be obtained in the lattice-matched $Ga_xAl_{1-x}As$ heterostructures. Both AlAs and GaAs have good mechanical properties. There are no miscibility gaps and the good alloy epilayers can be grown on GaAs substrates with all compositions. Over much of the composition range the $Al_xGa_{1-x}As$ alloy has direct bandgap and the heterostructure $Al_xGa_{1-x}As/GaAs$ has Type I band alignment (band alignments are shown later in Fig. 1.8). First semiconductor LDs fabricated in 1962 were GaAs homojunction devices. These devices were not very useful because the threshold current density was very high, ≥ 50 kA/cm^2. Room temperature operation of these lasers was not possible [16]. It was only in 1969 that room temperature operation of a *heterostructure* semiconductor laser was demonstrated (a comprehensive account of the early developments of LDs and references to original papers is given in [16]). However attempts to fabricate HBTs and LDs using $Ga_xAl_{1-x}As/GaAs$ heterostructures did not produce greatly improved devices until late 1970s mainly because of the limitations of the growth technology. Advances made in the technology (MBE and MOCVD growth technologies are described in chapter 2) of epitaxial growth enabled the development of high-performance $Ga_xAl_{1-x}As$ HBTs, lasers and other heterostructure devices [17]. GaAs quantum wells cladded with AlAs or $Al_xGa_{1-x}As$ are easily formed. Two dimensional electron gas in the modulation doped GaAs quantum wells formed with AlGaAs barriers shows very high mobilities (see Table 1.1). The quantum wells reduce the threshold current and enhance performance of lasers. To summarize, $Al_xGa_{1-x}As/GaAs$ heterostructures have played a dominant role in manufacturing optoelectronic devices and high frequency transistors.

$Al_xGa_{1-x}As$ has some problems. Al has high affinity for oxygen. Oxygen at the interface reduces mobility and generates trapping centers. It may be due to these reasons that AlGaAs has a high surface recombination velocity, of the order of 10^6cm/s. If the concentration of Al is more than 30%, deep levels, known as DX centers are formed. These centers cause bias-dependent distortion of the $I - V$ characteristics of the Heterostructure Field-Effect Transistors (HFET).

1.2. CONVENTIONAL III-V-BASED HETEROSTRUCTURES

Table 1.1: Low temperature mobilities (cm^2/Vs) in semiconductors and Modulation Doped (Mod.) AlGaAs/GaAs [12]. The values for Ge and Si are for both electrons and holes. The values for GaAs at 50 K and Mod. AlGaAs/GaAs at liquid helium temperature (LHeT) are for electrons.

Ge(4 K)	Si(4 K)	GaAs(50 K)	Mod. AlGaAs/GaAs (LHeT)
2,000,000	500,000	300,000	10,000,000

They also give rise to the photoconductivity at low temperatures which persists for long periods of time and interferes with the performance of the HFETs [48, p 291].

Considerable research has also been done on the ternary $Ga_xIn_{1-x}As$ and the quaternary $Ga_xIn_{1-x}As_yP_{1-y}$ alloys grown on InP substrates. Since Al is absent in these systems, they are less sensitive to the environments. The ternary alloy $Ga_{0.47}In_{0.53}As$ is lattice matched to InP. The quaternary alloy $Ga_xIn_{1-x}As_yP_{1-y}$ can be lattice matched to InP for several pairs of values of x and y. The mobility of $Ga_xIn_{1-x}As$ increases substantially with the increase of In concentration (even in the unstrained alloys). In the InGaAs alloys and Phosphide containing alloys the structure of the conduction band also offers an advantage. The separation between the upper and the lower valleys in GaAs is only 0.3 eV as compared to 0.69 eV in InP and 1.1 eV in InAs [48, 49]. In GaAs FETs the electrons near the drain acquire high energies and scatter into the upper valley where the electron mass is larger. The performance of the device degrades. As the In concentration in the $Ga_xIn_{1-x}As$ alloy increases, the separation between the two valleys increases and performance of the FETs improves [49, 50]. Similar advantage is also obtained with Phosphide containing alloys. The quaternary lasers emit at 1.3 to 1.5 μm range which is very well suited for optical communication. The surface recombination velocity of the quaternary $Ga_xIn_{1-x}As_yP_{1-y}$ is 10^3 cm/s, three orders of magnitude better than that of the $Al_xGa_{1-x}As$. However the technology of fabricating the InP substrates and the quaternary epilayers is involved because of the large vapor pressure of P. InP is weaker than GaAs and breakage rates are higher. InP begins to evaporate as InP molecules at 360°C. Heterostructures fabricated using GaInAsSb alloys have also been investigated [50] because they have direct bandgaps in the range 1.43 to 0.18 eV. The semiconductors GaSb, InAs, and InSb have extremely large electron and hole mobilities at room temperature. However if the bandgaps of the alloys are small, thermal generation of carriers and dark currents create problems.

1.2.2 Strained heterostructures

Use of pseudomorphic InGaAs layers grown on GaAs substrates has enhanced the performance and range of applications of GaAs based devices. Highly improved LDs and Modulation Doped FETs (MODFETs, also known as High Mobility Transistors or HEMTs) have been made and are in production. The performance InGaAs-HEMTs improve because of high electron mobility and large inter-valley separation in the conduction band of InAs. Both theory and experiments show that InAs mole fraction required for best performance is high. Devices have been fabricated with 30 to 60% mole fraction of InAs. At this high concentrations critical thickness of InGaAs layers become very small and stable strained layer devices can not be fabricated. Recently unstrained high In content HEMTs have been fabricated on GaAs substrates. These devices, known as Metamorphic HEMTs (MM-HEMTs) are discussed in chapter 8.

Strain makes the valence band sharp and reduces the hole effective mass much more than the electron effective mass [51]. If the hole effective mass decreases and becomes roughly equal to electron effective mass, the lasing condition (population inversion) is achieved at lower carrier concentration. This combined with the suppression of Auger recombination by strain has far reaching beneficial effects. Threshold current is reduced, lasers work at higher temperatures and lifetime of the laser increases. Threshold current densities as low as 50 A/cm^2 have been observed. Contrary to the general feeling which existed in 1980s that strain will degrade the lasers, InGaAs lasers showed better high temperature performance and longer lifetimes. In InGaAs and GaSb based semiconductor epilayers, Auger recombination affect adversely the performance of the optoelectronic devices [22, 42]. Energy and momentum must be conserved during the Auger recombination process. Using strain the structure of the conduction band and separation of the sub-valence bands can be changed in a manner that it becomes difficult to conserve momentum and energy. Therefore Auger recombination in suitably designed strained layers is suppressed. In Sb based narrow bandgap semiconductors the suppression of Auger recombination is by up to two orders of magnitude. This has made it possible to fabricate the mid-IR lasers with significantly improved performance [22]. Optical absorption by transition of an electron from the spin-orbit split off to the heavy hole band competes with the laser gain. Strain can be used to suppress this parasitic absorption [51].

InGaN/GaAs strained layer technology is now quite mature. Good reviews have been written on the InGaAs strained layers and devices [12, 42]. The properties of InGaAs have been reviewed by us recently [52]. Recent work has been concentrated on Sb containing strained layer mid-IR lasers. These lasers are discussed in chapter 7.

1.3 III-Nitrides

1.3.1 Properties

Wide bandgap semiconductors extend the areas of application of semiconductor electronics where conventional materials Si and III-V semiconductors fail. The bandgaps and lattice constants of III-Nitrides and related materials are shown

1.3. III-NITRIDES

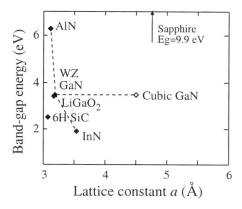

Figure 1.2: Bandgaps and lattice constants of III-Nitrides and substrate materials.

in Fig. 1.2. GaN based semiconductors have wide bandgaps and at the same time have other properties which are necessary for making high performance semiconductor devices. AlN, GaN and InN, and their alloys are direct bandgap semiconductors. Their bandgap values are 1.9 eV for InN, 3.4 eV for GaN and 6.2 eV for AlN. Bandgaps and emission wave lengths can be tuned by making alloys of GaN with the other two nitrides. III-Nitrides are suitable for short-wavelength optoelectronic devices. They are stable because the bonds between nitrogen and group III elements are very strong. We can dope GaN and its alloys with shallow donors to create electrons and with shallow acceptors to create holes. The electrons and holes have reasonable mobilities [54]. Selected relevant properties of GaN are compared with those of Si and GaAs in Table 1.2. The highest reported room temperature electron mobility in GaN epilayer is 900 cm^2/Vs [55]. Theoretically calculated mobilities in GaN without background compensating impurities are much higher [57]-[60]. Electron mobility as high as 1700 cm^2/Vs has been reported in AlGaN/GaN heterostructures [57]. The Monte Carlo simulated peak velocity in GaN is close to 3×10^7 cm/s and the saturation velocity is 1.5×10^7 cm/s [60]. Both values are considerably higher than the values for Si and GaAs. III-Nitrides have good thermal conductivity [54]. III-Nitrides and their alloys support the heterostructure technology. In-GaN quantum wells are indispensable for light emitting devices. AlGaN is used for cladding and barrier layers in both the light emitters and the FETs.

Nitride semiconductors have ideal properties for high temperature and high power electronics. Compared to Si and conventional III-V devices, III-Nitride devices can work at much higher temperatures and in hostile environments. Binari and Dietrich [55] and Duboz and Khan [61] have enumerated properties of III-Nitrides which make them useful for high temperature and high power device applications. The intrinsic carrier density decreases exponentially with the bandgap. Therefore it is very small in the wide bandgap semiconductors. The temperature at which the carrier density reaches 10^{15} cm^{-3} is 300°C for Si,

Table 1.2: Comparison of 300 K material properties of Si, GaAs, 4HSiC and GaN. The data have been taken from references [55] except those indicated in the footnotes at the end of the Table.

Property	Si	GaAs	4H SiC	GaN
Breakdown field (10^5 V/cm)	2	4	30	30
Electron mobility (cm^2/Vs)	1400	5000^a	800	900^b
Electron mobility (cm^2/Vs)	–	8000^c	–	1700^d
Maximum velocity (10^7 cm/s)	1	2	2	3
Thermal conductivity (W/cm K)	1.5	0.5	4.9	1.3

^{a} Ref. [56].
$^b n = 10^{16}$ cm^{-3}. See chapter 6 for values at other concentrations and temperatures.
^{c}For an AlGaAs/GaAs heterostructure [56].
^{d}For an AlGaN/GaN heterostructure [57].

500°C for GaAs, and 1300°C for GaN. Therefore the dark saturation current in devices fabricated using wide bandgap semiconductors remains small at high temperatures. In processing of Si and GaAs devices, special care has to be taken to keep the temperature low within the specified limits. This involves high cost processing steps. These steps can be avoided in the processing of GaN based devices. GaN has high breakdown fields. The breakdown field for Si is 2×10^5 V/cm, for GaAs 4×10^5 V/cm, and for GaN it is 3×10^6 V/cm. High breakdown voltage is a very desirable property for high power devices.

1.3.2 Historical perspective

Pankove [62] and Monemar [63] have described the early work on the development of GaN material and devices. Juza and Hahn [64] synthesized small needles and platelets of GaN by passing ammonia over hot gallium in 1938. Grimmeiss et al. [65] used the same method to produce small crystals of GaN and measured their PL spectra in 1959. The first large area GaN layer was deposited on sapphire by Maruska and Tietjen [66] in 1969. The success of Maruska and Tietjen [66] in depositing large area GaN layers created wide spread interest in GaN research. First blue LED using Zn doped GaN was fabricated by Pankove et al. [67] in 1972. It was a metal/insulating GaN:Zn/n-GaN (m-i-n or MIS) structure. The color of the emitted light depended on the concentration of Zn in the insulating layer and could be tuned from the blue to the green regions of the spectrum [62]. A number of other important discoveries were made, e.g. negative electron affinity [68], surface acoustic wave generation [69], and solar blind UV detectors (unpublished work of Pankove quoted in Ref. [62]). However further research on GaN was hampered because it could not be doped p type. The quality of GaN was also not very good. Making low resistance ohmic contacts was another problem.

1.3. III-NITRIDES

Break-throughs results in the growth technology were obtained in 1980s [70]-[73]. Akasaki, Amano and co-workers in late 1980s [71, 72] showed that high quality GaN layers can be grown by MOCVD on a sapphire substrate if the layer is grown in two steps. In the first step a thin buffer layer of AlN is grown on the sapphire substrate at a low temperature of $\sim 500°C$. The main GaN layer is then grown on the buffer layer at a much higher temperature. The second breakthrough was made by Amano et al. [71] when they obtained p-type conductivity in GaN for the first time. They found that p-type conductivity can be obtained by low energy electron beam irradiation (LEEBI) of Mg doped GaN. In 1992, Nakamura et al. [74] discovered that annealing the Mg doped GaN layers at ≥ 750 °C in nitrogen or vacuum also activated the Mg acceptors.

First AlGaN layered structures were grown by Khan et al. [75] and Itoh et al. in 1990 [76] in 1991. First InGaN/GaN layered structures were grown by Nakamura et al. [77] in 1993. First GaN based FETs were fabricated by Khan et al. [78]. In 1994 Pankove et al. [79] reported successful fabrication of a HBT using p-type 6H SiC as the base and n-type GaN as the emitter. Rapid improvements in the performance of the FETs were made. Recent emphasis has been to fabricate these devices on SiC substrates. Because of high thermal conductivity of SiC substrate, the transistors fabricated on SiC show better performance particularly at high temperatures.

Akasaki and co-workers have done extensive work on GaN-based light emitters [54, 80, 81]. They made the first p-n junction GaN LED in 1989 [72]. The success of bringing the blue and green LEDs to a level that they could be commercialized is due to Nakamura and his colleagues at Nichia Laboratories. The work done by Nakamura et al. is very extensive. It is described in the recent book by Nakamura and Fasol [45]. Nakamura has also written several reviews on the GaN-based LEDs and LDs [82]–[85]. Onset of stimulated emission by current injection was first observed by Akasaki and collaborators in 1995 (see the review of Akasaki and Amano [81]). Nitride based injection laser was fabricated by Nakamura and co-workers in 1996. Rapid improvements in the performance of these lasers were made by Nakamura's group. Akasaki et al. [86] fabricated an LD emitting at 376 nm also in 1996. Room temperature CW laser emitting at ~ 4000 nm with more than 6000 hour lifetime has been demonstrated by Nakamura [85].

Above discussion shows that in recent years the evolution and rise of III-Nitride semiconductors has taken place at an extraordinary pace. Several conference proceedings [87]–[89], special issues [90], authored book [45] and edited books [46, 47, 91] have been published on these semiconductors and devices fabricated using them. Reviews on growth and on light emitting devices have been written by Akasaki and co-workers [54, 80, 81] and by Nakamura and co-workers [82, 83, 84, 85, 92]. Pearton [93] has reviewed the work on device processing. The work on Field Effect Transistors (FETs) has been reviewed by Binari [94], Binari and Dietrich [55] and Khan and co-workers [57, 61, 95, 96]. Morkoç and co-workers have written comprehensive reviews covering several aspects of the nitride semiconductors [97]–[100].

Table 1.3: Properties of ZnSe, CdTe, III-V and Si crystals. The abbreviations are: MP (°C) is melting point, K_{th} (mW cm^{-1} K^{-1}) is thermal conductivity, H_v (kg mm^{-2}) is Vickers microhardness, EPD (cm^{-2}) is etch-pit density, FWHM (arcsec) is x-ray rocking curve half-width, α_{th} (10^{-6} K^{-1}) is the coefficient of thermal expansion and VP (MPa) is vapor pressure of the melt at 1600°C for ZnSe and at the melting point for other semiconductors. Value of vapor pressure of ZnSe melt is taken from [107] and for other semiconductors from [108]. The other data in this table were compiled by Triboulet [109] in 1991.

Semiconductor	MP	K_{th}	H_v	EPD	FWHM	α_{th}	VP
ZnSe	1520	139	92	10^5	10	7.57[1]	0.8
CdTe	1092	55	40	10^5	60	5.31	0.07
InP	1054	660	406	10^4	11	3.9	–
GaAs	1237	500	360	10^3	14	6.7[1]	0.1
Si	1412	1235	1250	0	7	2.6	6.5×10^{-8}

[1]Average values of $\alpha_{th} \times 10^6$ between 300 K and 600 K are 8.7 for ZnSe and 6.2 for GaAs (see Table I in [326]).

1.4 Wide bandgap II-VI semiconductors

1.4.1 Crystal structure, phase diagram and growth

Numerous attempts to grow the bulk crystals of ZnSe and CdTe have been made [101]–[106] but there have been only sporadic reports of good quality crystals because of their very difficult material parameters. Physical properties of ZnSe and ZnTe relevant for crystal growth are summarized[1] in Tables 1.3 and 1.4. The FWHM of the rocking curves and etch pit density (EPD) are also included in Table 1.3. Data for InP, GaAs and Si are given for comparison. II-VI semiconductors have high ionicity (see Table 1.4), high congruent melting points

Table 1.4: Ionicity and structure of II-VI semiconductors quoted by Triboulet [110]. In several cases the structures are different at high temperatures, see text and Table A in Appendix A.

Semiconductor	Ionicity	structure
CdS	0.74	Hexagonal (wurtzite)
CdSe	0.64	Hexagonal (wurtzite)
CdTe	0.55	Cubic (zincblende), twinned
ZnS	0.77	Cubic (zincblende), twinned
ZnSe	0.63	Cubic (zincblende), twinned
ZnTe	0.49	Cubic (zincblende),

and they are highly volatile. Ionicity of III-V semiconductors is < 0.5 and that

---
[1]Melting points and vapor pressures of other II-VI semiconductors are given in [104].

1.4. WIDE BANDGAP II-VI SEMICONDUCTORS

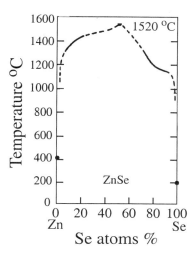

Figure 1.3: Phase diagram of ZnSe [111].

of Si is zero [104]. Because of the large ionicity, the thermal conductivity of II-VI semiconductors is small (see Table 1.3). Thermal conductivity of ZnSe quoted in [107] is only 0.2 W cm^{-1} K^{-1} near the melting point. Thermal conductivity of CdTe melt is 0.01 W cm^{-1} K^{-1} as compared to 0.18 W cm^{-1} K^{-1} of the GaAs melt [107]. Poor thermal conductivity of the melt makes it very difficult to use Czochralski pulling technique for the growth of the II-VI crystals. It also makes it difficult to control the temperature at the liquid-solid interface in the Bridgman method. The stacking fault energy in the II-VI semiconductors is low. It is only 10.84×10^{-7} J cm^{-2} for ZnSe. Therefore II-VI semiconductors tend to have a large density of twins and stacking faults [112, page 21]. Critical resolved shear stress for plastic deformation is low and dislocations are easily created [112]. Tendency of twinning is high in the II-VI semiconductors. Some workers have reported that addition of Zn to CdTe and Mn to ZnSe suppresses twining [109].

Several authors have reported the hexagonal structure of CdTe at high temperatures. At low temperatures the structure becomes zincblende [110, 113]. ZnSe [107, 110] and ZnS [104] show a reversible phase transition from the high temperature wurtzite structure to the low temperature zincblende structure. The transition temperatures are 1425°C and 1020°C respectively. A value 1420°C for the phase transformation temperature of ZnSe has also been reported in the early literature (see references in [114]). Recently Okada *et al.* [114] have determined a value 1411 ± 2 °C for the phase transformation temperature by differential thermal analysis. As grown crystals from the melt pass through the transition temperature during the cooling process. At the transition temperature the zincblende structure nucleates spontaneously at several points. When these nuclei grow and meet they give rise to twin boundaries.

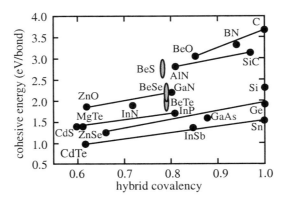

Figure 1.4: Bond energy versus hybrid covalency for III-V and II-VI nitrides [116].

Phase diagram of ZnSe is shown in Fig. 1.3. A recent determination [114] of the phase diagram is qualitatively similar but the congruent melting point is 1522 ± 2 °C. The values of the congruent melting point of ZnSe given in the literature vary from 1515 to 1526 °C [114]. Phase diagram (both solidus and liquidus) of CdTe are given in a recent paper by Rudolph et al. [106]. Phase diagrams of most other wide bandgap II-VI semiconductors are similar. They are quite different from those of the III-V semiconductors. The liquidus curves for the III-V semiconductors is broad while for the II-VI semiconductors it is sharply pointed at the maximum melting point. The shape of the liquidus curve of II-VI semiconductors makes it necessary to control the thermal conditions very precisely during the crystal growth. A small change in temperature changes the phase equilibrium significantly, gives rise to uncontrolled growth, and produces crystals of poor quality.

Zn partial pressure over the melt of ZnSe is about twice as large as that of Se. This results in preferential evaporation of Zn and a Se rich-melt (by 0.5 to 1.5 at.% depending upon the melt surface temperature). Similarly Cd evaporates preferentially from the CdTe melt and produces Te rich melt. This change of composition of the melt gives rise to constitutional supercooling and large deviations from stoichiometry in both ZnSe [107, 110, 115] and CdTe [105, 108] crystals. Se and Te inclusions and precipitates in the ZnSe and CdTe crystals are formed [105, 107, 108]. This also gives rise to the native defects (e.g. Zn vacancies in ZnSe and Cd vacancies in CdTe) which form stable clusters with impurities making purification of the crystals difficult. The bond strength of III-V, III-Nitrides and II-VI semiconductors are compared in Fig. 1.4. II-VI semiconductors have weak bond strength and energy of formation of defects is low. Therefore strain relaxation by introduction of defects is a problem in the epilayers of these materials [116]. Interdiffusion at the interface of II-VI semiconductors epilayer and GaAs substrate causes doping of the epilayers [117, 118, 119] and makes the interface rough.

1.4. WIDE BANDGAP II-VI SEMICONDUCTORS

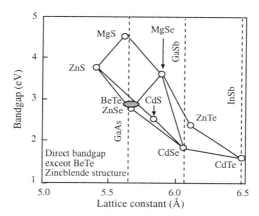

Figure 1.5: Bandgaps versus lattice constants of II-VI semiconductors. The lightly filled ellipse shows the uncertainty in the values for BeTe given in Refs. [116, 120].

1.4.2 Historical perspective

ZnSe is probably the most important and most extensively studied II-VI semiconductor. It is an essential constituent in most short wavelength light emitting diodes and lasers fabricated so far. However large single crystals of ZnSe crystals are not yet available because of the difficulties mentioned in the previous subsection. Most II-VI devices fabricated so far are grown on GaAs substrates. Lattice constants and bandgaps of wide-bandgap II-VI semiconductors are shown in Fig. 1.5 and in Table A given in Appendix A. ZnSe has a small lattice mismatch with GaAs (values of lattice mismatch used in different papers are slightly different, e.g. 0.265% in [121], 0.27% in [122] and 0.272% in [123]). The mismatch of most other II-VI semiconductors (except MgS) with GaAs is large. Both lattice constant and bandgap of ZnSe can be adjusted by alloying it with ZnS, CdS, CdSe, MgS and MgSe. Until a few years ago it was difficult to grow high quality epilayers of II-VI semiconductors and their alloys. Extremely thin epilayers with high quality can now be grown using Molecular Beam Epitaxy (MBE) [124]. Thickness of the MBE layers can be controlled on the scale of a single atomic layer. Quantum wells and superlattices of II-VI semiconductors grown on GaAs are strained. Extensive theoretical and experimental work has been done on strain, strain relaxation and critical thickness of II-VI semiconductors [125].

ZnSe has a low refractive index and is transparent to a large part of the spectrum where GaAs is opaque. These features are useful for fabrication of devices such as electron beam- and photo-pumped lasers, photodetectors and waveguides [101, 109]. Early report of emission of blue light from ZnSe appeared in 1985 [126]. Stimulated emission from optically pumped ZnSe based quantum wells was also reported in 1985 (quoted in [127]). Glass *et al.* [128] observed stimulated emission from optically pumped $Cd_{0.25}Zn_{0.75}Te/ZnTe$ quantum wells at room-temperature in 1988. Though crystal quality of the II-VI semiconductors improved dramatically in late 1980s, problems with p-type doping persisted. In the early work p-type conductivity could be obtained with Li doping. But control of doping profile was difficult because diffusion coefficient of Li is large [129]. First efficient p-type doping of ZnSe was accomplished by using radio frequency plasma sources of nitrogen by Park et al. [130] in 1990 and independently by Ohkawa et al. [131, 132] in 1991. The first ZnSe based blue-green injection lasers were fabricated in 1991 [133, 134]. Important advances were made and improved devices were reported in subsequent years [43]. Degradation of the II-VI semiconductor devices and short working life time continue to be a problem. Considerable work is being done to improve the growth techniques and fabricate better layers so that the degradation of the lasers can be avoided [135, 136, 137]. Maximum lifetime of the room temperature CW lasers ($\lambda \leq 550$ nm) is now ~ 400 h (Ref. [137]). However a lifetime of several thousand hours must be achieved before the devices can be commercialized. As mentioned earlier, the weak bond strength of the direct bandgap II-VI semiconductors is probably responsible for the ease with which defects are created [138] in ZnSe based materials. Beryllium Chalcogenides have indirect bandgap but high bond strength (shown in Fig. 1.4) comparable to that of GaN. Because of the indirect bandgap BeTe can be used only in small concentrations in the active layer. Lasing action (at $\lambda = 509$ nm) in a structure with BeZnCdSe active layer has been demonstrated [138]. However longer lifetime lasers using Be compounds has not yet been reported.

ZnTe is used as a buffer layer in the growth of HgCdTe on GaAs, in solar cells, and in X-ray detectors [139, and references given in this review]. Applications of ZnTe in wave guides, optical detectors, optical switches and frequency doublers are being considered [140]. Good ohmic contacts can be made on p-type ZnTe and therefore it is useful as a contact layers for wide-gap II-VI layers [139]. CdTe crystals or mixed II-VI crystals containing high concentration of CdTe are used in fabricating electro-optic modulators, windows, prisms and x- and γ-ray detectors [102, 112]. CdTe is an important materials for fabricating high efficiency solar cells [112].

1.5 Material parameters

1.5.1 Lattice constants and bandgaps

Two fundamental quantities of the semiconductors most relevant for device fabrication are the lattice constants and the minimum bandgaps [3, 12]. The bandgaps and lattice constants of important compound semiconductors have already been shown in Figs. 1.1, 1.2 and 1.5 respectively. BeTe and other Be chalcogenides have been studied only recently [116, 120]. Their material parameters are not known with accuracy. Kalpana et al. [120] have calculated band structure and lattice parameters of Be chalcogenides. Tables and figures of the presently available material parameters are given in Refs. [116, 120]. Be con-

1.5. MATERIAL PARAMETERS

taining compounds are interesting because their bonds are more covalent than the bonds of other II-VI semiconductors as shown in Fig. 1.5. Extensive work is being done on designing and fabricating devices by mixing BeTe with other II-VI semiconductors to fabricate devices. BeTe is an indirect bandgap semiconductor. It can be used in the active layer only in small concentrations. However it can be used in cladding layers to make them harder without any problem. The lattice mismatch between BeTe and ZnSe (and also GaAs) is small. This makes BeTe attractive for fabricating epilayers nearly lattice matched to ZnSe and GaAs.

The emission and absorption wavelengths in optoelectronic devices depend on the bandgaps. The thermal ionization rates and the dark currents in the electrical devices are large if the bandgaps are small. On the other hand the electron hole mobilities are large if the bandgaps are small and they decrease rapidly as the bandgaps increase (see Fig. 1.6 and 1.7) in the next subsection.

1.5.2 Effective masses and mobilities

The effective mass normalized by the free electron mass, m^*, is given by:

$$\frac{q\tau}{2m^*} = 2q\tau \left(\frac{\pi}{h}\right)^2 \frac{d^2E}{dk^2}, \tag{1.1}$$

where h is the Plank's constant, τ is the average time between the collisions, and the second derivative of the energy E is evaluated in the direction in which the charge carrier is moving in the k-space. The mobility μ of the charge carriers is given by

$$\mu = \frac{\text{velocity}}{\mathcal{E}} = \frac{q\tau}{2m^*}, \tag{1.2}$$

where $\mathcal{E}$ is the electric field. This equation shows that the mobility depends inversely on the effective mass. For the effective mass to be low and the mobility to be high the bandstructure of a semiconductor should have sharply curved peaks and valleys. Figs. 1.6 and 1.7 show that the mobilities decrease as the bandgap increases. If interactions between neighboring band edges is taken into account, it is found that the bands broaden and the effective mass increases as the minimum bandgap increases [50]. This effect is large in the valence band resulting in a high effective hole mass. This explains qualitatively the dependence of mobility on the bandgap shown in Figs. 1.6 and 1.7 and 1.7 and also the fact that the hole mobilities are much also the fact that the hole mobilities are much smaller than the electron mobilities.

1.5.3 Band alignments

To obtain population inversion, it is necessary to localize large concentrations of electrons and holes (or excitons) in a small region of the semiconductor. Localization is also necessary for the formation of 2D electron or hole gas. The 2D gas has high carrier mobility and is used for fabricating MODFETs. Localization is achieved by joining together semiconductors of different bandgaps and forming

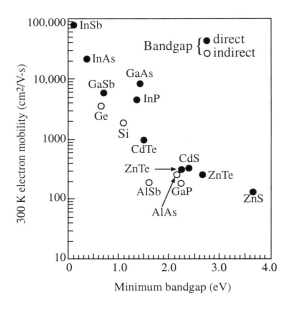

Figure 1.6: 300 K electron mobility of semiconductors [50].

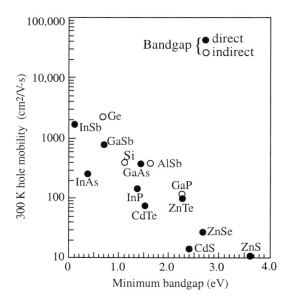

Figure 1.7: 300 K hole mobility of semiconductors [50].

1.5. MATERIAL PARAMETERS

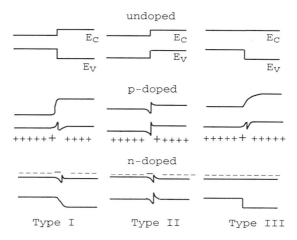

Figure 1.8: Three types of band alignments in heterostructures. + symbols show the holes and − symbols show and electrons in the p- and n-type semiconductors respectively. A missing symbol shows the depletion of the charges and a large size symbol shows the accumulation.

quantum wells. When two semiconductors of differing bandgaps are brought together, their band edges align in one of the three possible ways shown in Fig. 1.8. Depending on the band alignment and the doping, the equilibrium concentration of charges may produce regions of depletion and accumulation with dips and spikes in the band edges as shown in the figure. In the lightly doped regions the widths of the spikes are large and tunneling is not possible. The spikes behave as barriers. The dips behave as two dimensional charge wells. A type I alignment is needed to form a quantum well whereas a type III alignment is favored for HBTs. The magnitudes of the band offsets can be controlled by strain providing another degree of freedom for optimizing the device design. In the AlGaAs/GaAs HFET, Al concentration must be kept below 22% to avoid the problem of DX centers. However low Al concentration reduces the conduction band-offset and the performance of the HFET degrades. By using the strained layer AlGaAs/GaInAs structure sufficient conduction band edge discontinuity is obtained without the problem caused by the DX centers [49].

Before concluding this section we should mention that strain can only be used for modifying the bandstructure and electronic to suit a particular requirement. There are other non-electronic properties which also play significant role in the design, fabrication and performance of the device. Notable among them are the mechanical strength, thermal conductivity and thermal expansion. Bean has tabulated these properties for important semiconductors [50, see Table I on p 23].

1.6 Scope and organization of this book

The number of publications on strained layers has grown very rapidly. A survey of the literature reveals that several thousand papers have been published during the last few years on strained layers and devices. It is not possible to review all this work in a single monograph. Therefore the scope of this book must be defined. Since InGaAs strained layer technology is mature and has been reviewed, we discuss these layers briefly. However new developments on InGaAs based high electron mobility transistors are described in details. Most published work in the recent literature is devoted to ZnSe- and GaN-based layers and devices. These layers and devices are discussed in greater details. For any specific topic, the space devoted to the II-VI family of semiconductors is not necessarily equal to the space devoted to the III-Nitride semiconductors. The relative space devoted to a topic reflects the importance of the topic and the extent to which the topic has been studied. As an example strain relaxation and mechanical properties have been studied more extensively in II-VI semiconductors than in III-Nitrides.

We start with the discussion of growth and characterization of epilayers in chapter 2. Quality of interfaces in quantum wells and superlattices are also discussed in this chapter. Because of the mismatch in the lattice constants and thermal expansion coefficients of the epilayers and the substrates, the epilayers are strained. Theoretical and experimental studies of strain and critical thickness are discussed in chapter 3. The layers are grown at relatively low temperatures and experiments show that in most cases they are metastable. Therefore strain can produce defects in the epilayers. Even if the layers are of good quality when they are grown, dislocations can be introduced during the fabrication process or during the working life of the devices. The mechanical structure and stability of the strained layers are therefore of vital importance. Indeed a large majority of the papers continue to be devoted to the mechanical structure and strain relaxation in the layers. Relaxation of strain by introduction of misfit dislocations is discussed in chapter 4. Band structure of both zincblende (ZB) and wurtzite (WZ) semiconductors is described in chapter 5. Effects of strain on the band structure and a discussion of band offsets in strained heterostructures are discussed in detail. Photoluminescence and other optical properties are important in determining the quality of the epilayers. Strong excitonic band edge luminescence is obtained with good quality epilayers. Deep level luminescence indicates poor quality. Exact positions of the excitonic lines are used to determine the valence band structure and state of strain in the epilayer. Optical properties of the strained layers and superlattices are included in chapter 5. Electrical and magnetic properties are described in chapter 6. Doping methods, carrier concentrations and mobilities are discussed. Piezoelectric effects, quantum confined Stark effect, and magnetic polarons are also discussed. Chapter 7 is devoted to optoelectronic devices . Strained layer transistors are discussed in chapter 8. Finally a summary of important results is given in the last chapter, chapter 9. Values of lattice constants, elastic constants, bandgaps, refractive indices, and deformation potentials are given in Tables A to K in Appendix A.

Chapter 2

Characterization and growth

2.1 Methods of characterization

2.1.1 Electron Microscopy and X-ray diffraction

A large number of techniques have been used to characterize the epilayers. Discussion of all the techniques is beyond the scope of this book. The most extensively used methods are electron microscopy, X-ray diffraction and rocking curves, reflection high energy electron diffraction (RHEED), and optical methods. Recently micro-Raman and magnetic methods have also been used. Micro-Raman method has been reviewed by us recently [52, 141, 142]. We discuss briefly recent developments related to these methods. We also discuss other methods briefly. However we give references to books, reviews and papers for additional information.

Transmission Electron Microscopy (TEM) and High Resolution Electron Microscopy (HREM) are the standard techniques for determining defect-and dislocation-structure in the epilayers (see for example [143]). Recently TEM has also been used for measurement of localized stresses in epilayers. Vanhellemont et al. [144] have shown that convergent beam electron diffraction (CBED) and electron diffraction contrast imaging can be used to measure the localized stresses with a resolution on the nm scale. However the method is destructive and stress values can change appreciably during the specimen preparation. Possibilities and limitations of these techniques have been discussed by Vanhellemont et al. [144]. Bierwolf et al. [145] have developed a method which allows direct measurement of local displacement within a strained layer using high resolution electron microscopy (HREM). The experimental micrograph of the strained lattice is superimposed on the unperturbed lattice of the same crystal and the resulting moiré structure is analyzed. The analysis yields the displacement of the perturbed lattice with respect to the unperturbed lattice.

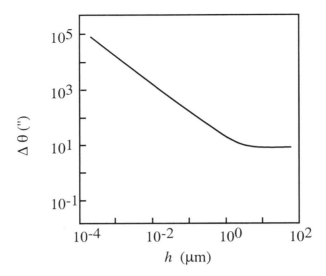

Figure 2.1: The calculated intrinsic reflection width $\Delta\theta$ versus layer thickness h for the (004) reflection from a ZnSe layer. Cu Kα radiation was assumed in the calculations [146].

This method is complementary to the method of Vanhellemont et al. [144] as it does not involve contrast interpretations. Interface abruptness, layer thickness, layer uniformity, and composition variation can also be determined using this technique [145].

X-ray diffraction is a powerful tool for characterization of semiconductor layers [146]. Diffraction measurements (rocking curves and reciprocal space mapping) yield information about the structural perfection. They can be used to determine the lattice constants, layer thickness, composition and strain. If the crystal quality is high, the width of the x-ray diffraction curves is intrinsic reflection width. The intrinsic width $\Delta\theta$ depends on the wavelength of the x-rays used, composition of the layer, angle of reflection and layer thickness h. Finite resolving power of the diffractometer also contributes to the width. Dislocations, grain boundaries and other defects broaden the curves. The broadening is a measure of the quality of the layers. To determine the defect induced broadening, intrinsic width expected from a perfect layer must be calculated and subtracted from the observed width. The intrinsic reflection width $\Delta\theta$ can be calculated using the dynamical theories (see references given in [146]). The calculated intrinsic width $\Delta\theta$ for the 004 reflection from a ZnSe layer is shown in Fig. 2.1. Fig. 2.1 shows that the intrinsic width of a thin layer is large. For thin layers estimation of crystal quality by measuring the width becomes less reliable. The broadening due to finite resolving power of the diffractometer can be taken into account by using the relation [146]:

$$\Delta\omega = \sqrt{\Delta\theta^2 + \Delta\epsilon^2} \qquad (2.1)$$

2.1. METHODS OF CHARACTERIZATION

where $\Delta\omega$ is the total width and $\Delta\epsilon$ is the width that depends on the resolving power of the diffractometer and on the Bragg angle. If the intrinsic width $\Delta\theta$ is small, $\Delta\epsilon$ becomes important.

The x-ray techniques are also used to determine the quality of the interfaces, evolution of strain relaxation and diffusion across the interfaces [147]. An excellent review on x-ray diffraction methods for characterizing semiconductors has been published by Fewster [148]. The capabilities and limitations of double-crystal diffractometer are discussed. Advantages of 'reciprocal space mapping' with a multiple crystal diffractometer are outlined. It is emphasized that care should be taken to avoid pitfalls while determining the thickness and composition of the layers. It is difficult to measure the lattice constant of layers with high resolution x-ray diffraction (HRXRD) if the layers are only a few ML thick. The review of Fewster [148] should be consulted for details of x-ray characterization methods.

2.1.2 Characterization by RHEED technique

RHEED is an invaluable tool for studying the real-time MBE growth of semiconductor epilayers. If the crystalline quality of the layers is good, well defined streaks in the RHEED pattern are observed. In the case of 3-D growth or if the crystalline quality is poor, the pattern becomes spotty. Stacking faults and twins give rise to diagonal streaks between the spots in the RHEED pattern [124]. RHEED surface reconstruction pattern depends on whether the surface is cation terminated or anion terminated. It is therefore useful in determining and optimizing the flux ratio of the cations and the anions for growth of good quality epilayers. The spacing between the streaks is used to determine the surface lattice constants [149]. If the streak pattern is observed from the beginning of the growth, the interface quality is likely to be good and the growth is 2D from the very beginning. RHEED patterns provide accurate measure of growth rates. These patterns are also used to determine the surface reconstruction during the cleaning of the substrate and during the nucleation of the epilayer. Ion gauges are used to measure the beam fluxes. Doping calibrations for different growth rates and temperatures are done on separate thicker layers. Growth rates can be controlled by RHEED oscillations to an accuracy of $\pm 1\%$ ML.

RHEED intensity oscillations are utilized for *in-situ* characterization of the layer [150]. RHEED oscillations are helpful in controlling the thickness of the semiconductor epilayers and in ascertaining that the growth is 2-dimensional. RHEED oscillations observed during the MBE growth of $AlAs_{0.16}Sb_{0.84}/InAs$ superlattices [151] are shown in Fig. 2.2 and observed during the MOMBE growth of ZnSe quantum wells and ZnMgSSe barriers [150] in Fig. 2.3. One period of intensity oscillation corresponds to the growth of one molecular monolayer. If the substrate temperature and flux ratio are not optimum and the growth is not 2D i.e. layer by layer, the RHEED intensity oscillations are absent. If the initial growth is 2D and becomes 3D after the growth of a few layers, the oscillations are observed in the beginning but decay as soon as the growth becomes 3D. RHEED intensity oscillations observed during the growth

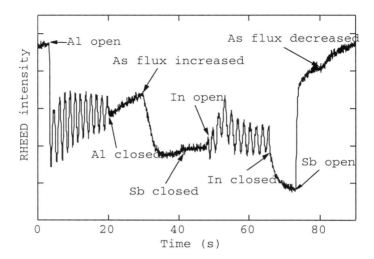

Figure 2.2: RHEED intensity oscillations observed during the growth of a AlAs$_{0.16}$Sb$_{0.84}$/InAs superlattices. The labels indicate when the different effusion cells were opened or closed and when the As$_2$ beam equivalent pressure (BEP) was changed [151].

of AlGaN are shown later in Fig. 3.21.

2.1.3 Optical and magnetic methods

A good review of optical methods useful for characterizing semiconductor epilayers has been published by Aspnes [155]. Layer thickness and compositions can be obtained with bulk oriented optical probes such as ellipsometry and reflectometry. Kinetics and chemistry of growth can be obtained by surface oriented methods such as laser light scattering, surface photoabsorption and reflectance-difference spectroscopy [155, 156]. The optical methods have not been used as extensively as the x-ray methods because photons interact with crystals weakly and interpretation of the signals obtained by the probes is not easy. The penetration depth of the visible-near-ultraviolet light in semiconductors is about 10 nm. One Å thick surface region will contribute to the signal using only about 1% of this light. Longer wavelengths penetrate several microns and therefore they give only some kind of average over the large volume of the layer. However optical methods also have some advantages over the other methods. In reactive ambients or where UHV can not be used, it becomes difficult to use RHEED because atmosphere attenuates the electron beam. Optical methods have the advantage that they can be used in any transparent ambient. New approaches to achieve high sensitivity with the optical methods are being developed [155]. Since RHEED can not be used in atmospheric pressure organometallic vapor

2.1. METHODS OF CHARACTERIZATION

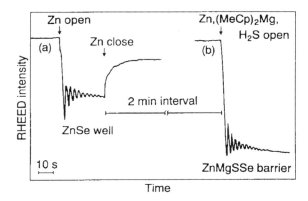

Figure 2.3: RHEED oscillations during the MOMBE growth of ZnSe quantum wells and ZnMgSSe barriers. The structure was grown for fabricating a photo-pumped laser [150].

phase epitaxy (OMVPE)[1], other in-situ techniques for characterization of epilayers are required. Recently an optical technique, known as reflection-difference spectroscopy (RDS) has been developed [157]. This technique can be used both in high vacuum and under atmospheric pressure. In this technique reflectivity with monochromatic light polarized in two orthogonal {110} directions is measured as a function of the wavelength of light. The difference in the reflectivity spectra measured with the two polarizations depend uniquely on the type of reconstructed (001) surface of the semiconductor, e.g. of GaAs. RDS measurements show that the surface reconstruction of GaAs under Ga and As rich conditions is the same at atmospheric pressure as that under high vacuum [157, see p 516].

Photoluminescence (PL) is the most frequently used technique to characterize the semiconductor layers. It does not suffer with the disadvantages mentioned in the previous paragraph. Most intrinsic defects (e.g. Zn vacancies in ZnSe or N vacancies in GaN) give rise to deep levels which show PL lines known as deep level peaks. Dislocations and extended defects also give known PL bands. Donor-acceptor pairs give rise to specific PL peaks. Peak positions for several defects in II-VI, conventional III-V and III-Nitride semiconductors are well documented[2]. Therefore PL provides information about the presence or absence of the defects. Point defects also cause broadening of the PL line. If the defects are absent, the PL is dominated by narrow lines of the near band-edge spectra. (Near band-edge spectra consist of free exciton and bound exciton PL lines.) The shift of the near band-edge PL peak position due to strain can be used to determine the magnitude of the strain in the layer [52]. The shift

---

[1] Also known as metalorganic vapor phase epitaxy (MOVPE) and as metalorganic chemical vapor deposition (MOCVD).

[2] The defect structure is not known with certainty in many cases.

can also be used to determine the nature of strain i.e. whether the strain is compressive or tensile. Optical studies of Olego et al. [121] for characterizing ZnSe/GaAs interfaces are discussed in chapter 4.

In heterostructures containing diluted magnetic alloys, an enhanced Zeeman splitting is observed due to diffusion across the interface. Diffusion of Mn produces layers of diluted magnetic alloy and therefore antiferromagnetic interactions become less effective in blocking the spins. The diffused interface layers have a larger magnetic susceptibility than the susceptibility of the more concentrated barriers. This enhances the Zeeman splitting of the carriers. The Zeeman splitting is measured before and after the heat treatment and the difference in the splitting is used to characterize the interfaces of the Mn containing heterostructures [158]. An example of characterizing the interface of CdTe-$Cd_{0.8}Mn0.2Te$ quantum wells is given in section 2.4.5. At the interfaces of a superlattice consisting of non-magnetic wells and diluted magnetic alloy barriers magnetic polarons are formed. Magnetic polarons are discussed in chapter 6.

RHEED can not be used in MOVPE reactor because RHEED can be used only in high vacuum. A reflected laser beam from the surface of a growing epilayer shows intensity oscillations due to interference effects. Nakamura [153] and more recently Mesrine et al. [154] have used the oscillations to measure the thickness of the film. The method can be used in both the MOVPE and MBE. The layers were grown by MOVPE in the first case and by GSMBE in the second case. The oscillations look similar to the oscillations of intensity of specularly reflected RHEED beam. As the thickness increases, the amplitude of the oscillations decreases and finally they disappear. The oscillations disappear because the fluctuations in the thickness become larger in the thicker films. The intensity of the reflected beam both in the oscillation regime [153] and beyond the oscillation regime [154] is used to determine the thickness of the epilayer.

Crystal field produces a fine structure in the electron paramagnetic resonance (EPR) spectra of paramagnetic ions present in the crystal. Atomic positions are rearranged and the crystal field is modified by strain. This produces a change in the crystal field splitting of the EPR lines. Furdyna et al. [159] have investigated the EPR spectra of ZnTe/MnTe, ZnTe/MnSe, and CdTe/MnTe superlattices and have determined the strain using the observed epr spectra. This work is discussed in section 4.3.

2.2 Epitaxial growth methods

2.2.1 Substrates for the growth of II-VI and III-V semiconductors

Epitaxial layers (abbreviated as epilayers or simply layers) can be grown with high purity, good crystalline quality and low concentration of defects. The epilayers can be grown with highly uniform structure and uniform doping. The doping profile and composition can be changed with depth in a controlled manner. Therefore devices are fabricated in the epilayers. In the early days epilayers

2.2. EPITAXIAL GROWTH METHODS

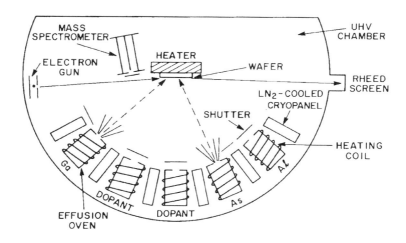

Figure 2.4: Schematic diagram of a typical MBE growth chamber for the growth of GaAs or AlGaAs, quoted in [49].

were grown by the Liquid Phase Epitaxy (LPE). In LPE the thickness of the layers can not be controlled so accurately, the interfaces are not abrupt, surface morphology is rough and the layer uniformity is not good [50]. LPE is cheap and is still used when it is not necessary that the quality of the layers must be very high. Generally molecular beam epitaxy (MBE) (see Refs. [99, 124, 127] and [160]–[162]) and MOVPE (see Refs. [45, 139, 157] and [163]–[166]) are used for growing the epilayers.

GaAs substrates have been used for the growth of epilayers of III-V and II-VI semiconductors by MBE [124] as well as by MOVPE [163]. GaAs is favored because it has zincblende structure, it is commercially available and the cost of the good quality substrates is relatively low. The (100) orientation is preferred because GaAs can be easily cleaved in the vertical (110) planes [163, page 43]. The substrate temperature is 275 °C to 400° C for II-VI semiconductors and 550 °C for GaAs. Other substrates on which compound semiconductor epilayers have been grown are discussed along with the growth of the individual semiconductors in sections 2.3 and 2.4. Sapphire and SiC are the most frequently used substrates for the growth of III-Nitrides. The substrates used for the growth of III-Nitrides are discussed in more details in section 2.5.

2.2.2 MBE

References to reviews of the MBE technique were given in the previous subsection (additional references are given in the recent review of Herman and Sitter [160]). We discuss only the main features of the technique here. Reviews cited above should be consulted for more details. A schematic diagram of the MBE chamber for the growth of III-V semiconductors is shown in Fig. 2.4. Chambers used for the growth of III-Nitrides [99] and II-VI semiconductors [43, 124] are essentially similar. Atomic or molecular beams of the constituents of the epilayer are produced by heating high purity solid sources of the constituents in crucibles (known as Knudsen cells or K-cells) placed in a high vacuum ($\sim 10^{-10}$ Torr) growth chamber as shown in the figure. The shutters (the doors between the K-cells and the substrate are known as the shutters) allow switching of the dopants and Al in or out of the molecular beam. If elemental sources are used, ratio of the fluxes of the cations and the anions is critical in determining the crystalline quality and stoichiometry of the epilayers. The ratio is related to the BEP which can be determined by a quartz crystal-monitor placed within the growth chamber. The monitor is placed as closely as possible to the growth position of the layer. Since the quality of the layers is extremely sensitive to the ratio, flux calibration experiments are performed to control the flux ratio more precisely [124]. When compound sources are used, a crucible with small diameter aperture over the opening of the crucible is employed. The small orifice prevents excessive loss of charge during thermal processing.

At a chamber pressure of $P = 10^{-5}$ Torr mean free path of air like molecules is $L = 50$ meters [50]. This mean free path exceeds the dimensions of the growth chamber. Therefore the beams go straight on to the substrate without any collisions. The substrate is heated radiatively or inductively. Vapor pressures of the semiconductors and dopants are low (except P, Hg and S, see Fig. 2.11. More extensive data on vapor pressures is given in Fig. 13 of [50])[3]. The beams condense on the substrate because it is kept at a lower temperature. Unlike the processes that occur in MOVPE (discussed in the next subsection), decomposition of precursors is not required and therefore the growth temperature is substantially lower. Interdiffusion across the interfaces and diffusion of the impurity from one layer to another are minimized. The most important feature of MBE is its capability of producing extremely thin layers of several different semiconductors or their alloys forming a superlattice. The thickness of each layer can be controlled on atomic scale. Since in the MBE technique the layers are grown in ultrahigh vacuum the contamination is reduced to low values [167, 168]. Doping sources used in MBE growth are discussed in chapter 6.

During the MBE growth two processes compete. In the first process, the impinging atoms attach themselves at the surface steps and the terraces propagate. In the second process, random nucleation takes place and 2D islands grow. At the high growth temperatures, the first process dominates and high quality layers grow in the 2D mode. In the second process which dominates at low growth temperatures, the nucleation can take place on uncompleted layers and the growth front becomes rough which ultimately results in the 3D growth mode. Surface at the growth front is never perfectly flat and smooth. Local enhancement of the attachment of the impinging atoms takes place at threading dislocations and other defects which roughens the growth front. Under opti-

---

[3]Since vapor pressure of phosphorus is high and is difficult to control, it is difficult to grow epilayers containing phosphorus as one of the constituents using MBE growth technique.

2.2. EPITAXIAL GROWTH METHODS

mum growth conditions surface migration is used to flatten and smoothen the surface. This effect may dominate in the Migration Enhanced MBE (MEMBE) (MEMBE is discussed in section 2.2.5. On the other hand under non-ideal conditions in the conventional MBE, the rapid migration can give rise to facetting, formation of hillocks and 3D growth. The low temperature of growth generally used in MEMBE also promotes 2D growth by slowing down the islanding process.

Stoichiometry of the epilayers is very important. Even a small (parts per million) deviation from 1:1 ratio of the constituents of a compound semiconductor (e.g. Ga and As in the case of GaAs) degrades the electrical properties of the semiconductor. The arrival rates of the atoms can not be controlled to an accuracy of parts per million. However epilayers of the compound semiconductors with good stoichiometry (deviations less than parts per million) are grown with MBE as a routine. The solution comes from the fact that the strength of the bond between two As atoms is much weaker than that of the bond between As and Ga. At an appropriate substrate temperature (which has to be determined experimentally for each semiconductor and which is around 550°C for GaAs), As adheres only to a Ga or Al layer and not to an As layer. Stoichiometry is therefore assured as long as arrival rate of As is slightly larger than the arrival rate of Ga. The sticking coefficient of Group III elements is high. The arrival rate of Ga atoms on the surface determines the growth rate. The Ga atoms and As atoms migrate short distance before they are stabilized on the surface. Epilayers of stoichiometric GaAs grow in the 2D (layer by layer) mode. The variation of doping concentrations between different runs is $\pm 5\%$. Since the source materials are highly pure and growth environments are clean, undoped GaAs has a low carrier (hole) concentration of $10^{13} - 10^{15}$ cm^{-3}. The residual impurity which gives rise to p doping is generally C. The main drawback of the MBE has been the low throughput of the wafers. Only one wafer is processed per run. However efforts are being made to develop multi-wafer processing equipment [49, 50].

2.2.3 MOVPE

MOVPE has a high degree of flexibility in selecting and designing precursor molecules. Like MBE, MOVPE growth takes place under conditions far from equilibrium. Stable and metastable compounds can be grown on any crystalline substrate. Epilayers on large area substrates (up to quarter square meter) can be grown. Uniform epilayers can be grown over a large number of closely stacked wafers. These are very strong points in favor of the MOVPE. MOVPE needs large volumes of metalorganic materials and other materials. The question of cost and safety (due to use of hydrogen carrier gas and problems of toxic waste disposal) becomes important. Therefore efforts are being made to develop new precursors (defined below) which have lower toxicity, are easier to decompose and have appropriate vapor pressures and reactivities. A schematic diagram of a typical MOVPE equipment used for the growth of semiconductor epilayers is shown in Fig. 2.5. Sources shown are for the growth of GaAs and AlGaAs

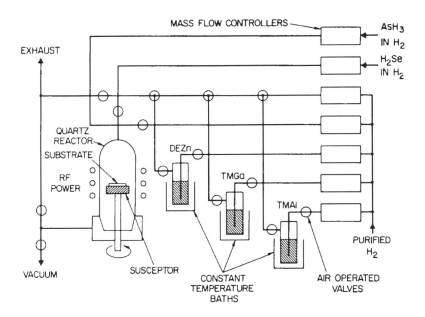

Figure 2.5: Schematic diagram of an atmospheric-pressure metal-organic CVD system used for the growth of doped GaAs [50, and references given therein].

epilayers. Gas handling system forms a major part of the MOVPE system. The system allows switching of the sources. The substrate is loaded on to a graphite susceptor and rotated to improve the uniformity of the epilayers. High precision electronic mass-flow controllers and fast switching valves are used for the control of composition and growth rate of the epilayer. The desired semiconductor components are released from the gaseous compounds, and organometallic compounds by pyrolysis. In the case of III-Nitrides, ammonia is generally used as the nitrogen source. These gaseous chemical sources are known as precursors. Pyrolysis takes place either in a cracker or the precursors decompose during the reaction at the growth surface. A detailed discussion of the pyrolysis of several precursors is given in [157]. The cracked or uncracked precursors are transported by a carrier gas, usually hydrogen. Due to the high temperature near the substrate a 'boundary layer' is formed. The concentration of the precursors in the lower part of the boundary layer near the growth surface is low because they are consumed at the surface during the growth of the epilayer. The concentration in the upper part of the boundary layer is large and the precursors or their cracked products are transported to the growing surface by diffusion through the thickness of the boundary layer. Three types of reactors are used in the MOVPE [163]. The reactors are HE (horizontal even i.e. not tapered), HT (horizontal tapered) and vertical. SiC coated graphite susceptors are generally used to support the substrate. Both IR and RF heating are employed. The temperature profile of the susceptor and growth rates depend on the position of

2.2. EPITAXIAL GROWTH METHODS

the susceptor.

The flux of the arriving atoms is given by [50],

$$F(\text{Mol.}/\text{cm}^2\text{s}) = 3.5 \times 10^{22} \; P(\text{Torr})/\sqrt{m(\text{g})T(\text{K})}. \tag{2.2}$$

The growth rate is equal to this flux multiplied by the probability of reaction and condensation. The precursor molecules must release the desired species at the substrate temperature. The rejected species must go out of the system before affecting adversely the quality of the epilayer. The substrate temperature should not be too high. If the substrate temperature is too high, autodoping and intermixing at the interface occur. GaAs is grown at a temperature around 650°C and AlGaAs at 675 to 700°C. It is due to the high growth temperature employed in the MOVPE that sharp interfaces are difficult to obtain. In Low Pressure Chemical Vapor Deposition (LPCVD) the growth temperature is lower. The abruptness of the interfaces is better in the LPCVD. Interface mixing is more of a problem in the II-VI semiconductors. No interface mixing seems to occur in the III-Nitrides.

The purity of precursors is of crucial importance in determining the quality of the epilayers. Certain contamination can render the epilayers useless. The layers may also be contaminated if the intermediate species are formed in the gas phase and are incorporated in the epilayer. However the incorporation of impurity such as H and C is determined by the probability of their attachment to the growing surface. If Al is a part of the III-V semiconductor alloy, carbon contamination is difficult to avoid because the bond strength between Al and C is large.

Commonly used Group III precursors are trimethylmetal (TMX or $(CH_3)_3X$), and triethylmetal (TEX or $(C_2H_5)_3X$) where metal and X stand for Al, Ga, or In. TMAl and TMGa have been used in Fig. 2.5. Arsine (H_3) is used as a source for arsenic. The required source chemicals are mixed with H_2 gas and the mixture flows over the substrate. The substrate is heated either inductively or radiatively. Dopants generally used in III-V semiconductors are Si (from disilane) for n-type layer and Zn (from DEZn or TMZn) for p-type layer. Dopant sources for the growth of II-VI and III-Nitrides semiconductors are discussed in chapter 6. Other precursors which are receiving attention are dimethylethylaminealane, trimethylaminealane, methyl and ethyl arsenic sources, tertiarybutylarsine and phosphine [157, 166]. TMAl gives high C contamination in the epilayers. Group III precursors for the growth of III-Nitrides are the same as for III-V semiconductors. Nitrogen sources are discussed later in this chapter.

Several different metalorganic precursors have been used in the MOVPE growth of II-VI epilayers: dimethylzinc $(CH_3)_2Zn$ abbreviated as Me_2Zn and commonly known as DMZn, diethylzinc $(C_2H_5)_2Zn$ abbreviated as Et_2Zn and commonly known as DEZn, dimethylzinc-triethylamine $(CH_3)_2Zn$-$N(C_2H_5)_3$ abbreviated as Me_2Zn-Et_3N and commonly known as DMZn-TEN, diethyltelluride $(C_2H_5)_2Te$ abbreviated as Et_2Te and commonly known as DETe, diisopropyltelluride $((CH_3)_3CH)_2Te$ abbreviated as iPr_2Te and commonly known as DIPTe, and methylallyltelluride $C_3H_5TeCH_3$ abbreviated as MeTe-allyl and commonly known as MATe. Precursors for the growth of other II-VI epilayers are similar and are discussed in Refs. [139, 157, 163, 164]. Tong *et al.* [169] have studied the pyrolysis of precursors DEZn, DipS, and t-BuSH. Zn was obtained with 98% efficiency at 950°C, the major by-product being ethyl radicals. With the low pressure injector design used in the experiments, H_2S could not be cracked below 1300°C. DipS cracked between 600 and 900°C producing ip-S. At higher temperatures ip-S was dissociated into S and propyl and ethyl radicals. The efficiency of S production was 95% at 1100°C. The precursor t-BuSH started decomposing at 700°C. The production efficiency of SH was 98% at 950°C, the by-product being t-butyl radicals. Reaction mechanisms during the growth of ZnSe by Chemical Beam Epitaxy have been investigated in Ref. [170]. DMZ decomposes readily but the sticking probability of Zn on GaAs or on ZnSe is small. The sticking probability of Zn increases dramatically in the presence of Se and efficient growth of ZnSe takes place.

The conditions for optimum growth are determined by the properties of the precursors also. The relevant properties are vapor pressure and decomposition energies. The growth rate is limited by the slowest reaction step which depends on the pyrolysis rate and also on the surface reactions. In many combinations of cation and anion containing precursors, strong interaction in pyrolysis is observed. For example Te containing alkyls Me_2Te and Et_2Te are relatively stable but their decomposition is enhanced by co-pyrolysis with Et_2Zn. In addition to pyrolysis, mass transport and surface reaction kinetics play a significant role in MOVPE. The roles played by these factors are interdependent. Modeling of the growth kinetics is very difficult. However the authors of Ref. [163] have done the modeling with considerable success.

For the growth of III-Nitrides, the precursors used as sources for Ga, In and Al are the same as used in conventional III-V semiconductors [157]. A detailed discussion of Nitrogen sources and additional discussion of Group III sources is given in section 2.5.

2.2.4 MOMBE and related methods

Industry requires improved yields, higher throughputs and short time between laboratory R & D and commercial production. Uniform growth over larger wafers is also necessary. To meet the above requirements, newer techniques such as MOMBE have been developed. MOMBE is a hybrid technique which combines the high vacuum clean environment of MBE and flexibility of the gaseous sources including the metal-organic sources. A schematic diagram of the MOMBE growth equipment is shown in Fig. 2.6. MOMBE uses long-lived sources of MOVPE and 'line of sight' flow regime of MBE [171]. In several cases solid sources can not be used conveniently. In the growth of S containing alloys, composition of the alloy can not be controlled accurately because vapor pressure of S is very large. Solid source of Mg also presents difficulties because it is easily oxidized. MOMBE provides the flexible choice of sources, accurate control of composition and precise control of layer thickness and interfaces. However the MOMBE growth equipment is more complex than the equipment used for either

2.2. EPITAXIAL GROWTH METHODS

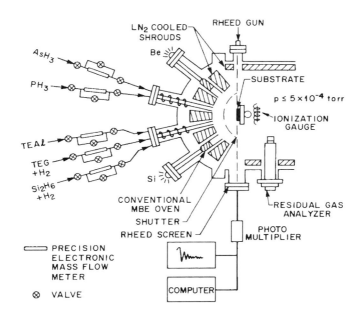

Figure 2.6: Schematic diagram of MOMBE equipment [49].

MBE or MOVPE [172]. Other similar growth techniques are gas-source MBE (GSMBE) and chemical MBE (CBE). If all sources are gaseous, the method is generally known as CBE. However in the literature the demarcation between GSMBE, CBE and MOMBE is not precise. Similar situation exists for the Atomic Layer Epitaxy and Migration Enhanced Epitaxy (to be discussed in the next subsection). The two techniques are practically identical but two different names have been used for this technique in the literature.

In the MBE the growth has to be interrupted to replenish the source ovens but in MOMBE it is not necessary because the source chemicals are placed outside the growth chamber. The cation alkyl molecules and Group V dimer species travel directly to the heated substrate surface as molecular beams. If the temperature is low, the alkyls may not dissociate or dissociate only partially and will re-evaporate. At sufficiently high temperatures the alkyls dissociate leaving the Group III atoms on the surface. The growth rate is determined by the rate at which the Group III alkyls arrive at the substrate, a result similar to that obtained with MBE. There is no chemical reaction in the gas phase. Molecular beam fluxes and substrate temperature can be controlled independently. Large area beams are easily obtained. If the substrate temperature can be controlled tightly, large area epilayers with highly uniform composition thickness and composition can be grown [172]. Very low surface defect density is achieved and different materials including phosphorous can be used. Since the vacuum is of the order 5×10^{-4} Torr, RHEED oscillations can be used to monitor the thickness and therefore abrupt interfaces can be achieved.

2.2.5 ALE, MEMBE, and MMBE

In the atomic layer epitaxy (ALE) and migration enhanced MBE (MEMBE) one atomic layer of each of the constituent elements is alternately deposited to form one molecular layer of the compound semiconductor in each cycle [173]–[180] The MEMBE technique has been developed independently by different groups and has been designated differently as Migration Enhanced MBE (MEMBE) and Modulated Molecular Beam Epitaxy (MMBE) [175]. We will use the name MEMBE for both the techniques. The layers grown by the ALE and MEMBE techniques have superior planarity of surfaces and interfaces. The growth temperature is considerably reduced but the high quality of the layers as determined by optical properties is maintained. We discuss both these techniques in this section.

In the compound semiconductors a layer by layer deposition is made possible by the fact that the bonds between cations and anions are stronger than those between cations-cations or anions-anions. Consider the growth of II-VI semiconductors. If the temperature is sufficiently high, only one layer of atoms of Group II semiconductor is deposited on the existing layer of Group VI semiconductor. Because of the lower binding energy, the second layer of the Group II semiconductor can not be deposited, it evaporates. The Group II semiconductor is purged out of the chamber and a molecular or precursor beam of Group VI semiconductor is allowed to enter. Again only one atomic layer of group VI semiconductor is formed on the existing layer of group II semiconductor. Thus by evaporating the two elements alternately (for sufficiently long time to complete the layer) one atomic layer of each element is formed in one cycle. The film thickness is equal to number of cycles completed multiplied by the thickness of one layer of the compound, the thickness being $a/2$ for (100) orientation and $a\sqrt{2}/2$ for (110) orientation (as shown later in Fig. 2.10.). The quality of the layers grown by ALE or MEMBE is very superior. The layers can be grown at a lower temperature, defect density in the layer is reduced and surfaces and interfaces become planar. At present it is not clear whether the improvements are caused because the growth is strictly two dimensional or due to the enhancement of the surface migration of the molecular species when supplied separately [175].

The ALE technique is somewhat slower because after the completion of one monolayer of an element, this element must be totally purged out before the deposition of the other element [175]. Furthermore it is difficult to grow the III-V semiconductors by the ALE technique with MBE. In MOVPE ALE the layer by layer growth is controlled by an intrinsic mechanism. Once a monolayer of one specie is deposited, the subsequent deposition of the same species is inhibited. Once the layer is completed, there are no more reactive sites available for the deposition of the same element. Such a mechanism is not available in MBE. At standard growth temperature Group V elements grow only on a layer of Group III elements and behave in ALE like manner. The sticking probability of a Group V element on an existing layer of the same element is practically zero. This is not true for the Group III elements. Group III layers keep growing on the

Group III layers in the absence of Group V elements. The MEMBE avoids this difficulty in the following manner. Group III dose per cycle is carefully adjusted to complete just one atomic layer. This can be done using the RHEED technique to monitor the growth of the layer. Group III source is then switched off and Group V source is switched on.

The quality of the layers grown by ALE or MEMBE is generally superior. The surfaces and interfaces are planar, defects are suppressed and the layers can be grown at lower temperatures. At present it is not clear whether the improvements are caused because the growth becomes strictly two-dimensional or due to the enhancement of the surface migration of the molecular species when supplied separately [175].

Briones *et al.* [175] have used several modifications of the MEMBE technique. In one modification, designated as ALMBE, Group III flux is supplied continuously but Group V flux is in pulses which synchronize with the completion of a monolayer of Group III elements. The quality of the layers grown by the ALMBE technique was found to be very good.

2.3 Growth of conventional III-V semiconductors

2.3.1 Growth by MBE, MOVPE and other techniques

Numerous reviews, conference proceedings and books exist on the growth of III-V epilayers by the conventional MBE [160, 157, and references given there in]. We discuss some recent work on the growth of these semiconductors. We also discuss the growth in nitrogen carrier gas in the next subsection.

Donnelly em et al. [172] have discussed kinetic models for the growth of GaAs by CBE or MOMBE. They have also reviewed the growth and growth rates of GaAs by these techniques. They compared the published observed growth rates with the predictions of a new model developed by them. The experimental and calculated GaAs MOMBE growth rates versus growth temperature are shown in Fig. 2.7. The precursors used in the experiments were TEGa and arsine. The maxima observed in the curves is due to two competitive processes, desorption and decomposition of DEGa. Before the work of Donnelly *et al.* [172] was published the increase in growth rate with temperature at lower temperatures was attributed to increased dissociation of TEGa into DEGa and C_2H_5 radical. The decrease at higher temperatures was attributed to desorption of DEGa. The work of Donnelly *et al.* [172] showed that low temperature behavior is more likely to be due to the blocking of adsorption sites by the C_2H_5 radicals and adsorption of TEGa without dissociation. Above 660 °C the growth rate drops rapidly due to desorption of the Ga atoms. Small amounts of In reduce the growth of GaAs drastically. It is believed that Ga bonded to In atoms desorbs more easily. This makes it difficult to grow GaInAs using MOMBE. It has been suggested that the use of precursors such as trimethylamine-galane might solve this problem [157].

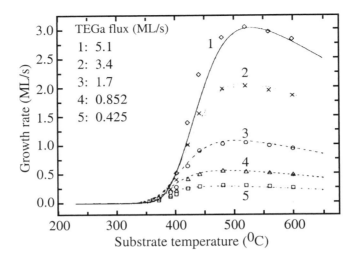

Figure 2.7: CBE growth rate of GaAs versus growth temperature [172]. TEGa beam fluxes (expressed in GaAs monolayers/s) starting from the top curve are: 5.1, 3.4, 1.7, 0.852, and 0.425 respectively. These fluxes correspond to the measured sticking coefficients if multiplied by a correction factor on 0.79.

Briones et al. [175] have grown several III-V semiconductor and alloy epilayers and superlattices using ALMBE. High quality GaAs and GaAs/AlAs superlattices were grown at a substrate temperature $< 400°C$. Surface and interfaces were flat, defect free and smooth. Modulation doped $(GaAs)_4(GaAs)_4$ superlattices were also grown at a substrate temperature $T = 380°C$. Only two central atomic planes in GaAs quantum wells were doped with Si. A carrier density of $n = 2 \times 10^{18}$ cm^{-3} with good mobility ($\mu = 10^3$ cm^2/Vs) were obtained. The activation energy of the donors was 3 meV. It is generally difficult to grow III-V alloy layers containing both As and P. During the MBE growth As displaces P. This makes it very difficult to obtain a desired ratio of As and phosphorus in the alloy. This problem can be solved by growing alternate single monolayers using ALMBE. Briones et al. [175] have grown successfully high quality $(GaAs)_7(GaP)_2$ superlattices on a GaAs substrate using this method.

InAs and AlSb are nearly lattice matched. InAs/AlSb superlattices have been grown by MBE in which solid elemental Al, Sb and In sources were used but for arsenic, the gas source AsH$_3$ was used [178]. Undoped InAs and Te-doped GaSb substrates were used. Both gas-source MBE and gas-source MEMBE were used for the growth. However atomically controlled interfaces could not be obtained in this system. Recently surface morphology of AlP epilayers grown by ALE has been examined by atomic force microscopy [180]. The layers were grown in a MOVPE system with a horizontal quartz reactor. The sources used were dimethylaluminum for Al and PH$_3$ for P. It was found that Group III gas feeding time was important in determining the surface smoothness on atomic scale. Shortening the gas feeding time improved the surface morphology.

2.3.2 Highly mismatched layers

If the lattice mismatch between the epilayer and the substrate is large, the growth front becomes rough and 3D SK growth begins after a few monolayers. The 3D growth and islanding of InGaAs has been reviewed in the book by Jain [3]. If short period $(InAs)_m(GaAs)_n$ superlattices are grown (at 520°C with As-stabilized surface) on GaAs, a progressive deterioration of the RHEED pattern occurs for n or m more than 2. If $n = m = 3$, the RHEED pattern becomes spotty and the growth becomes three-dimensional when the thickness of the superlattice is 90 nm [176]. Low temperature PL shows a broad peak with FWHM = 30 meV to 60 meV. The width is still around 60 meV at a growth temperature of 600°C. Chen et al. [177] have found similar results with the $In_xGa_{1-x}As$ layers (with $x > 0.2$) grown on GaAs substrates. In the layers grown with low V/III ratio and high substrate temperatures, RHEED oscillations terminated abruptly after a few monolayers. At very high temperatures the surface becomes cation stabilized and droplets of Ga are formed. By increasing the V/III ratio the oscillations persisted somewhat longer. However they are by no means comparable to those obtained with low In containing layers.

InAs layers using ALMBE have been grown between 300 and 400°C [175]. The layers were almost completely relaxed after a thickness of 7 MLs. However even 2.5 μm thick layers had flat surfaces with good morphology. Very good quality symmetrically strained $(InAs)_m(AlAs)_n$ ($3 < m < 17$, $1 < n < 10$) were grown on GaAs substrate. Intense network of dislocation was created near the superlattice/substrate interface which de-coupled the superlattice from the substrate. RHEED measurements showed that after the second period the inplane lattice constant became constant. Strain induced shift of LO Raman phonons showed that the superlattice was symmetrically strained and pseudomorphic [175]. The MEMBE growth of $(InAs)_m(GaAs)_n$ (at 370°) short period superlattices on InP substrates [176] gave similar results. RHEED pattern remained streaky during the growth of a 130 nm thick $(InAs)_4(GaAs)_3$ superlattice. This shows that unlike the results obtained with conventional MBE, the MEMBE growth remains two-dimensional during the growth of the whole superlattice. High quality of the superlattices was confirmed by x-ray diffraction and TEM studies. PL peak related to the SL was at 0.75 eV with FWHM equal to 10 meV. This is the smallest FWHM ever reported for InAs/GaAs SL grown on InP substrate. Similar improvements were observed in the MEMBE growth of $In_xGa_{1-x}As$ layers (with $x > 0.2$) grown on GaAs substrates. RHEED oscillations decayed rapidly with the MBE growth. When the layers were grown by MEMBE, the oscillations lasted much longer. FWHM of the low temperature PL from the MEMBE grown $In_xGa_{1-x}As$ quantum wells was very narrow. MODFETs fabricated with MEMBE layers showed much better performance.

2.3.3 Growth in nitrogen carrier gas

In recent years hydrogen has been used almost exclusively as the carrier gas in the MOVPE technique. However recent results of growth with nitrogen carrier

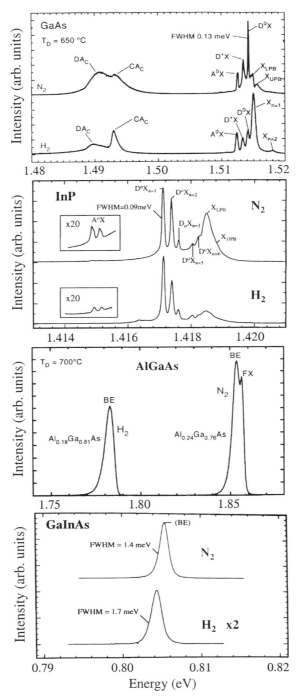

Figure 2.8: PL spectra of III-V epilayers grown in H_2 and N_2 respectively [166].

2.3. GROWTH OF CONVENTIONAL III-V SEMICONDUCTORS

gas (shown in Fig. 2.8) are very encouraging. Al-C bond is very strong and the reactivity of Al organics towards oxygen is high. The carbon and oxygen impurities are therefore a problem in the growth of AlGaAs by MOVPE. Group V precursors AsH_3 and PH3 are highly volatile and toxic. This combined with large volumes of highly explosive hydrogen gas raises the safety issues in the MOVPE technique. Molecular weight of nitrogen is 28 g/mol and that of hydrogen it is 2 g/mol. Nitrogen has higher efficiency of energy transfer during collision with the source molecule and is more efficient in decomposing the compounds because of its higher weight. Use of nitrogen avoids the necessity of costly hydrogen detecting system. Nitrogen is inexpensive and can be obtained with very high purity. However it is difficult to obtain laminar flow with the heavier nitrogen. Thermal conductivity of nitrogen is much lower and transfer of heat from the substrate to the compounds is not as efficient. Hardtdegen *et al.* [166] examined whether nitrogen carrier gas instead of hydrogen can be used in low pressure MOVPE. Despite the difficulties mentioned above, Hardtdegen and co-workers found that indeed high quality III-V epilayers can be grown by LPCVD using nitrogen carrier gas. They have grown epilayers of GaAs/GaAs, $Al_xGa_{1-x}As$/GaAs, InP/InP and $Ga_{047}In_{0.53}As$/InP using both the hydrogen and the nitrogen as the carrier gases. The epilayers were grown at a pressure of 20 mbar. The growth temperature was 650°C for GaAs and InP, and 700°C for $Al_xGa_{1-x}As$ and $Ga_{0.47}In_{0.53}$. 2 K PL spectra of the epilayers are shown in Fig. 2.8. For GaAs and InP the crystal quality and purity of the epilayers grown in nitrogen is comparable to those of the epilayers grown in hydrogen. Halfwidth of the bound exciton peaks is similar for both carrier gases and compare well with the state of the art of these layers.

In some respects the layers grown in nitrogen carrier gas were superior. Deep levels are generally a problem in the epilayers, particularly if the layers contain Al. Carbon impurity and deep levels in the $Al_{0.29}Ga_{0.71}As$ epilayers grown in nitrogen carrier gas (not shown in the figure) was much smaller when the layers were deposited with the nitrogen carrier gas. The improvement is believed to be due to better decomposition of the Al-organic with nitrogen carrier gas. The better decomposition of AsH_3 gives higher concentration of atomic hydrogen which in turn suppresses the incorporation of oxygen in the epilayer [166]. $Al_xGa_{1-x}As$/GaAs epilayers were grown in nitrogen carrier gas using the less stable triethylgalllium (TEGa) and dimethylethylaminealane (DMEAAl $(CH_3)_2C_2H_5N$-AlH_3) group III precursors and AsH_3 as the arsenic source. The PL spectrum for $x = 0.24$ is shown in Fig. 2.8. BE and FX (free exciton) transitions are resolved. Such high purity and crystal quality layers with high Al concentration and using the less stable precursors had not been reported earlier. PL spectra of $Ga_{0.47}In_{0.53}As$/InP epilayers grown in nitrogen also shown in Fig. 2.8 show that the quality of these epilayers is also extremely good. The FWHM of the bound exciton (BE) line is 1.4 meV for the nitrogen grown layers and 1.7 meV for the hydrogen grown layers. For an electron concentration of 1.6×10^{15} cm^{-3} the mobility was 11000 cm^2/Vs at 300 K and 99,000 cm^2/Vs at 77 K. These values compare favorably with the best reported for this system in the literature [166]. Nitrogen grown epilayers also showed better composition and thickness homogeneity.

2.4 Growth of II-VI semiconductors

2.4.1 Growth of ZnSe

Several efforts have been made to grow ZnSe layers on ZnSe substrates [132, 181, 182]. However large ZnSe wafers are not commercially available at a reasonable price. As mentioned earlier II-VI epilayers are grown on GaAs substrates. RHEED studies show that the surface of a GaAs substrate prepared for the growth of II-VI epilayers is deficient in As. The deficiency can be removed by impingement of an As beam. The growth temperature used for ZnSe is generally in the range 250 − 400 °C. At a given substrate temperature, the flux ratio is varied and the surface stoichiometry is monitored using RHEED. The flux ratio at which transition from cation rich to anion rich (or vice versa) surface takes place is determined. The flux ratio which gives good stoichiometry at a given substrate temperature is close to this transition.

Upon nucleation of ZnSe on arsenic stabilized GaAs surface (which has (2×4) reconstruction), a mixed layer of As and Zn is formed. In the mixed layer the number of Zn–As bonds is roughly equal to the number of Se–Ga bonds. Under these conditions the interface instability and disorder are suppressed and the growth becomes 2-D. X-ray rocking curves measured with thick ZnSe layers show FWHM of about 126 arcsec. The width is large because misfit and threading dislocations are created in a thick layer. The thinner pseudomorphic layers also show large FWHM because now the width is determined by the thickness of the layers. However PL measurements confirm that the quality of the layers grown on As stabilized surfaces is high.

Fujita *et al.* [150] have compared the quality of ZnSe epilayers grown by MOMBE and MBE. Effect of using different sources was also investigated. Activated nitrogen for The figure shows that when an alkyl group precursor is used the I_1^d line due to Zn vacancy-related defects dominates. It appears that the alkyl containing precursor is not completely decomposed which results in defect formation and C impurity incorporation in the layers. This shows that the window for growth and doping conditions is narrower in MOMBE compared to that in MBE. The efficiency of p-doping is also lower in MOMBE. The DLTS (deep level transient spectroscopy) signals of the ZnSe layers grown by MBE and MOMBE were also compared. A low temperature ($\sim 120°C$) trap, designated as trap 1, was observed in MBE and MOMBE layers grown under Zn-rich conditions and was attributed to Se vacancies V_{Se}. In the layers grown under Se rich conditions, a high temperature ($\sim 300°C$) trap was observed and was attributed to Zn vacancies V_{Zn}. A third trap was observed at $\sim 190°C$ only in the MOMBE layers. The origin of this trap is not known. These experiments seem to suggest that ZnSe layers grown by MOMBE are not as good as those grown by MBE.

High quality films of ZnS and ZnSe have been grown by MBE-ALE by Nelson

2.4. GROWTH OF II-VI SEMICONDUCTORS

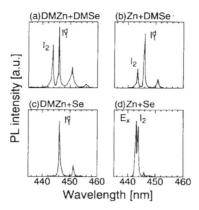

Figure 2.9: PL spectra measured at 4.2 for undoped ZnSe layers grown by different source combinations. Figs. (a), (b) and (c) are for MOMBE growth and (d) is for conventional MBE growth. The sources used are shown in each case [150].

[173]. The films were grown on both the (100) and (110) surfaces of GaAs. The measured film thickness as a function of number of cycles is shown by symbols in Fig. 2.10. The thickness calculated assuming ideal ALE growth is shown by the solid straight line. The good agreement between the observed and the calculated thicknesses show that the growth mode was ALE. Nelson [173] characterized the films by ellipsometry and determined the refractive index which agreed with the known values for ZnSe. X-ray diffraction studies showed high crystalline equality of the films. XRD and Laue x-ray backscattering data indicated that the films grew in the same orientation as the GaAs substrate. PL spectra showed strong near band-edge features confirming the high optical quality of the layers. Monte Carlo simulations of ALE growth rates of ZnSe have also been made [183] and good agreement between the calculated and the experimental values of the growth rates has been found. The model used is similar to that used for MBE growth of ZnSe [184]. However ALE does not always give a 2D growth. Three-dimensional growth of ZnSe on GaAs substrates has been observed in ALE by Yao and Takeda [174]. Recently ZnSe layers have been grown by atmospheric MOVPE ALE by Lee et al. [179]. The PL of 800 nm thick layer showed sharp excitonic luminescence around 2.8 eV. Donor acceptor peak at $\sim$ 2.78 eV, broad emission due to Cu at $\sim$ 2.35 eV and other defect related peaks were not detected. Phonon replica of the free exciton peak was observed at 2.774 eV. These PL results confirmed the high quality of the epilayers. The free exciton peak in the 100 nm thick layer is shifted from its bulk value to higher energy by 4.4 meV [179]. Calculations showed that the compressive inplane strain of 0.27% gives a shift which agrees with this value. As the thickness increases, the free exciton peak gradually returns to the bulk value. For a 800 nm thick layer the peak position becomes identical to

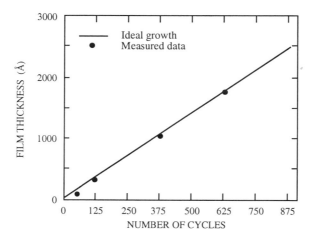

Figure 2.10: Film thickness versus number of cycles for ZnSe layers grown on GaAs (100) by atomic layer epitaxy [173].

that observed in the bulk crystal indicating that the strain is now completely relaxed by the creation of the misfit dislocations.

2.4.2 Growth of CdTe and ZnTe

Despite the 14.6% mismatch between CdTe and GaAs (see Table A in appendix A) high quality CdTe layers have been grown on GaAs substrates by MBE by Gunshor et al. [124]. Both (100) and (111) oriented epilayers were obtained. The orientation depended on the cleanliness of the interface and on the surface reconstruction of GaAs surface on which ZnTe nucleated. Since lattice mismatch between CdTe and InSb is very small, CdTe epilayers have been grown on InSb substrates [124]. The quality of the layers was improved by growing an InSb homo-epilayer on the InSb substrate before growing the CdTe layer. Very thick and high quality CdTe epilayers were grown without dislocations.

Extensive work on the growth of ZnTe layers has been done by MBE and by MOVPE on (100) GaAs substrate [163, 164, and references given therein]. For Te rich conditions in MBE growth RHEED studies show a streaky (1×2) reconstruction along the [110] azimuth. A streaky two fold $c(2 \times 2)$ reconstruction along [001] azimuth is observed under Zn rich conditions. A small region of intermixing of both patterns appear when the beam pressure ratio Te/Zn is close to unity. The ratio at which this region appears depends on the growth temperature. The ratio is between 1.5 and 2 for 350°C and between 1.2 and 1.9 for 300° C. The crystal quality of the layers is best when they are grown under conditions when the two patterns co-exist. In MOVPE growth, if partial pressure of one component becomes larger than its equilibrium vapor pressure, nuclei of the element may be formed and incorporated as precipitates in the epilayer. Excess element can evaporate from the hot surface and nuclei of the

2.4. GROWTH OF II-VI SEMICONDUCTORS

element can be formed in the adjacent colder gas phase. Kuhn *et al.* [163, 164] have studied the growth of ZnTe at 350 °C with partial pressure of DIPTe equal to 10 Pa. The partial pressure of DEZn was varied in the range 5 Pa to 35 Pa. High concentration of particles was observed at low and high Zn concentrations. There was only a small window near DEZn pressure of 13 Pa where the particles were absent and the surface morphology was good. SEM pictures showed that the particles as big as 5 μm were present in the layers grown with DEZn partial pressure of 33 Pa. With DETe and DEZn precursors, growth of whiskers was observed. These results emphasize again that the quality of layers depend on the precursors used, their partial pressures, and temperature of the substrate in a very complicated way. Good quality ZnTe crystals are generally grown when input partial pressures are equal. A very detailed review of growth of ZnTe by MOVPE has been published recently in three parts by Kuhn *et al.* [139, 163, 164]. Reference should be made to this review for more details.

Since lattice constants of ZnTe and GaSb are closely matched (see Table A in Appendix A), growth of ZnTe on GaSb substrates by MBE has been studied [185]. Nucleation of ZnTe directly on GaSb resulted in a spotty RHEED pattern characteristic of 3D growth. However as the growth continued, streaky RHEED patterns were evolved. Two dimensional growth of ZnTe layers can be obtained right from the beginning if GaSb or AlSb buffer epilayers are first deposited and ZnTe layers are grown on these buffers. RHEED studies showed that in the initial stages of the growth the III-V epilayer surfaces were Sb-stabilized with (1×3) reconstructed surfaces. After the nucleation of ZnTe the surface was Te-stabilized as shown by the (2×1) surface reconstruction.

2.4.3 Growth of ZnS

ZnS has a bandgap of 3.7 eV and among the II-VI family it is a primary candidate for ultraviolet light emitting devices. Good quality of ZnS epilayers have been grown by MBE [186]. It has been demonstrated that in spite of its very high vapor pressure (vapor pressures of Mg, Zn, Se and S [113] are shown in Fig. 2.11), elemental source of sulfur along with the solid-source of Zn can be used in the MBE growth. This is accomplished by using a specially designed effuser cell with a *post-heating* zone. The temperature of the effuser cell is kept low, 120–150°C and the temperature of the post heating zone is 200–400°C. RHEED pattern of the ZnSe layer grown in this manner was spotty in the beginning but became streaky after 30 minutes. PL at 4 K showed peaks due to excitons bound to neutral acceptors and neutral donors. Recently growth of ZnS layers by CBE and MOMBE [169] has been investigated. Pyrolysis of several precursors (DEZn, DipS, t-BuSH, H_2S) was studied in detail [169]. The highest growth rate was obtained by CBE using uncracked DEZn and H_2S sources. The growth rate increased with temperature up to 240°C and then started decreasing slowly as shown in Fig. 2.12. Initial increase in the growth rate is due to increase in the rate of reaction between DEZn and H_2S at higher temperatures. The growth rate decreases beyond 240°C presumably because now the desorption rate of the precursor and S-species became large. The growth with cracked

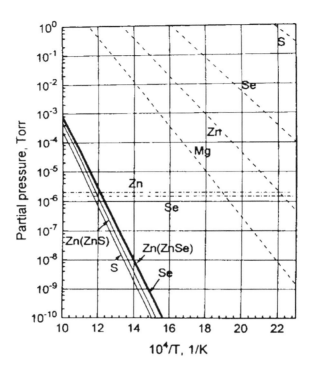

Figure 2.11: Vapor pressure of selected II-VI compounds and elements [113].

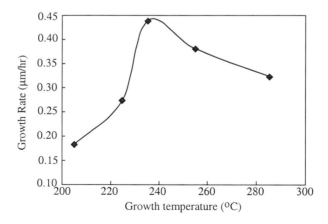

Figure 2.12: CBE growth rate of ZnS as a function of temperature using uncracked DEZn and H_2S [169].

2.4. GROWTH OF II-VI SEMICONDUCTORS

DEZn and t-BuSH showed a similar behavior. Hydrogen plays an important role in removing the alkyl radical from the growth sites and promotes better growth rates. The growth rate with cracked DEZn + DipS was rather small. The alkyl radicals produced by cracking by-products of DEZn and DipS block the growth sites as in this case hydrogen was not available to remove the radicals. Use of Zn with cracked DipS or t-BuSH also does not give appreciable growth rates. This result might appear surprising because it is known that good MBE growth rates are obtained using Zn and S_2. It is possible that the cracking product of DipS blocked the growing sites and inhibit the growth rate. It is also possible that the monomer S does not enable the growth at the temperatures used. In the case of t-BuSH, the SH molecule must react with a surface attached organic radical in order to free a sulfur atom to bond to Zn. It is the radical obtained by decomposition of DEZn which reacts with the SH and releases S. X-ray measurements showed that using DEZn and H_2S single crystal layers of ZnS could be grown at a low temperature of 140 °C. However RHEED pattern showed micro-twinning in the layers.

MOVPE has the ability to handle high vapor pressure elements phosphorus and S [187]. Good quality ZnS layers have also been grown by MOVPE [188, 187, and references given therein]. Heuken *et al.* [187] grew ZnSSe/ZnSe superlattices using atmospheric MOVPE. Both diethylsulphide (DES) and H_2S were used as sources of sulfur. Pd-purified H_2 was used as carrier gas. PL measurements and x-ray diffraction were used to characterize the layers. Both methods confirmed very high quality of the epilayers. Abounadi *et al.* [188] have also grown ZnS layers of high quality by MOVPE. More details of the growth of ZnS by MOVPE are given in the review of Fujita and Fujita [189].

2.4.4 Photo-assisted growth of II-VI semiconductors

Several authors have investigated the effect of above bandgap photo-irradiation on the growth of ZnSe layers [150, 190, 191]. Effect of ultraviolet (UV) light (325 nm, 3.81 eV) from a He-Cd laser on the MBE growth of ZnSe was investigated by Ohishi *et al.* [190]. The power density of the beam was ~ 40 mW/cm^2 and the flux ratio was Se:Zn = 4:1. For the unirradiated samples, the RHEED pattern became spotty upon nucleation of ZnSe. The streaked pattern observed with the GaAs surface continued with the nucleation and growth of ZnSe under UV irradiation. The growth rate under UV irradiation was 2–2.5 times faster. Presumably the Se_2 molecules (with a dissociation energy of 3.55 eV) were dissociated under the UV irradiation and this resulted in better quality ZnSe epilayers with enhanced growth rates. The growth temperature in MOVPE and MOMBE can be lowered using the above-bandgap photo-irradiation [150, 191]. Photo-irradiation also reduces the formation of defects (which compensate the acceptors) and improves the *p*-type conductivity. Photo-irradiation enhances the decomposition of zincalkyl with very high (1–10%) efficiency. Fujita *et al.* [150] have grown ZnSe layers by MOMBE with and without photo-irradiation. A high pressure mercury lamp with an infrared filter (to eliminate heat) was used as a source if light. In the layers grown with photo-irradiation during the supply of DMZn, the I_1^d PL peak (attributed to Zn vacancies) was suppressed. PL lines due to free excitons and bound excitons became dominant. This improvement is partly due to better decomposition of DMZn and partly due to desorption of Se from the growing surface. In both cases the stoichiometry of the layer improves. Gokhale *et al.* [191] observed similar results in the OMVPE growth of ZnSe, using the same sources i.e. DMZn and DMSe. They found that the surface morphology of the layers grown with photo-irradiation also improves. However they found that there is temperature dependent threshold in the light intensity below which the improvement does not occur. The photo-irradiation effects are different for different source combinations. The improvements in the quality of the layers are remarkable only for certain source combinations [150].

2.4.5 Interfaces of II-VI heterostructures

Highly abrupt and smooth interfaces can be obtained in the III-V semiconductor heterostructures. This is not the case with the II-VI epilayers grown on GaAs substrate. In addition to intermixing, the interfaces become rough. The interfaces show lateral fluctuations at different scales. Different techniques of characterization probe the roughness at different lateral scales. X-ray diffraction gives an average of the roughness over large distances. Excitonic photoluminescence line shape is sensitive to fluctuations at the scale of excitonic radius. It is therefore necessary to use different techniques to characterize an interface.

Olego [121] has investigated the interfaces of ZnSe/GaAs heterostructures using Raman and PL techniques. ZnSe layers were grown by MBE at 350°C. Six samples with ZnSe layer thicknesses $h = 50, 80, 180, 350, 1300$ and 5000 nm were fabricated. The GaAs substrate was n-type with 1.3×10^{18} electrons cm^{-3}. At the bare surface the Fermi level is pinned in the gap and the bands bend which give rise to a depletion width of about 25 nm. Raman spectra of the heterostructure were measured in the backscattering geometry with the 488.0 nm light from an Ar$^+$ ion laser. ZnSe is transparent to this light. First-order Raman spectra of three samples are shown in Fig. 2.13. The high energy LO$_{GaAs}$ peak at ~ 292 is due to the GaAs LO phonons in the depletion layer. Intensity of this mode is sensitive to the band bending, it varies as E_s^2 where E_s is the electric field in the depletion layer [121]. Since band bending depends upon the surface states due to dangling bonds, the intensity of this mode can be used to monitor the density of surface states and the dangling bonds. The L$_{GaAs}^-$ mode is due to the lower branch of the LO phonon-plasmon coupled modes. Its intensity is independent of band bending and it can be used as a reference to compare the intensity of the LO$_{GaAs}$ mode in different samples. The LO$_{ZnSe}$ line is due to ZnSe LO phonons. It is seen from Fig. 2.13 that the intensity of the LO$_{GaAs}$ line decreases sharply for the 80 nm ZnSe sample but starts increasing again for thicker samples. More detailed measurements show that the intensity remains low up to $h = h_c = 150$ nm (see Table 3.1). As the thickness goes past this value, the intensity starts increasing and regains its full strength in the 5000 nm sample. These results suggest that since pseudomorphic ZnSe layer has contiguous lattice planes, the density of dangling bonds is decreased.

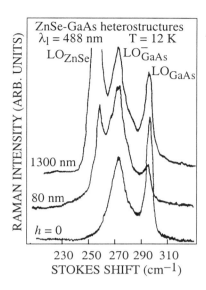

Figure 2.13: First-order Raman spectra of ZnSe-GaAs heterostructures for selected values of the ZnSe layer thickness h. Note the strong reduction in the intensity of the LO$_{GaAs}$ line for $h = 80$ nm as compared to the bare surface case, i.e. $h = 0$ [121].

This decreases the band bending and also the electric field E_s in the depletion layer suppressing the intensity of the of the LO$_{GaAs}$ line. For thicker layers misfit dislocations are produced and the dangling bonds are created again. This increases the band bending and also the strength of the of the LO$_{GaAs}$ line. Since absence of dangling bonds and surface states reduce the surface recombination velocity, a corresponding increase in the PL intensity is also observed in the heterostructures. As the layer thickness increases to more than h_c the PL intensity starts decreasing. Pagés et al. [192] have made similar studies using semi-insulating GaAs substrate.

Faschinger et al. [147] studied strain relaxation and interface structure in ZnSe and CdSe layers and in superlattices which had a period thickness varying from (2 ML ZnSe)/(2 ML CdSe) up to (6 ML ZnSe)/(6 ML CdSe). Total thickness of each superlattice was 1500 monolayers. RHEED measurements showed that complete relaxation of CdSe on relaxed ZnSe buffers (and of ZnSe on relaxed CdSe buffers) occurs when the layer thickness is 10 MLs. The total thickness of the superlattice is so large that the superlattice as a whole must be relaxed by generation of misfit dislocations in the layers close to the relaxed buffer. HRXRD spectra were simulated for different values of lattice constants of the CdSe and ZnSe layers varying between the values of completely relaxed and completely (symmetrically) strained layers. It was found that relaxation shifts the zero order superlattice peak and changes the relative intensities of the satellite. The first effect depends on the sum of the vertical lattice constants

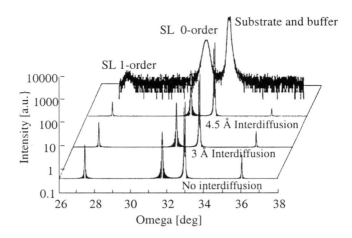

Figure 2.14: Observed HRXRD spectrum of 2ML/2ML CdSe/ZnSe superlattice (top curve). Calculated curves using a dynamical theory are also shown. A $Cd_{0.5}Zn_{0.5}Se$ layer of variable thickness at each interface was assumed to model the interdiffusion. A good fit with the experiment was achieved for a 4.5 Å interface layer [147].

of both layers in a period and the second effect depends in the vertical lattice constants of individual layers. A fit of the observed spectra with the simulations can be used to determine the individual vertical lattice constants of each layer. Tilt between the SLS and substrate can be determined by measuring each spectrum in a second azimuth rotated by 180 degrees. The tilt determined in this manner can be taken into account in the simulations. Interface mixing can be studied by the x-ray diffraction technique. A typical observed HRXRD spectrum of 2ML/2ML CdSe/ZnSe superlattice is shown by the top curve in Fig. 2.14. The other curves are obtained by calculations using a dynamical theory [147] and assuming interfacial layers of ZnCdTeSe alloy with half-half concentration of each binary compound. It is seen that the calculated and observed curves agree quite well for a 4.5 Å interfacial layer.

Grieshaber et al. [158] have studied the interfaces in $CdTe$-$Cd_{0.8}Mn_{0.2}Te$ quantum wells using photoluminescence and Zeeman splitting. Diffusion of Mn across the interface produces layers of diluted magnetic alloy. Therefore the diffused interface layers have a larger magnetic susceptibility than in the more concentrated $Cd_{0.8}Mn0.2Te$ barriers. In the interfacial layers antiferromagnetic interactions are less effective in blocking the spins. This gives rise to an enhanced Zeeman splitting. The enhanced Zeeman effect probes the magnetization locally due to Mn spins at and close to the interface. Grieshaber et al. [158] found that as the growth temperature increased from 220°C to 310°C, dilution of Mn at the interface measured by the increased Zeeman effect also increased suggesting that the interdiffusion of Mn ions was taking place at the interface[4]. In contrast

---

[4]A detailed discussion of formation of magnetic polarons at the interfaces between diluted-

the PL line width decreased. This line width measures the effective roughness on the scale of the exciton radius. In structures grown at lower temperatures (220°C) interfaces were rough on this scale. The interface probably became smooth on annealing.

2.5 Growth of III-nitride epilayers

2.5.1 Substrates for the growth of III-Nitrides

Several methods have been used to grow III-Nitrides [100, 179, 193]. MOVPE is the most extensively used technique for growing III-Nitride epilayers and heterostructures. MBE has also been used by many groups to grow the III-Nitride epilayers. Sapphire and SiC are widely used substrates for the growth of the epilayers [99, 194]. The lattice constants and bandgaps of GaN and related materials are shown in Fig. 1.2 and also in Table J given in Appendix A. There is a large lattice mismatch between the III-Nitrides and the substrate materials. Thermal coefficients of expansion of the Nitrides are also considerably different from those of the substrates. The layers are therefore strained. A large concentrations of defects is created in the epilayer to relieve the strain and lower the energy of the system. A few attempts have been made to grow the epilayers of GaN on (mm size) GaN substrates [195]. Large size GaN wafers are not yet available. Large area good quality wafers of sapphire are available at a low price. Sapphire is transparent and stable at high temperatures. Wurtzite GaN is hexagonal-close-packed (hcp) structure. Sapphire also has an hcp structure but it is slightly distorted [196].

Considerable work has also been on growth of III-Nitrides on SiC [99, 194]. SiC has several advantages over sapphire. Its lattice mismatch with GaN is small, only 3.5%, and with AlN, the mismatch is very small (see Table J in Appendix A). SiC has good electrical conductivity. A substrate bias can be applied to reduce the ion energy while using nitrogen ions for the growth of the Nitrides [198]. Large good quality SiC substrates are available commercially [99]. For the surface preparation, the SiC wafer is dipped in HF and then treated with a hydrogen plasma. This reduces the oxygen-carbon bonds to very small numbers. Low temperature GaN or AlN buffer layers are also grown on the SiC substrates before depositing the Nitrides. The stress in the films grown on SiC is smaller than in the films grown on sapphire.

III-Nitride epilayers have also been grown on Si substrate [199]-[202] Reviews of this work have been written by Abernathy [194] and Popovici *et al.* [99]. A low temperature buffer layer of about 30 nm GaN is generally grown on Si before the growth of the main epilayer. GaN grown on (100) Si is predominantly cubic. The epilayer contain a large number of defects such as dislocations, twines and stacking faults. The layers grown on (111) Si are predominantly wurtzite with localized inclusions of cubic phase. Better quality wurtzite GaN has been obtained on (111) Si by ECR MBE on Si surface containing specially

---

magnetic alloy and nonmagnetic layers is given in section 6.8.

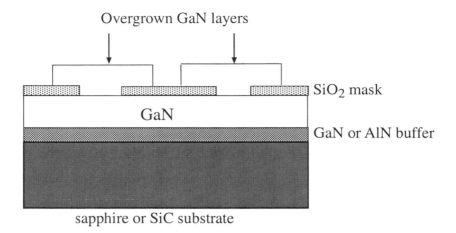

Figure 2.15: Design of an ELOG substrate. The top surface consists of patterned SiO$_2$ mask and window stripes. GaN grows first in the windows and then laterally on the SiO$_2$ stripes.

prepared atomically flat terraces [99]. These layers exhibited free exciton PL at low temperatures.

Several other materials have been used as substrates. Some of them are GaAs [100, 203], MgAl$_2$O$_4$ [99, 204], glass [205], quartz glass [206], and ZnO [207]. Properties of several substrate materials have been discussed in Refs. [99, 194]. Very high quality GaN films have been grown on ZnO substrates by MBE [99]. The width of the exciton PL line was only 8 meV, comparable to the best results obtained by MBE on sapphire substrates. Deep level yellow PL was absent.

2.5.2 Laterally Epitaxially Overgrown (LEO)- and Pendeo (PE)-epitaxial layers

Several authors have grown and studied Laterally Epitaxially Overgrown (LEO) GaN epilayers on the ELOG (Epitaxial Lateral Over Growth) substrates [208]–[212]. To fabricate an ELOG substrate, selective growth of 0.2 μm GaN layer is performed on a (0001) c-face of a sapphire substrate with a low temperature buffer. A 0.1 μm thick SiO$_2$ layer mask is patterned to form 4-μm wide stripe windows separated by 7 to 8 μm wide SiO$_2$ stripes (see Fig. 2.15). A ten μm thick layer of GaN layer is grown on the mask. The epilayer first grows on the buffer through the windows. As the thickness of the epilayer increases, it grows laterally on the SiO$_2$ stripes. At about ten μm thickness the epilayers that grow at the two edges of the SiO$_2$ stripes coalesce and a continuous flat GaN layer is formed on the patterned substrate. The dimensions and thickness of the layer at which the layer begins to grow laterally may vary somewhat from experiment to experiment. The LEO epilayers have been grown by MOVPE [208]-[210]

2.5. GROWTH OF III-NITRIDE EPILAYERS

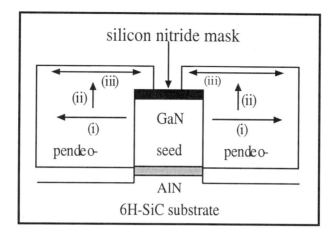

Figure 2.16: Schematic of GaN pendeo-epitaxial growth process steps. The Fig. is taken from the paper of Linthicum *et al.* [26].

and by HVPE [211, 213]. TEM and SEM have been used to determine the defect structure [208]-[211] and Raman, PL and CL techniques have been used to determine the optical properties [210]. Raman measurements showed that there was a general improvement in the quality of the laterally overgrown layers. Dislocation density in the GaN layer grown on the SiO_2 stripes was small, it was only 2×10^7 cm^{-2}. The lifetime of the InGaN/GaN/AlGaN-based laser diodes grown on this substrate increased to 6000 hours [85, 214].

In a LEO GaN layer, alternate stripes have high and low dislocation density. Since the device is fabricated on the GaN stripes with low density of defects, the size of the device is limited by the width of these stripes. Also, the device fabrication requires careful alignment of the device structure with the underlying mask stripe. Therefore it is desirable to have a large area layer with a low and uniform density of defects. Nam *et al.* [215] were able to produce such a layer by repeating the LEO process a second time. In the second LEO step, the mask stripes are placed on the openings of the first mask i.e. they cover the GaN stripes with high density of defects. The GaN layers in the second LEO step are therefore seeded on the good GaN stripes. They grow laterally on the second oxide masks. Thus a large area GaN layer with low defect density is obtained. More recently [26, 212] large area GaN films with low density of defects have been fabricated by the so-called Pendeo-Epitaxy (PE). The PE growth process using SiC substrates is illustrated in Fig. 2.16. GaN does not nucleate on the etched surface of SiC. It nucleates only on the side walls of the columns grown on AlN buffer layer as shown in Fig. 2.16. GaN grows laterally away from the opposite side walls on which it nucleates. Growing fronts of the layers approach each other and eventually meet (see Fig. 2.17a). A (0001) top surface is created. Now GaN begins to grow vertically on the (0001) surface. When it reaches the top level of the silicon nitride masks, it grows laterally over the mask. Since no

50   CHAPTER 2. CHARACTERIZATION AND GROWTH

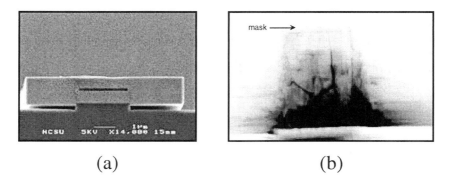

(a)   (b)

Figure 2.17: (a) Cross-sectional SEM of a GaN/AlGaN pendeo-epitaxial growth structure showing coalescence over the seed mask. (b)) Cross-sectional TEM of a GaN pendeo-epitaxial growth structure showing confinement of threading dislocation under the seed mask, and a drastic reduction of defects in the regrown areas (Linthicum et al. [26]).

nucleation takes place at the substrate or on GaN with large defects, the density of dislocations remains low in the whole layer. A large area of the low defect density GaN is produced [26]. TEM picture of the grown structure is shown in Fig. 2.17b. The dislocations are confined to the column under the mask. A drastic reduction of the dislocations takes place in the regrown layer.

2.5.3 Growth of GaN and AlN by MOVPE

The source materials ammonia, Trimethylgallium (TMGa), trimethylaluminum (TMAl), and trimethylindium (TMIn) are generally used for nitrogen, Ga, Al, and In respectively [216]. Mg is widely used as a p-type dopant and Si as n-type dopant. Biscyclopentadienyl (Cp_2Mg) is used as a source of Mg [217] and methyl silane ($MeSiH_3$) as a source of Si. Amano, Hiramatsu, Akasaki and their collaborators showed that good epilayers of III-Nitrides on sapphire substrates can be obtained by MOVPE if the growth is performed in two steps (see Ref. [218] and the reviews by Akasaki and Amano [54] and by Hersee et al. [196]). In the first step a thin AlN buffer-layer is grown at a low temperature. The main epilayer is grown in the second step at a higher temperature. The buffer layer provides the high density of the nucleation centers and promotes the lateral growth of the main epilayer. The quality of the main epilayer is best if the buffer-layer is annealed for about 20 minutes. PL, TEM, and X-ray measurements show that the layers grown by the two step method are of high quality. Nakamura [219] used GaN layers (instead of AlN layers) as buffers. Nakamura also introduced the so-called two-flow method. In this method, in addition to the main gas flow parallel to the substrate, another flow, designated as the 'subflow' is used. The subflow transports the mixture perpendicular to the substrate. The subflow changes the direction of the main flow of the inactive

2.5. GROWTH OF III-NITRIDE EPILAYERS

H_2 and N_2 gases to bring the reactant gas in contact with the substrate more effectively. Nakamura found that subflow suppresses the 3D growth, promotes the 2D growth and improves significantly the quality of GaN epilayers. The two step method is also used for the growth of conventional III-V semiconductor layers directly on Si substrates [52]. The observed FWHM of the GaN (002) x-ray diffraction is large. It varies from 2 to 5 arcmin for good quality films and to over 35 arcmin for poor quality films [99, see Fig. 2.14]. The calculated FWHM is 33.2 arcsec. The difference arises because of the large number of defects present in the layers.

The growth of GaN over (1120) oriented (A-face) sapphire substrate by MOVPE has been achieved using an InN buffer layer [220]. The 20-30 nm InN buffer layer was deposited at 600°C. As compared to the GaN layer grown on GaN buffer the density of dislocations in the layer was lower by a factor 10. Hall effect measurements showed that the maximum electron mobility was close to 500 cm^2/Vs in the GaN layer grown on an InN buffer. The maximum mobility with GaN buffer was lower, $\sim$ 330 cm^2/Vs.

2.5.4 Growth of GaN and AlN by MBE and related methods

The group III sources for the growth of the III-Nitrides are the same as for the growth of conventional III-V semiconductors. These sources and their sticking coefficients have been discussed earlier.

As compared to MOVPE, development of MBE for growth of the Nitrides has been slow mainly because ammonia, common source of nitrogen, is stable at low temperatures of growth used in MBE. Experiments show that incorporation of ammonia and growth of GaN begins at 450°C [154]. The incorporation efficiency is 0.5% at 500°C, rises rapidly and becomes 3.8% at 700°C. At higher temperatures it increases slowly to 4% at 830°C. (DMHy)[$(CH_3)_2NNH_2)$] decomposes at lower temperatures (below 600°C). However it introduces carbon impurity in the epilayer [221]. Reactive species of N_2 obtained by electron-cyclotron resonance (ECR) or radio frequency (RF) plasmas[5] are generally used to avoid this difficulty. The energy of nitrogen species in radio frequency plasma is large and can be detrimental to the quality of the epilayer. Electron cyclotron resonance (ECR) microwave plasma produces reactive nitrogen beams at lower energies. The ECR beam has a higher fraction of reactive nitrogen than the RF plasma beams. Early use of ECR to grow good quality GaN epilayers on (0001) sapphire substrate was made by Moustakas and Molnar [222]. The authors found that the surface was flat with low power plasma and the layers were n-type. With high power plasma, the layers were insulating and the surface was rough. Moustakas and Molnar [222] also deposited Si doped n-type and Mg doped p-type layers. Electrical properties of these films are discussed in chapter 6. Korakakis *et al.* [231] have grown $Al_xGa_{1-x}N$ on both (0001) sapphire and (0001) Si terminated

---
[5]The methods to create ECR and plasmas are described in detail in Refs. [13, 99, 194, 222, 223].

6H-SiC using an ASTEX compact ECR microwave plasma source to activate the molecular nitrogen gas. RHEED and XRD studies showed that the layers were of good quality.

Morkoç and co-workers [97]-[99] and several other groups [152, 162, 225] have grown GaN, AlGaN and InGaN epilayers on sapphire substrates using ammonia as the nitrogen source. In the method generally used (designated in Ref. [97] as the Reactive MBE or RMBE) ammonia decomposes at the surface of the substrate by pyrolysis. The two step method is generally used though some authors have grown layers without the buffer layer. At high ammonia to group III ratio, the stress generated in the film is low, and deep level luminescence is weak [99]. Maximum growth rate observed in RMBE by Kim et al. [97] was about 1.1 μm/h. In the early days the low growth rates were a problem but now growth rates comparable to MOVPE growth rates are easily obtained. Dislocation density determined from the TEM experiments is $\sim 10^9$ cm^{-2}. Typically the FWHM of the X-ray rocking curve is about 350 arcsec. RHEED pattern shows streaky-like pattern and the amplitude of the laser reflected intensity oscillations remains constants for growth of several microns. AlGaN epilayers have also been grown. Surface morphology of the AlGaN epilayers is not as good as that of the GaN epilayers. To avoid the generation of cracks in thick layers, AlGaN/GaN superlattices are used instead of thick AlGaN layers. Room temperature PL spectra of GaN shows the band edge emission at 365 nm and a broad yellow band[6] at 575 nm due to deep levels [221]. By increasing the V/III flux ratio by a factor 2, the yellow band disappears. Si doped layers do not show the broad yellow band.

Other methods of growth have also been used [229, 224]. Meng et al. [229] have grown AlN films on Si (111) and Si(001) substrates by ultrahigh vacuum reactive sputter deposition. Epilayers of hexagonal AlN have been grown on Si (100) substrate using supersonic jet molecular beam gas sources [202]. In this method a seeded jet accelerates precursors to kinetic energies which are one to two orders of magnitude higher than the average kinetic energies in the molecular beams of materials used as sources [230]. High energies enhance precursor adsorption and increase the growth rate. The low base pressure and high growth rates make this technique attractive. The focused jet exit stream generates a very high peak flux and increases further the growth rates. However, due to focusing of the flux, the deposition non-uniformity is quite large [230]. Rizzi et al. [228] have grown GaN and AlN on 6H SiC substrates. The lattice mismatch between AlN and SiC is small (in plane lattice constants are 3.08 Å for 6H-SiC and 3.11 Å for 2H-AlN, see Table J in Appendix A) and thermal coefficients of the two materials are not so different [227]. AlN buffer layer was grown at 620°C and GaN/AlN heterostructure was grown at 760°C [228]. The surface showed clear (1 × 1) LEED pattern. PL studies showed that the epilayers were of good optical quality. Recently MBE layers of III-Nitrides using activated nitrogen have been grown by several groups. Properties of these layers are discussed in chapters 5 and 6.

---

[6]Yellow PL band is also discussed in chapter 5.

2.5. GROWTH OF III-NITRIDE EPILAYERS

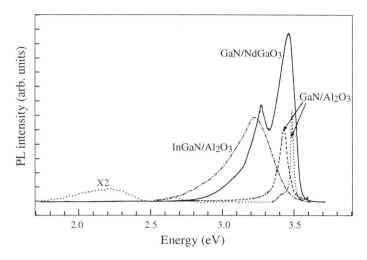

Figure 2.18: 5 K PL spectra of GaN and InGaN samples. Dotted curve is for the GaN/Al$_2$O$_3$ layer grown by ECR. The other three samples were grown by a novel plasma source [232].

2.5.5 Comparison of III-Nitride layers grown on different substrates and by different methods

Mamutin et al. [232] have grown GaN and InGaN (with 10% In) epilayers on NdGaO$_3$ as well as on sapphire substrates. The layers were grown by MBE using ECR and modified RF Plasma sources of nitrogen. The modified source was a compact magnetron source with an RF excitation of the discharge [232, and references given therein]. All samples showed streaky RHEED pattern confirming that the crystalline quality of the layers was good. The PL spectra of the four samples studied by Mamutin et al. [232] are shown in Fig. 2.18. The yellow band is absent in the three samples grown using the plasma source of nitrogen. PL intensity is maximum in the spectrum for the GaN/NdGaO$_3$ sample (solid line). In this spectrum the main peak at 3.46 eV is due to donor bound excitons. It was not possible to assign the lower energy peak at 3.26 eV to a specific transition unambiguously. The possible transitions responsible for this peak are donor bound excitons in cubic GaN inclusions or DAP transitions in hexagonal GaN. This peak disappears on warming the layer to 80 K. Since electron concentration in the sample was small, DAP transitions are less likely to be the cause for this peak. It is more probable that transitions in cubic nitride inclusions give rise to this peak. The width of the PL spectra in this sample is very large due to inhomogeneous broadening. The width of PL in the sample grown by ECR is the smallest, it is 20 meV. This suggests that the number of defects in the sample grown on NdGaO$_3$ substrate using the novel plasma source of nitrogen is large. As compared to the PL of the ECR grown sample on sapphire substrate (dotted curve in Fig. 2.18), the PL spectra of the other two samples grown by the novel plasma source are also broad. However the PL spectrum of the ECR grown sample shows a yellow band at 2.3 eV and an unidentified weak band at 2.65 eV. The stresses in the sample were measured by the radius of curvature method. The stress was minimum in the GaN sample grown on $NdGaO_3$ substrate. The number of defects in this sample was larger than in the sample grown on sapphire substrate. It is possible that the strain was relaxed by the creation of misfit dislocations.

Johnson et al. [31] compared the quality of GaN epilayers grown by MOVPE and by MBE. The motivation of the work was to determine why MBE has not been successful in fabricating advanced optoelectronic devices. The MOVPE layers were grown by the conventional two-step method on 5-cm sapphire substrates. The growth temperature for GaN was 1060°C to 1130°C and for InGaN 725°C to 800°C. For doping with Si and Mg, silane and bis(cyclopentadienyl) magnesium were used as sources. MBE growth was performed using elemental Group II and dopant sources. Active nitrogen was obtained from a RF plasma source. Growth temperature was in the range 750°C to 900°C for GaN and 670°C to 700°C for InGaN layers. The growth rates were 1-2 μm/hr in MOVPE and 0.4-2 μm/hr in MBE.

The surface of the MBE undoped GaN layers had a "wormy" structure but that of the MOVPE layers was smooth and uniform. The surface of the Mg doped MBE GaN showed a faceted structure. Again Mg doped MOVPE layers had smooth structure. On annealing the MOVPE layers show characteristic blue emission. Annealed MBE grown GaN:Mg layers do not show this effect. The authors suggested that abundant supply of hydrogen in the case of MOVPE growth is responsible for the improved quality of the layers. Results obtained with InGaN are discussed in the next subsection.

2.5.6 Growth of InGaN and AlGaN alloys

Yoshida [221] deposited good quality InGaN epilayers by MBE on sapphire substrates using uncracked ammonia. InGaN layers with 0 to 20% In were grown. However good quality layer with mirror like surfaces could be grown only in a narrow temperature range in the neighborhood of 780°C. In the films grown at a temperature above 800°C the concentration of In was very low. At a temperature lower than 760, droplets of Ga were formed on the surface due to insufficient decomposition of ammonia. Epilayers with more than 20% In could not be grown. PL of InGaN epilayers did not show the deep level peak. There was band edge peak at 420 nm and a peak at 365 nm due to the buffer layer. Grandjean and Massies [152] have grown $In_xGa_{1-x}N$ layers with high In concentrations (up to $x = 0.46$) by GSMBE. Nitrogen precursor was 50 sccm ammonia. A 2 μm thick GaN layer was first deposited on (0001) sapphire at 500°C after nitridating the substrate at 850°C [162]. $In_xGa_{1-x}N$ layers were grown on the buffer layer at temperatures in the range 500 to 600°C. At high temperatures growth rate was 1.2 μm/h. At lower temperatures it was reduced to 0.2–0.4 μm/h. The effect of growth temperature on the incorporation of In was studied. The indium incorporation rate was smaller at higher growth

2.5. GROWTH OF III-NITRIDE EPILAYERS

temperatures. A similar result was also obtained by Yoshida [221]. Increasing the Ga flux had only a small effect on the In incorporation rate. By increasing the In flux, up to 46% In could be incorporated in the epilayers. Epilayers of InGaN alloys with entire composition range have been grown by MBE by Blant et al. [199] and by Singh et al. [233]. Blant et al. [199] have studied the MBE growth of both the AlGaN and InGaN alloys on clean and oxidized Si surfaces using an RF plasma source of nitrogen. The growth temperature was in the range of 400 to 750°C depending on the composition of the alloys. X-ray measurements showed that alloys of (InGa)N of controlled composition can be grown on Si substrate using a plasma enhanced Molecular Beam Epitaxy over the entire composition range. Films grown on oxidized surfaces of Si were polycrystalline or amorphous. The growth on chemically cleaned Si substrates show the usual columnar structure common to the growth of Group III-Nitrides [195]. Singh et al. [233] were also able to grow InGaN alloys over practically the entire composition range at the growth temperature of GaN (700-800°C) by MBE. These authors found that if the thickness of the films is > 0.3 μm, incorporation of more than about 30% indium results in spinodal decomposition. However such phase separation is absent in thin InGaN films (< 600 Å) grown as GaN/InGaN/GaN heterostructures. In such configurations up to 81% In could be incorporated.

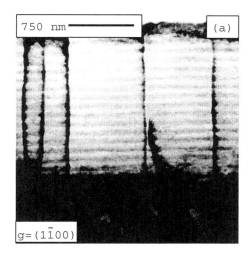 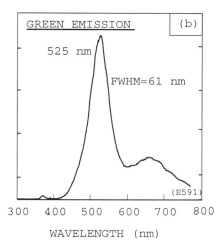

Figure 2.19: MBE grown InGaN MQW structure. (a) Cross sectional TEM Image (200 kV) and (b) Room temperature PL spectrum [31].

Akasaki and Amino [54] have grown InGaN layers by MOVPE and examined quality of InGaN/GaN MQWLs using X-ray and PL techniques. A $2\theta/\theta$ mode x-ray diffraction profile from a GaInN/GaN multilayered structure was measured. The calculated results using a kinematical theory agreed well with the experiment. There were four fingers between the 0th and -1st order peaks of the multi-layer structure in both the experimental and the theoretical curves. This

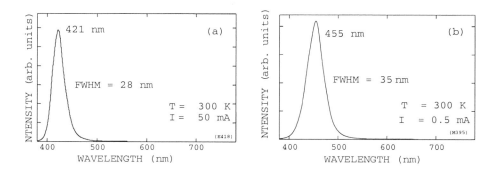

Figure 2.20: Spectrum from InGaN quantum well LED grown by (a) MBE and (b) MOVPE [31].

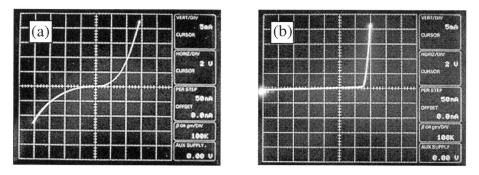

Figure 2.21: I/V curve for LEDs grown by (a) MBE and (b) MOVPE [31].

shows that the interfaces were smooth, and the composition and thickness are well controlled. At room temperature the PL intensity of the multi-quantum wells grown on the low temperature buffer is 2.5 orders of magnitude higher than in the structure grown without the low temperature buffer. At 77 K, the intensity is 3 orders of magnitude higher [54]. Good quality InGaN/GaN multiquantum wells have been grown by atmospheric MOVPE on c-plane (0001) sapphire substrate by Minsky et al. [234]. A 19 nm thick GaN nucleation layer was deposited at 600°C. A 1.8 μm thick GaN layer, doped with 1×10^{18} cm^{-3} Si atoms was grown at 1080°C on top of the buffer layer. The quantum well structure was grown on the GaN layer. The barriers consisted of 7 nm GaN layers doped with 1×10^{18} cm^{-3} Si atoms. The quantum wells were 2 nm layers of In$_{0.2}$Ga$_{0.8}$N with back-ground n-type doping of $3-5 \times 10^{17}$ cm^{-3}. The structure was capped with a 100 nm Al$_{0.06}$Ga$_{0.94}$N which also had an n-type back ground doping of $3-5 \times 10^{17}$ cm^{-3}. Three structures were grown, with one, five and ten quantum wells respectively. Time resolved PL measurements were made. The decay of PL was rapid in the single quantum well. The decay rate decreased with the number of quantum wells, being slowest in the ten quantum

2.5. GROWTH OF III-NITRIDE EPILAYERS

well structure. The FWHM of the ten quantum well structure PL was 98 meV at room temperature and 58 meV at 4 K. This shows that the quality of the epilayers was good. The FWHM on the single quantum well PL was larger. Minsky et al. [234] suggested that the gettering of defects may become more effective as the number of interfaces increases and therefore the quality of the 10 quantum well structure is better.

Recent emphasis is on the growth of InGaN quantum well layers with up to 50% In concentration. Johnson et al. [31] have grown InGaN multiple quantum well structure by the modulation beam technique which suppresses the formation of Ga droplets. Cross section TEM image of the multiple quantum well grown by MBE is shown in Fig. 2.19(a). Room temperature PL spectrum of the same structure is shown in Fig. 2.19(b). Using MOVPE high InGaN layers have been grown and blue and green LEDs have been commercialized (see chapter 7). However these are probably the best MBE results for such high In concentration. To compare further the quality of the MOVPE and MBE layers single quantum well LED test structures were fabricated. The doping concentrations in the two test structures were similar. The emission spectra of the LEDs are shown in Fig. 2.20(a) for the MBE frown layers and in 2.20(b) for the MOVPE grown layers. The LEDs emit light in the violet and the blue regions of the spectrum. The current for the comparable emission of light is two orders of magnitude smaller in the MOVPE grown LEDs. The current voltage characteristics of the two diodes are shown in Fig. 2.21(a) for the MBE frown layers and in 2.21(b) for the MOVPE grown layers respectively. The reverse break-down voltage in the MBE diodes is low. On the whole the behavior of the MOVPE diodes is considerably superior.

Chapter 3

Strain and critical thickness

3.1 Strain and energies of epilayers

3.1.1 Misfit strain

Consider an epilayer of a semiconductor with lattice constant a_l grown on a thick substrate with lattice constant a_{sub} (illustrated in Fig. 3.1). The lattice mismatch is measured by the misfit parameter f_m defined below [3, 235]:

$$f_m = \frac{a_l - a_{sub}}{a_{sub}}. \tag{3.1}$$

If both misfit parameter f_m and thickness h of the epilayer are small the layer remains pseudomorphic. Misfit between the two semiconductors is accommodated by the tetragonal strain in the epilayer as shown in Fig. 3.1(a). The strain is homogeneous and is known as the 'misfit strain'. The in-plane homogeneous strain in the pseudomorphic layer is given by [3, 235]

$$\epsilon_\| = -f_m. \tag{3.2}$$

If thickness h of the epilayer exceeds a certain thickness, known as critical thickness h_c, misfit dislocations are created. Now the strain is partly accommodated by the misfit dislocations as shown in Fig. 3.1(b). The value of h_c decreases as f_m increases. If the epilayer contains dislocations with average inter-dislocation space p, each dislocation relaxes the strain by an amount b_1. b_1 is equal to the active component of the Burgers vector b. Expression for the strain becomes

$$\epsilon_\| = -\left(f_m + \frac{b_1}{p}\right), \tag{3.3}$$

where $b_1 = -b\sin\alpha\sin\beta$ and b is the Burgers vector. The value of b is 3.84 Å for silicon ~ 4.0 Å for compound semiconductors. Semiconductors. For 60° dislocations,

$$\alpha = \operatorname{archtan}\frac{1}{\sqrt{2}}, \quad \beta = \frac{\pi}{3}. \tag{3.4}$$

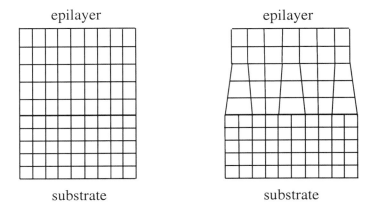

Figure 3.1: Structure of an epilayer. The layer is under biaxial compression. (a) The layer is pseudomorphic layer and (b) the layer is relaxed by creation of misfit dislocations.

Some authors have used $(f_m - b_1/p)$ instead of $(f_m + b_1/p)$ in Eq. (3.3). The +ve or the −ve sign depends on the sign convention used for the Burgers vector. Numerically the value of b_1/p is always subtracted from f_m in this equation. If the lattice constant of the layer is smaller than that of the substrate, f_m is negative, the strain is tensile and $\epsilon_\parallel$ is positive. According to convention tensile strain and stress are positive and compressive strain and stress are negative. Strain in superlattices [325] is discussed in a later section.

The normal stress σ_0 in a pseudomorphic layer grown on a (100) surface is given by

$$\sigma_0 = -2\mu \frac{\nu+1}{\nu-1} f_m = \frac{E}{1-\nu} \epsilon, \qquad (3.5)$$

where E and μ are the Young's and shear moduli of elasticity and ν is Poisson's ratio. Assuming $f_m = 0.02$, Young's modulus $E = 163100$ MPa and Poison's ratio $\nu = 0.28$ for the substrate and $E = 147200$ MPa and $\nu = 0.28$ for the alloy layer, σ_0 comes out to be − 4069 MPa.

3.1.2 Dislocation and strain energies: Periodic arrays of dislocations

The energy of a single isolated dislocation is given by (see e.g. [236, 235])

$$E_D^\infty = \frac{\mu b^2}{4\pi(1-\nu)} \left[(1 - \nu \cos^2 \beta) \ln \frac{\rho_c h}{q} \right]. \qquad (3.6)$$

The parameter ρ_c is introduced to account for the non-elastic part of the core energy of the dislocations[1] and q is the core radius of the dislocation line, usually

---
[1] Values of ρ_c are not known with any certainty. Different authors have used different values of this parameter [237].

3.1. STRAIN AND ENERGIES OF EPILAYERS

taken to be equal to b [3, 235].

We now consider two perpendicular arrays of interacting dislocations with a periodic distribution. Their strain field consists of an average homogeneous part and a fluctuating part with average equal to 0. The energy of the arrays contained in a unit area of the layer is given by [235],

$$E_D^{array} = Bh\left(\frac{b_1}{p}\right)^2 + \frac{2}{p}E_{DS}, \qquad (3.7)$$

where

$$B = 2\mu\frac{1+\nu}{1-\nu}. \qquad (3.8)$$

The first term in (3.7) is the energy associated with the average homogeneous strain and the second term is the energy due to the fluctuating part of the strain of dislocations in the arrays. E_{DS} is the energy per unit length of a dislocation line (due to the fluctuating strain). A part of the energy (due to average strain caused by the dislocation network) is included as $Bh(b_1/p)^2$ in Eq. (3.7). Therefore $E_{DS} < E_D^\infty$, the difference $E_D^\infty - E_{DS}$ is large for small values of p. The explicit expression for E_{DS} is given in Refs. [238, 3, 235].

The homogeneous strain energy consists of Bhf_m^2 due to misfit strain, the energy $Bh(b_1/p)^2$ due to average strain of the dislocation arrays and the product of these two terms due to interaction between the two strains. The expression for the total homogeneous strain energy E_H is,

$$E_H = Bh\left(f_m + \frac{b_1}{p}\right)^2. \qquad (3.9)$$

Total energy E_T of a partially relaxed layer is the sum of the energy E_H and $\frac{2}{p}E_{DS}$ [235],

$$E_T = E_H + \frac{2}{p}E_{DS}. \qquad (3.10)$$

For large p, $E_{DS} \approx E_D^\infty$ and E_T can be written as [235]

$$E_T^m = E_H + \frac{2}{p}E_D^\infty. \qquad (3.11)$$

In Fig. 3.2, calculated values of E_T/μ of the strained epilayers with $f_m = 0.0042$ (calculated for 60° interacting dislocations using Eq. (3.10)) are plotted as a function of strain relaxation $|b_1/p|$ for 7 different values of the epilayer thickness h [238]. Since elastic constants of III-V, Si and Ge are not very different, the values of energy shown in Fig. 3.2 are also valid for GeSi/Si, InGaAs/GaAs and other epilayers as long as $f_m = 0.0042$. The calculated value of h_c is 236 Å (critical thickness h_c is discussed in section 3.3) for this composition. Each curve shows a minimum, the minimum becoming stronger and moving to larger values of $|b_1/p|$ as h increases. The position of the minimum tends to move to f_m as h approaches ∞. These results show that for each thickness, there is a definite concentration $1/p_s$ of dislocations per unit length needed for the

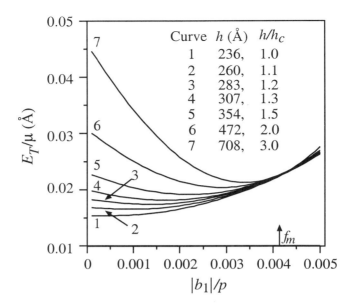

Figure 3.2: Total energy E_T/μ of the strained epilayers with $f_m = 0.0042$ is plotted as a function of strain relaxation $|b_1|/p$ for 7 different values of h [238].

epilayer to be stable (the corresponding strain relaxation is $|b_1/p_s|$). These minima in energy give pairs of values of h_e and $|b_1/p_s|$ for stable configurations. The values of $|b_1/p_s|$ at which minima occur for different values of h increase with h, first rapidly and then slowly. For any given thickness, the concentration $|b_1/p_s|$ of dislocations is smaller if interactions of dislocations are not properly taken into account and E_D^∞ instead of E_{DS} is used in the energy expressions. However the average concentration of dislocations decreases for a non-periodic distribution. The observed concentrations are always much smaller than the predicted values for a periodic distribution. The discrepancy arises partly due to the non-periodic distribution and partly due to the difficulty in nucleating the dislocations, see chapter 4.

If the layer under consideration has a thick cap-layer on the top, the above expressions for energies are not valid. If the cap is sufficiently thick, dislocation dipoles instead of dislocations are formed. A dipole consists of a pair of dislocations, one each at the upper and lower interfaces. Fig. 3.3 shows a schematic representation of a 60° array of dislocations in an uncapped layer and dipoles in a capped layer. Both dislocations of the dipole are of 60° type and α is the same as for the uncapped layers. The dislocations at the lower interface have a Burgers vector b and at the upper interface, a Burgers vector $-b$. The dipole spacing p is the spacing both in the upper and the lower array of dislocations (see Fig. 3.3). The angle θ is the angle between the line joining the two dislocations of a dipole and perpendicular to the interface.

For obtaining the total energy E_D^{pcap} of the arrays in a capped layer, E_{DS} in

3.1. STRAIN AND ENERGIES OF EPILAYERS

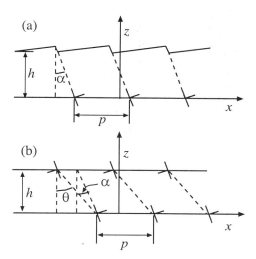

Figure 3.3: Schematic representation of (a) 60° dislocations in an uncapped layer and (b) 60° dipoles in a capped layer [235].

Eq. (3.10) is replaced by E_{DS}^{cap},

$$E_D^{pcap} = Bh\frac{b_1^2}{p^2} + \frac{2}{p}E_{DS}^{cap}. \qquad (3.12)$$

The total energy of the capped layer is given by,

$$E_T^{cap} = Bh\left(f_m + \frac{b_1}{p}\right)^2 + \frac{2}{p}E_{DS}^{cap}. \qquad (3.13)$$

The expression for E_{DS}^{cap} is given in Refs. [235, 239] for interacting as well as non-interacting dipoles. The energy of an array of dipoles in a capped layer is considerably larger than the energy of the array of dislocations in an uncapped layer. If interactions between the dipoles or between dislocations are neglected, the energy of a dipole in the capped layer is approximately twice as large as the energy of a dislocation in an uncapped layer.

3.1.3 Non-periodic arrays of dislocations

The energy of an array of dislocations is lowest if the distribution is periodic. However dislocations generally nucleate at heterogeneous sources (defects, impurities, ledges etc.) and therefore they are clustered. We therefore consider the energy of the arrays with irregular distribution of dislocations. The energy of interaction $E_I(h,p)$ (per unit length of the dislocation lines) of a pair of dislocations is given by [240],

$$E_I(h,p) = Ab_1^2\left[\ln(4(h/p)^2 + 1) + \frac{4(h/p)^2\left(4(h/p)^2 + 3\right)}{\left(4(h/p)^2 + 1\right)^2}\right]$$

$$+Ab_2^2 \left[\ln(4(h/p)^2+1) - \frac{4(h/p)^2\left(12(h/p)^2+1\right)}{\left(4(h/p)^2+1\right)^2}\right]$$

$$+Ab_3^2 \left[\ln(4(h/p)^2+1)(1-\nu)\right]. \tag{3.14}$$

Let the number of dislocations be $N+1$ with p_i as the inter-dislocation spacing between the ith and $(i+1)^{th}$ dislocations. The total interaction energy of the array is given by [240, 241],

$$E_I^{irr}(N,h,\{p\}) = \frac{2}{N\bar{p}}\left[\sum_{i=1}^{N} E_I(h,p_i) + \sum_{i=1}^{N-1} E_I(h, p_i+p_{i+1})\right.$$

$$+ \sum_{i=1}^{N-2} E_I(h, p_i+p_{i+1}+p_{i+2}) + \sum_{i=1}^{N-3} E_I(h, p_i+p_{i+1}+p_{i+2}+p_{i+3})$$

$$\left.+\cdots\cdots\cdots + \sum_{i=1}^{N-(N-1)} E_I(h, p_i+p_{i+1}\cdots+p_{i+(N-1)})\right]. \tag{3.15}$$

To obtain the energy of the two perpendicular arrays per unit area of the layer we must add to the interaction energy given above (1) the self-energy $(2/\bar{p})E_D^\infty$ ($\bar{p}$ is the average inter-dislocation distance in the array) of the dislocations and (2) the energy of interaction $E_{I\perp}^{\bar{p}}$ between the two perpendicular arrays. The energy of interaction $E_{I\perp}^{\bar{p}}$ is given by [240],

$$E_{I\perp}^p = \frac{2\mu h}{\bar{p}}\left[\frac{\nu b_1^2}{1-\nu} - \frac{b_3^2}{2}\right]. \tag{3.16}$$

It has been assumed in writing the above equation that this energy depends only on the average spacing $\bar{p}$ and not on the details of the distribution of spacings, p_i. Fig. 3.4 shows the numerical values of the energy of two orthogonal arrays of periodic and non-periodic interacting dislocations. We can see from this figure that interactions between dislocations farther than the next-near neighbors are important. It is not a good approximation to replace non-periodic arrays with average spacing $\bar{p}$ by periodic arrays with spacing $p = \bar{p}$. This is more clearly demonstrated by curves 3 and 4.

The total energy of an epilayer containing non-periodic arrays is obtained by adding together all the contributions discussed above. The energy values calculated in this manner for an epilayer with $f_m = 0.0042$ are plotted in Fig. 3.5. Curve 1 is for the periodic arrays. Curves 2 and 3 are for Gaussian distributions and curve 4 is for a uniform-random distribution. For a given misfit the number of dislocations is smaller if the distributions is irregular. The number is not sensitive to the details of distribution but it does depend on the standard deviation. The number decreases as the standard deviation σ increases.

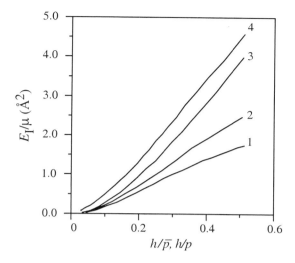

Figure 3.4: Interaction energy of two orthogonal periodic and non-periodic arrays of 90° dislocations. Curves 1, 2 and 3 are for the periodic arrays with p as inter-dislocation spacing. Curve 1 includes interactions between nearest neighbors only, curve 2 includes interactions between nearest and next nearest neighbors and curve 3 includes all interactions. Curve 4 is for 41 non-periodic dislocations. A Gaussian distribution of non-periodic dislocations with spacing $\bar{p} = p$ and standard deviation $\sigma = 0.44p$ is assumed for the non-periodic distribution [242].

3.2 Processes involved in dislocation generation

3.2.1 Propagation of dislocations

To understand the mechanism of strain relaxation by introduction of misfit dislocations, it is necessary to understand how dislocations nucleate, propagate and multiply. The energy for homogeneous nucleation is very large. Without the assistance of defects the dislocations can not nucleate even at high temperatures. The layers free from dislocations can be grown even if their thicknesses are larger than the critical thickness. Such layers are metastable. When the metastable layers are annealed for longer times at higher temperatures the misfit dislocations are introduced and the strain relaxes. Creation of misfit dislocations by Matthews and Blakeslee, by multiplication or by nucleation involves glide motion of dislocations. Fig. 3.6 shows how misfit dislocations are deposited by the nucleation and by propagation of threading dislocations. When dislocations come close together, motion of dislocation may be blocked by interaction with other neighboring dislocations. Propagation of dislocations is discussed in this section 3.2.1 and nucleation, multiplication and blocking in section 3.2.2. Using the results of these sections, annealing and strain relaxation are discussed later.

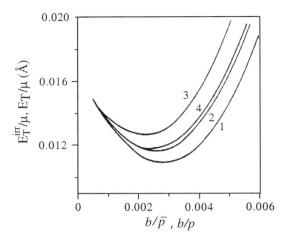

Figure 3.5: Plot of normalized total energy E_T^{irr}/μ vs $h/\overline{p}$ (non-periodic distribution) and E_T/μ vs h/p (periodic distribution). Curve 1 is for a periodic array, curve 2 is for a Gaussian distribution, and curve 4 is for a uniform-random distribution, both with the same standard deviation $\sigma = 0.44\overline{p}$. Curve 3 is for a Gaussian distribution with $\sigma = 0.7\overline{p}$. Thickness $h = 254$ Å and $f_m = 0.0042$ [241].

3.2.1.1 Excess stress

The dislocation velocity v_d can be represented by an equation of the type [243, 244],

$$v_d = v_0(\sigma_{exc})^m \exp(-E_v/kT), \qquad (3.17)$$

where v_0 is a constant, σ_{exc} is the excess stress and E_v is the energy of activation for the glide motion of the dislocation. Values of m equal to 1 [244, 245] and 2 [246, 247] in Eq. (3.17) have been used. The excess stress can be written as,

$$\sigma_{exc} = \sigma_e - \sigma_t. \qquad (3.18)$$

σ_e is the misfit stress and σ_t is the self stress associated with the line tension of the dislocation. The values of these stresses are well known [3]. Substituting these values one obtains the expression for the excess stress,

$$\sigma_{exc} = 2S\mu\epsilon\frac{1+\nu}{1-\nu} - \frac{\mu b\cos\alpha(1-\nu\cos^2\beta)}{4\pi h(1-\nu)}\ln\frac{\rho_c h}{b}. \qquad (3.19)$$

Here S is the Schmid factor and has a value $1/\sqrt{6}$. The self-energy of the dislocation line in a capped layer (i.e. a dipole) increases by a factor ~ 2 [235] and therefore σ_t is replaced by $2\sigma_t$ in (3.19). σ_{exc} and the velocity of the dislocation are reduced in the capped layers. Effect of dislocation interactions are neglected in the above expression for the excess stress. When the dislocations are clustered the spatial distribution of stress and dislocation velocities become

3.2. PROCESSES INVOLVED IN DISLOCATION GENERATION

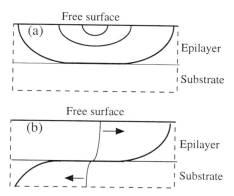

Figure 3.6: Generation of misfit dislocations (a) by nucleation and expansion of a surface half-loop and (b) by motion of an existing threading dislocations.

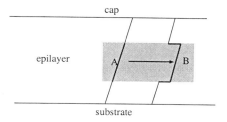

Figure 3.7: A small portion of the dislocation A moves to right to form a double kink B

nonuniform. The observed dislocation velocities in different regions of the same sample differ by a factor up to 3 [245]. Experiments discussed in section 4.2 also show that strain in the partially relaxed layers is highly nonuniform.

3.2.1.2 The kink model of dislocation propagation

Consider the motion of a dislocation line from A to B shown in Fig. 3.7. It can be easily shown that the energy for the whole dislocation line to glide in one step is excessively large and no glide motion by this mechanism is possible at ordinary temperatures. The double kink model provides a mechanism which avoids this difficulty [236, 245]. To start with, a small portion A of the line jumps forward (to the right in Fig. 3.7) forming a double kink as shown at B. The double kink is formed if the layer is capped or if the layer thickness is sufficiently large. In the case of a thin layer with free surface, a single kink is formed. Subsequently the two component of the double kink move apart along the dislocation line. In the case of a single kink, the kink moves in a manner such that the length of the dislocation line in the displaced position increases. A larger and larger portion of the dislocation line moves forward to the right and ultimately whole dislocation moves by one atomic spacing. If the dislocation line is sufficiently long, several kinks can be formed at the same time. The process is repeated and the line moves large distances.

Let $2F'_k$ be the energy required to nucleate an isolated double kink[2]. The average distance X_k between the double kinks is given by [245],

$$X_k = 2c \exp\left(\frac{F'_k}{kT}\right), \qquad (3.20)$$

where c is the kink jump distance (≈ 3.8 Å). The rate of nucleation J_k of the kinks is given by [245],

$$J_k = \frac{\nu_D b d_p \sigma_{exc}}{kT} \exp\left(-\frac{E_m + 2F'_k}{kT}\right), \qquad (3.21)$$

where d_p is the distance between Peierls valleys and has a value 3.3 Å, ν_D is the Debye frequency and E_m is the energy of activation for kink jumps along the line direction. The transverse dislocation velocity v_d is given by [245],

$$v_d = \frac{2\nu_D c b d_p^2 \sigma_{exc}}{kT} \exp\left(-\frac{E_m + F'_k}{kT}\right) \frac{L_d}{L_d + X_k}. \qquad (3.22)$$

Here L_d is the length of the dislocation line. If $L_d \gg X_k$, Eq. (3.22) reduces to

$$v_d \approx v_\infty = \frac{2\nu_D c b d_p^2 \sigma_{exc}}{kT} \exp\left(-\frac{E_m + F'_k}{kT}\right). \qquad (3.23)$$

In this case the velocity is independent of the length of the dislocation line. Eq. (3.23) is similar in form to Eq. (3.17). On the other hand if $L_d \ll X_k$, Eq. (3.22) becomes

$$v_d(L_d) = \frac{\nu_D b d_p^2 \sigma_{exc} L_d}{kT} \exp\left(-\frac{E_m + 2F'_k}{kT}\right). \qquad (3.24)$$

Now the velocity is proportional to the length of the dislocation line. This predicted behavior has been confirmed experimentally in bulk Ge and Si [248, and references there in]. The observed dislocation mobility increases with its length L_d up to a certain length L_0 and becomes independent of L_d for $L_d > L_0$. Measured values of L_0 are 0.2 to 0.4 μm in Si and 1.5 μm in Ge.

Louchet and co-workers [248] have modified the double kink model. They used statistical analysis of the double kink formation and derived an expression for the velocity of dislocations which is somewhat different. However the kink model is used more extensively to interpret the experiments.

Single kinks are formed only up to a certain depth s_0^* below the surface. The distance s_0^* is given by [236]

$$s_0^* = \left(\frac{\mu(1+\nu)bd_p}{8\sigma_a \pi(1-\nu)}\right)^{1/2}, \qquad (3.25)$$

---

[2]The prime on F indicates that it is modified by kink–kink interactions and by the work done by the applied stress if the stress is very high [236, 245].

3.2. PROCESSES INVOLVED IN DISLOCATION GENERATION

and in this model the glide velocity v_{skd} of the dislocations becomes,

$$v_{skd} = \frac{1}{4} J_{sk} s_0^* d_p, \qquad (3.26)$$

where J_{sk} is the single kink nucleation rate given by

$$J_k = \frac{\nu_D b d_p \sigma_{exc}}{kT} \exp\left(-\frac{E_m + F_k'}{kT}\right). \qquad (3.27)$$

This expression is similar to the nucleation rate of double kinks except that the energy of activation for the formation of the kink is now F_k' instead of $2F_k'$.

The rate of nucleation of double kinks increases with thickness and beyond a certain thickness, motion of dislocations by formation of double kinks dominates even for the uncapped layer. The thickness at which the transition from single kink to double kink nucleation takes place depends on the strain ϵ in the epilayer. It is ~ 1 μ for $\epsilon = 0.2\%$ and ~ 200 Å for $\epsilon = 1\%$ [235].

3.2.1.3 Measurements of dislocation velocity

Extensive measurements of dislocation velocity have been made on GeSi strained layers. The work on the compound semiconductors is not so extensive. Dislocation velocities in GaN have not been measured probably because they are too small and measurements are difficult. Estimated dislocation velocities in III-Nitrides are available and are discussed later in this chapter. We first discuss the work on GeSi layers because the work is very helpful in giving insight in to the processes that occur during the motion of dislocations.

Dislocation velocities measured by Tuppen and Gibbings [244] in capped GeSi layers are shown in Fig. 3.8. Calculated values are shown both for the large thickness limit and for the actual thickness of the layer used. Experimental values agree well with the calculated values. Curve (c) in Fig. 3.8 shows that the dislocations did not penetrate the cap i.e. the cap was sufficiently thick. The observed velocity v_∞ in $Ge_x Si_{1-x}$ strained layers for $L_d > L_0$ can be represented by [244],

$$v_\infty = B_0 \sigma_{exc} \exp\left(\frac{-(2.156 - 0.7x)}{kT}\right). \qquad (3.28)$$

Here $B_0 = 1.15 \times 10^{-3}$s m^{2}kg^{-1} and kT is in eV. Hull et al. [245] measured the glide velocity of a thin (1000 Å) uncapped $Ge_{0.15}Si0.85$ layer. The experimental results agreed with the single kink model as expected.

The dislocation velocities in $In_x Ga_{1-x} As$ have also been studied. 200 Å thick capped $In_x Ga_{1-x} As$ layers grown by MBE at 450°C have been investigated [249]. The dislocation velocity was ~ 1 μm s^{-1} at 450°C and tens of μm s^{-1} at 600°C. Capped layers of $In_x Ga_{1-x} As$ with different concentrations of In have been studied by Paine et al. [250]. They found that the dislocation velocity was 1.1 μm s^{-1} at 800°C. The concentration of Indium in this sample is not known. It was suggested that the velocity was low because of the interaction with other existing dislocations.

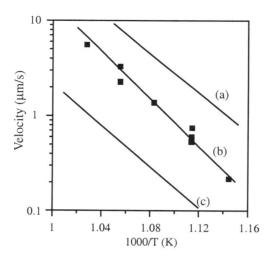

Figure 3.8: Dislocation velocity versus $1/kT$ for a 0.06 μm thick Ge$_{0.171}$Si$_{0.829}$ layer with 0.38 μm capping layer. Line (a) shows the values of V_∞, line (b) is the value corrected for small length of the dislocation line and solid square symbols show the experimental data. Line (c) represents the calculated values of the velocity if the dislocation is assumed to penetrate the capping layer up to the free surface [244].

Considerable work on the dislocation velocities in Si, GaAs and InP has been done. Sumino and Yonenaga [251] have published a review of this work. Measured velocities of dislocations in these semiconductors are plotted in Fig. 3.9. The velocity in III-V semiconductors is much larger than in Si. In GaAs the velocity of α dislocations is larger than that of β dislocations.

The bonding in the II-VI semiconductors is partly ionic. The metallic ions have some positive charge [252], the exact quantity of the charge is not known. Moving dislocations in these materials carry a large electric charge because they sweep the electrons from the point defects. The application of an electric field changes the flow stress. The materials also show a large photoplastic effect i.e. the change of flow stress on illumination or electron irradiation. A review of the work on the properties of the dislocations in electron- and photon-irradiated II-VI semiconductors is given in Refs. [252, 253]. The activation energy is small and it has the same value for the α, β and screw dislocations, a results strikingly different from that for other semiconductors. Dislocation velocity in unirradiated II-VI semiconductors has not been studied extensively.

Recent work of Sugiura et al. [254, 255] shows that in GaN the activation energy for propagation of dislocations is very large and in this material dislocations do not expand, move or multiply. Sugiura et al. [255] compiled the activation energy E_v for Si, Ge and several III-V compounds. The plots of E_v versus bandgap E_g are shown in Fig. 3.10. This figure shows that there is good linear relation between the activation energies and the bandgaps. Assuming

3.2. PROCESSES INVOLVED IN DISLOCATION GENERATION

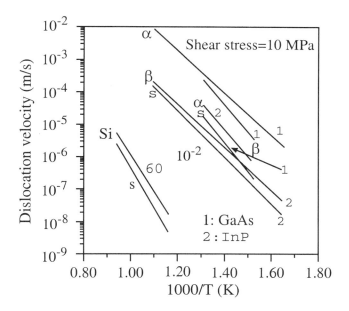

Figure 3.9: Comparison of dislocation velocities under a shear stress of 10 MPa in undoped silicon, GaAs and InP plotted against temperature. Symbols s, α, and β show screw, α, and β dislocations; 60 indicates 60° dislocations [251].

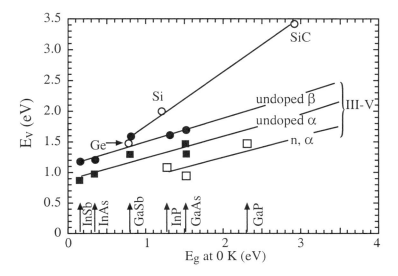

Figure 3.10: The activation energy E_v of dislocation glide motion as a function of the bandgap energy E_g at 0 K [255].

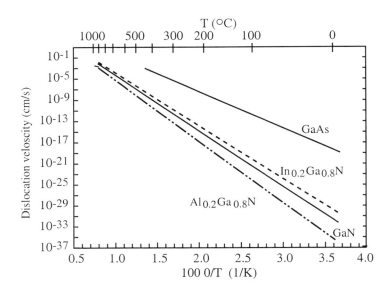

Figure 3.11: Comparison of dislocation velocity of α dislocations in GaAs and in GaN based semiconductors under the applied shear stress of 1 MN/m^2. Numerically mobility and velocity are equal under this applied stress. The figure is based on Fig. 5 of Ref. [255].

that the relation represented by the line for undoped α dislocations in the III-V semiconductors also holds for the III-Nitrides also, they calculated the values 2.1 eV, 2.0 eV and 2.3 eV of E_v for GaN, In$_{0.2}$Ga$_{0.8}$N, and Al$_{0.2}$Ga$_{0.8}$N respectively. Sugiura et al. [255] found that the pre-exponential factor v_0 in Eq. (3.17) were not very sensitive to the semiconductor type and the III-V semiconductors can be described approximately by one value, 1×10^6 cm/s. Using a value 1 MN/m^2 for the excess stress, the velocity calculated using Eq. (3.17) with $m = 1$ are plotted in Fig. 3.11. At room temperature the mobility of the III-nitrides is smaller by a factor 10^{-10} to 10^{-16}!

3.2.2 Nucleation, multiplication and blocking

3.2.2.1 Nucleation

Homogeneous nucleation of dislocation loops at the surface of semiconductor strained layers has been studied by many authors [256, 257, 258, extensive lists of references are given in these paper]. The total energy of the loop can be written as [2, 258],

$$E_{total} = E_{loop} - E_{strain} \pm E_{step}. \tag{3.29}$$

in this equation E_{loop} is the self-energy of the semicircular loop of radius R, E_{strain} is the reduction of the homogeneous strain energy due to strain relaxation and E_{step} is the energy of the surface step created or removed by the loop.

3.2. PROCESSES INVOLVED IN DISLOCATION GENERATION

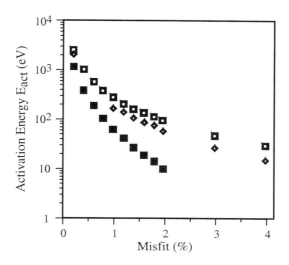

Figure 3.12: Energies E_{act} of a semicircular surface loop in a strained semiconductor layer are plotted as a function of misfit f_m. Filled squares: $\rho_c = 8/e^2 \approx 1$, filled diamonds: $\rho_c = 4$ and open squares: $\rho_c = 4$ but neglecting the surface energy term [258].

Expressions for these energies given in the literature [2, 257, 258] are somewhat different but they give practically the same numerical values. The energy of the loop increases with its radius R. It increases from 0 for $R = 0$ to a maximum value which is the energy of activation E_{act} for the nucleation, at $R = R_c$, the critical radius of the loop. For $R > R_c$, loop energy decreases and therefore it grows until it touches the interface. Now its threading arms move apart depositing the misfit dislocation at the interface (see Fig. 3.6). The rate of nucleation is generally assumed to be proportional to $\exp(-E_{act}/kT)$. The expression for E_{act} is [2]:

$$E_{act} = \frac{\mu b^2 R_c (2-\nu)}{8(1-\nu)} \left[\left\{ 1 + \ln\left(\frac{\rho_c R_c}{b}\right) \right\} - \frac{8\pi R_c \epsilon (1+\nu)}{3b(2-\nu)} + \frac{16\sigma_s}{\mu b} \frac{1-\nu}{2-\nu} \right], \quad (3.30)$$

where

$$R_c = \frac{3b(2-\nu)}{16\pi(1+\nu)} \left[\left\{ 2 + \ln\left(\frac{\rho_c R_c}{b}\right) \right\} + \frac{16\sigma_s}{\mu b} \frac{1-\nu}{2-\nu} \right]. \quad (3.31)$$

Here ρ_c is the core parameter already introduced in section 3.1 and σ_s is the surface tension of the layer. The value of σ_s commonly used is $\mu b/8$. The calculated values of E_{act} are shown in Fig. 3.12 [258]. Fig. 3.12 shows that (1) E_{act} decreases with increase in f_m and surface energy affects on E_{act} significantly. The core parameter ρ_c has a very large effect on the values E_{act}. Generally $\rho_c = 1$ is used for metals and $\rho_c = 4$, for covalent semiconductors

[236, 258]. For $f_m = 0.02$, $E_{act} \sim 10$ eV for $\rho_c = 8/e^2 \approx 1$ and > 100 eV for $\rho_c = 4$. If $\rho_c = 4$, the activation energy is so large that homogeneous nucleation is unlikely to occur. In most cases observed values of activation energies are much lower due to heterogeneous nucleation [237]. Interfacial structural defects, oxide or carbon contamination [259] and β-SiC particles [235] act as efficient sources of nucleation. Copper contamination of the GeSi strained layers even in trace amounts induces the dislocation nucleation and increases strain relaxation rate considerably [260]. The rate of strain relaxation in GeSi strained layers is enhanced if defects were introduced by ion implantation [261]. Jain et al. [237] have reviewed the work on homogeneous and heterogeneous nucleation in strained epilayers. An extensive list of references on the subject is given in the review. The defects as small as 1.5 nm in size make the nucleation possible at a mismatch of 1% [237]. 'Intrinsic' sources for the efficient generation of dislocations have also been observed. The source, designated as the diamond defect (because its shape resembles diamond) that can generate dislocations at low misfits in GeSi strained layers was observed by Eaglesham et al. [258].

Creation of dislocations involves both nucleation and propagation. It is difficult to extract the energy of both processes separately from the observations. Hull et al. [262] assumed a constant rate of nucleation for interpreting their annealing experiments. Since the activation is a function of strain Gosling et al. [256] assumed that the activation energy varies as $1/|\epsilon|$ to interpret the same experiments. The nucleation rate was assumed to be [256]:

$$J_{nuc} = J_0 e^{-\lambda/kT|\epsilon|}. \tag{3.32}$$

Here J_0 is a constant pre-exponential frequency factor per unit area and $\lambda/|\epsilon|$ is the activation energy.

3.2.2.2 Multiplication

The experimental evidence for the multiplication of dislocations is not clear cut. Hull et al. [257, 261, 262] found no evidence of multiplication in their experiments. While interpreting their results they neglected multiplication altogether. However, results of other authors provide evidence of the importance of dislocation multiplication [243, 244]. Relative role of multiplication in annealing and stress relaxation experiments depends on the experimental conditions and sample history.

Tuppen et al. [244] studied relaxation of strain and creation of dislocations in GeSi strained layers. Initially large number of dislocations produced on closely spaced {111} planes were observed. The local stress in these regions became small and therefore sources became ineffective. Further increase of concentration of dislocations occurred by multiplication. As the annealing continued, relaxation spread across the whole sample. Hagen-Strunk mechanism for the production of dislocations has been discussed in [235, 258]. This mechanism produces dislocations when two existing dislocations meet each other at right angles. Multiplication of dislocations by this mechanism has been observed in

3.2. PROCESSES INVOLVED IN DISLOCATION GENERATION

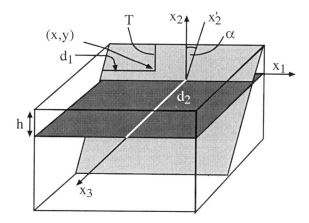

Figure 3.13: Schematic illustration of the motion of a threading dislocation T through a strained layer and deposition of the misfit dislocation d_1. The co-ordinate system used is shown. Also shown is a perpendicular misfit dislocation d_2 (shown in white) along the x_3-axis, with which the threading dislocation interacts [256].

both GeSi strained layers [258] and InGaAs strained layers [235, and references given therein].

The rate of increase of the number of dislocations by multiplication, J_{mult}, has been discussed by several authors. Dodson and Tsao [243] gave a phenomenological models of the multiplication rate. Gosling et al. [256] reviewed several multiplication mechanisms and showed that in general the multiplication rate can be written in the form

$$J_{mult} = \frac{1}{4} N_i J_0 e^{-\lambda/kT|\epsilon|} = \frac{1}{4} \left(\frac{\epsilon + f_m}{b_{eff}} \right)^2 J_0 e^{-\lambda'/kT|\epsilon|}, \qquad (3.33)$$

where $\lambda'/|\epsilon|$ represents an activation energy, and J_0 is a frequency factor per unit area [256].

3.2.2.3 Blocking

We now discuss the rate at which number of mobile dislocations decrease by blocking. The discussion follows the treatment of Gosling et al. [256] as summarized in Ref. [235]. Consider a misfit dislocation d_1 being deposited by a moving threading dislocation T in a strained layer as illustrated in Fig. 3.13. The free surface of the layer is at $x_2 = h$ and the substrate-layer interface is at $x_2 = 0$. The glide plane of the dislocation is inclined at an angle α to the x_2-axis. The coordinates of the bottom of the threading dislocation are $(x_1, x_2) = (x, y)$. It is depositing a misfit dislocation at a height y above the interface. A misfit dislocation shown running along the x_3-axis interacts and impedes the motion of the threading dislocation if the two dislocations have right kind of Burgers

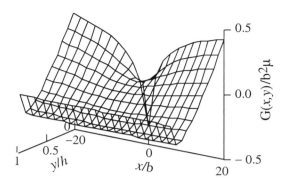

Figure 3.14: Surface plot of the driving force $G(x, y)$ on the threading dislocation T in a 110 Å thick strained layer (lattice-mismatched by 2% relative to its substrate) as it approaches a perpendicular misfit dislocation d_2, with which it interacts. The Burgers vectors of the two dislocations have been chosen so that the interaction is repulsive.

vectors [256]. There are four pairs of strain relieving Burgers vectors, only one of which causes significant blocking of the moving dislocation. For a given interaction, the probability that dislocation interaction can impede the motion of the threading dislocation is 1/4.

Let the Burgers vector of the perpendicular misfit dislocation be a, defined relative to a line direction chosen positive in the direction of the positive x_3-axis. The local net force $G(x, 0)$ acting on the threading dislocation, including the interaction with the orthogonal misfit dislocation, is given by the following expression to a very good approximation [256],

$$G(x, y) = -E^s(h - y) + \mu b \left(\frac{1+\nu}{1-\nu}\right) |\epsilon| (h - y)$$
$$+ \frac{2.5\mu b^2}{2\pi(1-\nu)} \left\{ \frac{1+\nu}{4} \ln \sqrt{\frac{x^2 + y^2}{x^2 + h^2}} \right.$$
$$\left. + \frac{1 - 3\nu}{2\sqrt{2}} \left[\tan^{-1}\left(\frac{x}{y}\right) - \tan^{-1}\left(\frac{x}{h}\right)\right] \right\}, \qquad (3.34)$$

Here $E^s(h-y)$ is the energy E_D^∞ defined in equation (3.6), with $h-y$ substituted for h. The term involving $2.5\mu b^2$ is the force due to the interaction with the perpendicular misfit dislocation. The force $G(x, y)$ calculated for $f_m = 0.02$ and thickness of the layer $h = 110$ Å is shown in Fig. 3.14. It is seen from the figure that there is a saddle-point in the plot. The saddle-point force $G^\star$ is a minimum with respect to x and a maximum with respect to y. In the case considered, $G^\star > 0$ and the threading dislocation will cross over depositing a misfit location at a height near the saddle point. If the $G^\star < 0$, blocking will occur provided Burgers vectors of the two dislocations are of the right type. If other dislocations are present in the layer, we replace the mismatch strain $-f_m$

in all equations by the mean strain in the layer, ϵ, given by (3.3).

The saddle-point values of x and y are given by [256]

$$y = \frac{(1-3\nu)\sqrt{2}x^2 - (1+\nu)xh}{(1-3\nu)\sqrt{2}h + (1+\nu)x}, \tag{3.35}$$

$$[b_e^2 + (1-\nu)b_s^2]\frac{(1-3\nu)\sqrt{2}h + (1+\nu)x}{(1-3\nu)\sqrt{2}(h^2-x^2) + 2(1+\nu)xh}$$

$$-4\pi b(1+\nu)|\epsilon| - 2.5b^2\frac{[(1-3\nu)\sqrt{2}h + (1+\nu)x]h}{2(x^2+h^2)x} = 0. \tag{3.36}$$

The smallest negative solution of (3.36) corresponds to the saddle point; this solution lies in the region of $x^\star = -2b$. The corresponding value of $y^\star$ is obtained from (3.35). Using these values of saddle-point coordinates, the sign of $G^\star$ is determined from (3.34). If $G^\star > 0$ then no blocking occurs for the given combination of ϵ and h. If $G^\star < 0$ then blocking will occur if the pair of Burgers vectors (**a**,**b**) are of right type.

The number of dislocations blocked per unit time, J_{block} is given by [256]:

$$J_{block} = -\frac{(\epsilon + f_m)N_m v_d}{b_{eff}}\frac{H(-G^\star(\epsilon,h))}{4}, \tag{3.37}$$

where the Heaviside step function

$$H(x) = \begin{cases} 1 & \text{if } x > 0 \\ 0 & \text{otherwise.} \end{cases}$$

If the propagating dislocation interacts with a closely bunched cluster of N misfit dislocations with identical Burgers vectors, the interaction force is multiplied by a factor N [256].

3.3 Critical thickness

3.3.1 Theory of critical thickness

We first discuss the layers with free surfaces. The principle of energy minimization states that at the critical thickness, the total energy of the layer must be minimum. Below the critical thickness, the energy of the epilayer should increase by the introduction of the misfit dislocations and beyond the critical thickness, the energy should decrease. By equating to zero the first derivative of the total energy E_T given by (3.11) with respect to $1/p$ and letting $p \to \infty$ in Eq. (3.11), we obtain the following expression for h_c:

$$h_c = \frac{b^2(1-\nu\cos^2\beta)}{8\pi f_m(1+\nu)b_1}\ln\frac{\rho_c h_c}{q}. \tag{3.38}$$

The error caused in the value of h_c by using Eq. (3.11) instead of Eq. (3.10) is small and can be neglected. However for calculation of strain relaxation in thick layer, the error caused by using Eq. (3.11) can be large because in this case p is not large [3].

Cohen-Solal et al. [263] have proposed a new model for calculating the critical thickness of strained layers. According to this model the critical thickness is the thickness at which the homogeneous strain energy and the energy of the totally relaxed layer are equal. The energies were calculated using Keating's valence force field approximation. They found that the critical layer thickness can be written as:

$$h_c = A^* f_m^{-3/2}, \qquad (3.39)$$

where A^* is an adjustable parameter. At $h = h_c$ there is a considerable strain in the layer and the arguments leading to this equation are not correct. However by adjusting the parameter A^*, fit of Eq. (3.39) (with $A^* = 0.45$ Å) with experiments on II-VI semiconductor epilayers (shown later by dotted line in Fig. 3.19) is reasonable.

The critical thickness of the capped layers can be calculated using an equation similar to Eq. (3.38) provided we minimize the total energy of a capped layer given by Eq. (3.13) instead of the energy of a layer with free surface.

3.3.2 Strain and critical thickness of superlattices

For the purpose of illustration consider a CdSe/CdTe superlattice. In this superlattices, each of the two layers consists of $CdTe_xSe_{1-x}$ alloy with two different values of the composition x. Let the lattice constants of the two layers (relaxed, without strain) of a period be a_1 and a_2 and the thicknesses, d_1 and d_2. The thin layers of the superlattice acquire an in-plane lattice constant equal to that of the substrate, designated as $a_\parallel$. The strains in the two layers are given by

$$\epsilon_1 = \frac{a_\parallel - a_1}{a_\parallel}. \qquad (3.40)$$

$$\epsilon_2 = \frac{a_\parallel - a_2}{a_\parallel}. \qquad (3.41)$$

In order for layers to remain pseudomorphic, their structure must satisfy two conditions. (1) Each layer of the superlattice must have a thickness less than its own critical thickness, given by

$$h_1, \ h_2 < h_{c1}(\epsilon_1), \ h_{c2}(\epsilon_2). \qquad (3.42)$$

The critical thicknesses h_{c1} and h_{c2} of the layers are somewhat larger than those given by Eq. (3.38) because these layers are sandwiched between two other layers [3]. (2) The total thickness of the superlattice as a whole (i.e., the sum of thicknesses of all the periods) must be less than the critical thickness h_c^{SL}, which is equal to the critical thickness of the alloy with the same average composition as that of the superlattice [3]. The critical thickness h_c^{SL} of the

3.3. CRITICAL THICKNESS

superlattice as a whole can now be calculated using Eq. (3.38) with the misfit parameter $f_m = f_m^{SL}$ defined by [3]

$$f_m^{SL} = \frac{a_s - a_{av}}{a_{av}} = 0.042 x_{av}, \qquad (3.43)$$

with x_{av} given by

$$x_{av} = \frac{x_1 d_1 + x_2 d_2}{d_1 + d_2}. \qquad (3.44)$$

The result of Eq. (3.43) implies that for the purpose of calculation of stability, the superlattice can be regarded as one single layer of $CdTe_x Se_{1-x}$ alloy with a value x_{av} equal to the average value given by Eq. (3.44).

3.3.3 Symmetrically strained superlattices

We consider a free-standing superlattice. Because a thick substrate is not used to force the in-plane lattice constant of the thin layers to acquire the value of the substrate, both layers of the superlattice will be strained to have a common in-plane lattice constant. The homogeneous energy stored in the two layers is proportional to $\epsilon_1^2 d_1$ and $\epsilon_2^2 d_2$, respectively. The condition of minimum energy in a period consisting of these two layers therefore requires [3]

$$\epsilon_1 d_1 = -\epsilon_2 d_2. \qquad (3.45)$$

The weighted average lattice constant of the two layers of a period of the superlattice is given by:

$$a_{av} = \frac{a_1 d_1 + a_2 d_2}{d_1 + d_2}. \qquad (3.46)$$

It can be easily shown that Eq. 3.45 leads to $a_\parallel = a_{av}$. If we now use a substrate with a lattice constant a_{av}, the in-plane lattice constant continues to be equal to a_{av} and the condition of minimum energy is satisfied. Eq. 3.43 shows that the misfit parameter of the superlattice as a whole is zero and therefore the second critical thickness defined earlier is infinitely large. However the thickness of each layer must be less than its critical layer thickness for the superlattice to be stable. The superlattice fabricated on a thick buffer layer whose lattice constant is equal to a_{av} satisfies this condition. The superlattice is known as a symmetrically strained superlattice or a free standing superlattice. If the superlattice is very thick, the top part of the superlattice can become symmetrically strained [264]. It is possible that the layers near the interface can partly relax by the generation of misfit dislocations and the thick superlattice can become symmetrically strained [265].

3.3.4 Experimental values of critical thickness

The observed values of h_c for several lattice mismatched III-V semiconductor layers are shown in Fig. 3.15. The scatter of the experimental points is large. The discrepancy in the calculated and the observed values is also very large,

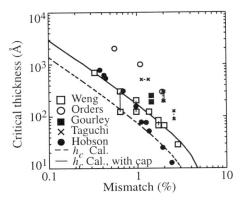

Figure 3.15: Critical thickness of III-V semiconductor lattice mismatched layers. Dashed and solid lines give the calculated values of h_c for the layers with free surface and for the capped layers respectively. The experimental data are taken from Hobson [266], Weng [267], Orders [268], Gurley [269] and Taguchi [270].

particularly for small values of mismatch. The values of h_c of Ge_xSi_{1-x}/Si strained layers determined experimentally by Bean *et al.* in 1984 [11] are shown in Fig. 3.16. It is seen that the observed critical thicknesses of GeSi strained layers are also much larger than the calculated values. The experimental data can be described by the following equation [11],

$$h_c \text{ (in Å)} = \left(\frac{1.9 \times 10^{-2}}{f_m^2}\right) \ln\left(\frac{h_c \text{ (in Å)}}{4}\right). \tag{3.47}$$

Curve 3 in Fig. 3.16 shows the plot of this equation. If we plot the data from other workers, the scatter in the points becomes very large. The discrepancy arises partly because the sensitivity of the x-ray and several other measuring techniques is small [10]. When h just exceeds h_c, strain relaxation is much smaller and is not detected by the insensitive techniques [10]. The other part of the discrepancy is due to the difficulty in nucleating misfit dislocations at relatively low growth temperatures [10]. Houghton *et al.* [271] used x-ray topography and Nomarski microscopy of defect etched surfaces to monitor the onset of misfit dislocations in Ge_xSi_{1-x} strained layers. These techniques are so sensitive that one dislocation in an area of 1 cm^2 can be detected. They found very good agreement between the theory and the experiment. In the case of $In_xGa_{1-x}As$ also, experimental values obtained using sensitive techniques such as photoluminescence spectroscopy and Hall measurements agree closely with the theoretical values [10, 235]. Experimental data for $In_xGa_{1-x}As$ quantum wells (capped layers) obtained using the sensitive techniques [272] are shown in Fig. 3.17. The solid line curve shows the predicted values using an equation derived by Matthews and Blakeslee [7, see Eq. (5) on p 124]. This equation was derived for a specific superlattice in which strain in the $In_xGa_{1-x}As$ was half of the misfit parameter. The equation therefore overestimates h_c. If correct

3.3. CRITICAL THICKNESS

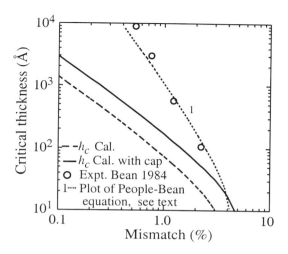

Figure 3.16: Critical thickness of Ge_xSi_{1-x} epilayers grown on Si substrate. Dashed line gives the calculated values of h_c for the layers with free surfaces, solid line is for the capped layers and dotted curve is the plot of Eq. (3.47). Experimental data are taken from [11].

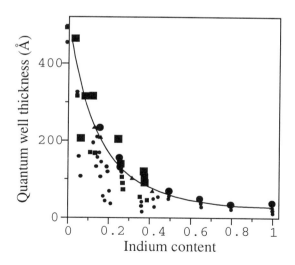

Figure 3.17: Thickness versus In content in $In_xGa_{1-x}As$ strained layers. Small circles and squares represent high quality layers with no strain relaxing defects. Triangles represent quantum wells with small number of defects and large symbols represent quantum wells with large concentration of defects. Calculated values are shown by the continuous curve (see text) [272].

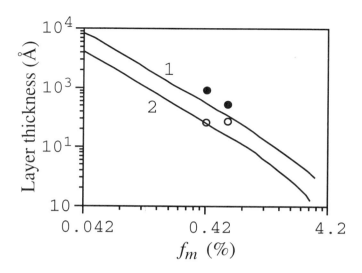

Figure 3.18: Critical thickness h_c (60° dislocations) of a capped layer (curve 1) and an uncapped semiconductor layer (curve 2) plotted as a function of misfit parameter f_m. The calculated values are valid for GeSi/Si and for InGaAs/GaAs layers. Experimental data [273] for the uncapped and capped Ge_xSi_{1-x} layer are shown by the open and the filled circles [239].

values of h_c for the capped layer are used the agreement between theory and the experimental values becomes better.

Careful measurements of the critical thickness of capped Ge_xSi_{1-x} layers have been made by Houghton et al. [273] and their results are shown in 3.18. The results for the uncapped Ge_xSi_{1-x} layers are also included in the figure. The calculated values of h_c are shown for both the capped and the uncapped layers. The critical thickness of a capped layer is approximately twice as large as that of the corresponding uncapped layer for low mismatch and is four times as large for large mismatch. For a given misfit the hollow circles represent the maximum thickness for which the layer was found to be dislocation-free and the solid circles represent the smallest thickness for which partial relaxation of strain by the introduction of the dislocations was detected [273]. Thus the hollow and solid circles provide experimental upper and lower bounds for the critical thickness. The theoretical curve for the capped layers falls within the bounds placed by experiment. The agreement between the calculated and observed values is quite satisfactory.

The value of h_c of ZnSe/GaAs has been determined by many workers [123, 122, 274]. Yokogawa et al. [123] measured the lattice constants of ZnSe layers grown on GaAs substrates by MOVPE. The calculated critical thickness is 21 nm as compared to the observed value of 150 nm. Other authors [122, 274] have also observed $h_c = 150$ nm for ZnSe layers grown on GaAs. Recently Horsburgh et al. [275] have measure the critical thickness of ZnSe grown on

3.3. CRITICAL THICKNESS

Table 3.1: Observed values of critical thickness. The -ve sign indicates that f_m is negative, the lattice constants of the epilayer is smaller and the in-plane strain $\epsilon_\parallel$ is positive, see Eqs. (3.1) and (3.2) [125].

	Sample, $f_m(\%)$	Method	h_c nm	Ref.
1	ZnSe/GaAs,	TO	110	[275]
2	ZnSe/GaAs 0.27	x, T	150	[122]
3	ZnSe/ZnS[1], 5	x	5	[123]
4	ZnTe/GaAs, 7.5	R	1.2	[149]
5	CdSe/ZnSe[1], 6.8	x, PL	0.91 (3ML)	[147]
6	ZnSe/CdSe[1]– 6.3	x, PL	0.28 (1ML)	[147]
7	CdTe/Cd$_{0.97}$Zn$_{0.03}$Te, 0.18	Ch	390	[277]
8	CdTe/ZnTe, 6	RO, PL, T	1.6 (5 ML)	[278]
9	CdTe/Cd$_{0.96}$Zn$_{0.04}$Te, 0.23	x	513[2]	[279]
10	CdTe/InSb, 0.04	x	2950[2]	[279]
11	CdTe/Cd$_{1-x}$Zn$_x$Te		see text	[263]
12	ZnTe/CdTe, –5.8	R	2.1(7ML)[3]	[280]

[1]Symmetrical SLS. [2]Values in (111) orientation were somewhat smaller.
[3]Relaxed buffer.
TO: Topography, x: x-ray diffraction, Ch: Channeling, R: RHEED, RO: RHEED Oscillations, and T: TEM.

GaAs substrate measured by synchrotron based x-ray topography. The value of the critical thickness was 100 nm, significantly lower than the value of 150 nm reported by earlier workers. On the other extreme the measured critical thickness of ZnSe on ZnS (mismatch 5%) buffer is 5 nm [123].

We have compiled experimental values of h_c of II-VI semiconductor and alloy epilayers in Table 3.1 [125].

The experimental data from Table 3.1 are plotted in Fig. 3.19. Using Burgers vector $b = 4.008$ Å and $\nu = 0.376$ we have calculated h_c for a layer with free surface using Eq. (3.38) and plotted it by the lowest dashed curve in Fig. 3.19. We have also calculated h_c for the capped layers, shown by the solid curve. Fig. 3.19 shows that the experimental values of h_c are larger than the calculated values in this case also. The discrepancy is very large for small values of f_m e.g. for the ZnSe layers. The uppermost dotted line is a plot of Eq. (3.39). This equation agrees with the experimental results for $A^* = 0.45$ Å.

Considering that the experimental values of h_c in Table 3.1 are for many different II-VI semiconductor and alloy layers and that the experiments were performed in different laboratories using different techniques of measurements, the scatter in the experimental values is small. The scatter in the values of h_c for InGaAs/GaAs and GeSi/Si layers discussed earlier is much larger. The main reason for the different behavior of II-VI semiconductors is the fact that the II-VI semiconductors have weak bond strength as shown earlier in Fig. 1.4 [138, 281]. The dislocation energy in these semiconductors is much smaller than

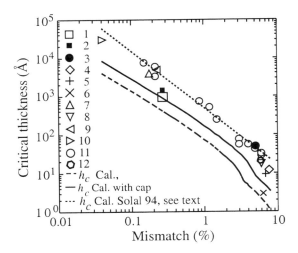

Figure 3.19: Critical thickness of II-VI semiconductor lattice mismatched layers. Dashed and solid lines give the calculated values of h_c for the layers with free surface and for the capped layers respectively. The lowest dashed line is a plot of Eq. (3.39) for $A^* = 0.45$ Å. The description of the samples for symbols 1 to 12 is given in Table 3.1 by the items 1 to 12 [125].

that in GeSi or III-V semiconductors [138, 281]. Therefore dislocations are more easily produced in these semiconductors.

Critical thickness of $In_xGa_{1-x}N$ and $Al_xGa_{1-x}N$ layers grown on relaxed GaN layers have been measured recently by Akasaki and Amano [81]. For $0.05 \leq x \leq 0.2$, the values were in the range 300 to 700 nm for $Al_xGa_{1-x}N$/GaN and ~ 40 nm for $In_xGa_{1-x}N$/GaN. Within the experimental errors the values were independent of the composition x. Critical thickness of III-Nitrides determined by islanding is given in the next subsection.

3.3.5 Critical thickness determined by islanding

For the growth of GeSi/Si and InGaAs/GaAs layers, if the mismatch between the epilayers and the substrate is more than 2 to 3%, the growth is in the Stranski-Krasnov (SK) mode. In this mode of growth, after the growth of first few monolayers (2 to 7, depending upon the mismatch) which grow layer by layer in the 2D mode, 3D islandic growth begins. In the initial stages, the relaxation occurs not by the introduction of the misfit dislocations but by the formation of islands. The 3D islandic growth provides another mechanism of strain relaxation in the lattice mismatched layers. Osten et al. [283] have shown that the critical thickness of highly strained semiconductor layers for relaxation of strain by generation of islands is smaller than by the generation of misfit dislocations. Islanding in InGaAs epilayers has been reviewed by us in Ref. [3]. Here we discuss recent results of experiments on III-Nitrides layers. Grandjean and

3.3. CRITICAL THICKNESS

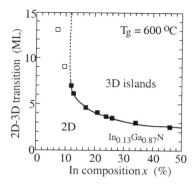

Figure 3.20: Critical thickness (filled squares) associated with 2D to 3D change in growth mode of highly strained $In_xGa_{1-x}N$ (for $x > 0.1$). Open squares correspond to the roughening of the growth front without change to real 3D growth mode [152].

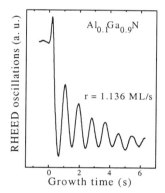

Figure 3.21: RHEED intensity oscillations recorded during AlGaN growth [29]. The value of growth rate r is given in the figure.

Massies [152] studied RHEED oscillations in MBE InGaN epilayers grown on 2 μm thick GaN layer/sapphire for several values of In concentrations. Grandjean and Massies [152] determined the critical thickness of $In_xGa_{1-x}N$ epilayers at which the growth of islands begins and the oscillations disappear for several values of x. The thickness at which the islanding begins is plotted in Fig. 3.20.

The critical thickness is 3 ML for $x = 0.3$ and 7 ML for $x = 0.12$. Below $x = 0.12$, islanding was not observed, only surface roughening was seen. The thickness at which roughening begins is shown by open squares in Fig. 3.20.

Using RHEED oscillations and lattice parameters of GaN on AlN and AlN on GaN have been measured recently [162, 29]. Epilayers of GaN on thick relaxed AlN buffers and of AlN on thick relaxed GaN buffers were grown on (0001) sapphire. The growth temperature was 600°C for GaN epilayers whereas it was 800°C the AlN layers. The change in the lattice parameter of GaN

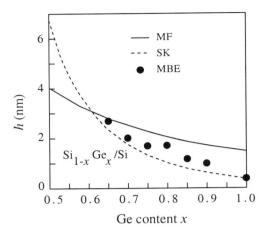

Figure 3.22: Observed and calculated critical thickness for SK growth of GeSi strained layers. Calculated critical thickness for breakdown of coherency by misfit dislocations is also shown by the solid curve MF [283].

layer due to relaxation of strain begins after the growth of two monolayers. The change in lattice parameter of AlN layer due to strain relaxation was also observed after 3 monolayers, a value similar to that for GaN grown on AlN. However during the growth of GaN, 5 RHEED oscillations were observed. After 5 MLs when the oscillations decay, the RHEED pattern changes from streaky to spotty, indicating the onset of a 3D islandic growth. 15 RHEED oscillations were observed and during the growth of AlN. The critical thickness in both cases is 2 to 3 monolayers. This value is smaller than that determined earlier by other experiments [162]. The value of critical thickness obtained by different workers are different. The critical thickness is 10 Å for AlN/GaN determined in Ref. [284] and 30 Å GaN on AlN determined in Refs. [285, 99].

In a more recent paper Grandjean et al. [29] have utilized the RHEED oscillations to determine the AlN mole fraction in the AlGaN epilayers. The authors found that if standardized reproducible growth conditions are used, the growth rate depends uniquely on the concentration of Al in the epilayers. The growth rates can be easily determined using RHEED oscillations. The RHEED oscillations for a growth rate of 1.136 ML/s are shown in Fig. 3.21. This growth rate corresponds to 10% Al concentration. The Al concentrations determined using RHEED oscillations agreed with those measured by the energy dispersive x-ray method.

Results of Osten et al. [283] on Ge_xSi_{1-x}/Si layers along with the theoretically predicted values of h_c for islanding are shown in Fig. 3.22. Critical thickness for SK growth was calculated using the following approximate expression for the critical thickness [283]:

$$h_c^{SK} = \frac{\text{constant}}{f_m^4}. \qquad (3.48)$$

3.3. CRITICAL THICKNESS

The constant in Eq. (3.48) was determined by using the observed critical thickness of 3 mono-layers for Ge layers grown on Si i.e. for $f_m = 0.04$. Fig. 3.22 shows that for $f_m > 0.02$ the critical thickness for generation of islands becomes smaller than that for misfit dislocation generation.

Chapter 4

Strain relaxation and defects

4.1 Strain in GeSi layers

We mentioned in chapter 3 that as grown thick strained semiconductor layers are generally metastable because sufficient number of dislocations are not generated at the growth temperature. On annealing at higher temperatures and/or for longer times, dislocations are produced and strain relaxes. Strain has been measured in pseudomorphic as well as in partially relaxed layers. Kinetics of strain relaxation have been studied in GeSi-, InGaAs- and II-VI-strained layers. Measurements of strain in pseudomorphic and partially relaxed layers and interpretation of the experimental results are discussed for several cases in this chapter. Strain relaxation has been measured using a very large number of techniques. Many different semiconductor heterostructures have been investigated. Additional measurements of strain are discussed along with the optical properties in chapter 5.

Early studies of strain relaxation in GeSi layers were made by Bean *et al.* [11] in 1984. Dodson and Tsao [243] formulated a phenomenological theory and fitted it with the data of Bean *et al.* [11]. Jain *et al.*[238] fitted the Dodson-Tsao theory with the same experimental data [11] taking into account the dislocation interactions. A good account of this work is given in the book by Jain [3]. The same group (Hull *et al.* [262]) made more extensive studies of strain relaxation on 350 Å $Ge_{0.25}Si_{0.75}/Si(100)$ strained layers in 1989. The layers were grown by MBE at 550°C at a rate of 3 Å s^{-1} and then successively annealed for 240 s at six temperatures in the range 550°C and 850°C. The threading dislocation concentration and average misfit dislocation spacing were measured. These results provided impetus for further theoretical work on strain relaxation. An improved theory of strain relaxation was given by Gosling *et al.* [256]. As the strain relaxes and number of dislocations increases, the total length L of the dislocations also increases. The total length L of misfit dislocations is given by the following differential equation [256],

$$\frac{dL}{dt} = N_m v_d. \tag{4.1}$$

where N_m is the number of mobile threading dislocations. In writing this equation it is assumed that the numbers of threading and misfit dislocations are equal. This is true when the threading dislocations are the substrate dislocations as shown in Fig. 3.6(b). If the dominant process is nucleation of surface loops as shown in Fig. 3.6(a), the number of mobile threading dislocations is twice as large as that of misfit dislocations and there will be a factor 2 on the right hand side of Eq. (4.1). The rate of change of the number of mobile threading dislocations is given by [256],

$$\frac{dN_m}{dt} = -J_{block} + J_{nuc} + J_{mult}. \tag{4.2}$$

Gosling et al. [256] showed that strain relaxation can be written as:

$$\frac{d\epsilon}{dt} = -\frac{b_1 N_m v_d}{2A}. \tag{4.3}$$

Expressions for v_d, J_{block}, J_{nuc} and J_{mult} are given in chapter 3. Eqs. (4.2) and (4.3) provide a pair of coupled first order differential equations that may be solved to calculate strain relaxation or the number of dislocations as a function of annealing time and temperature [256]. Gosling et al. [256] used the above equations to interpret annealing experiments of Hull et al. [262]. Gosling et al. [256] found that the experimental results can be reproduced only if it is assumed that blocking occurs from the very beginning and if misfit dislocations are present in clusters of two to three with identical Burgers vectors. Nucleation of dislocations occurs at fixed heterogeneous sources which generate several dislocations very close together [2, 235, 259]. Assuming these dislocations originate from the same source, they will have identical Burgers vectors [256]. Hull et al. [262] have mentioned that they there was no evidence for multiplication events, and therefore J_{mult} was neglected. J_0 and λ were used as adjustable parameters. Taking $\epsilon = -f_m = 0.105$ and the observed density of threading dislocations as 10^3 cm^{-2} at time $t = 0$, (4.2) and (4.3) were integrated [256] numerically. Gosling et al. [256] obtained a reasonable fit with experiment with $J_0/A = 1.3 \times 10^{10}$m^{-2}s^{-1} and $\lambda = 0.0032$ eV. This value of λ corresponds to a nucleation activation energy of 0.3 eV for $f_m = 1\%$. The activation energy is low and suggests that nucleation was heterogeneous. There is agreement between theory and experiment within a factor two or better over almost the whole temperature range. In view of the fact that the distribution of dislocations is highly irregular, that not all observations are made on the same sample and that there are inherent uncertainties associated with the measurements, this agreement is regarded as satisfactory.

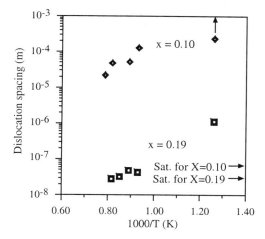

Figure 4.1: The average dislocation spacing versus $1/T$ where T is the temperature at which layers were annealed. The observed spacing immediately after growth at 480°C (marked by vertical arrow) is shown on the extreme right. Horizontal arrows indicate the saturation spacing for pure edge dislocations. For 60° dislocations this spacing will be reduced by a factor $1/2$ [250].

4.2 Strain in III-V semiconductor layers

Strain relaxation measurements in In_xGa_{1-x} strained layers have been made by Paine et al. [250]. The layers, capped, 236.4 nm thick and with 4 values of x, (= 0.05, 0.1, 0.19 and 0.22) were grown at 520°C. Two of the layers with $x = 0.1$ and $x = 0.19$ were annealed for 120 s at each of the 5 temperatures, 800, 850, 900, 950 and 1000°C respectively. After each anneal the spacing between the dislocations was measured by TEM. The measured values of spacing are shown in Fig. 4.1. Theoretical analysis of these results to extract values of nucleation activation energy and dislocation velocity has not yet been undertaken. Bonar et al. [249] measured the dislocation velocity and dislocation structure in capped In_xGa_{1-x} strained layers. For $x > 0.4$ pure edge dislocations with $b = (a/2)\langle 101\rangle$ gliding on $\{101\}$ plane were observed. High stresses due to the large In concentration and the large value of Schmid factor for the edge dislocations were probably responsible for this new glide system.

GaAs, GaP and InP epilayers have been grown directly on Si by several workers and extensive studies of their quality have been made (see the review [52]). These layers are deposited by the *two step* method described for the growth of GaN films in section 2.5. A thickness 30 nm to 0.2 μm buffer layer is first grown at a low temperature, in the range 300 to 450°C. The second layer, several microns thick, is grown on the buffer layer at higher temperatures (~ 600°C) generally used in homoepitaxy. Lattice mismatch is approximately 4% for GaAs/Si, 0.4% for GaP/Si, and 8% for InP/Si (see Table D in Appendix A). Initially the III-V layers are under compressive stress because their lattice constants are bigger than the lattice constant of Si. As discussed in section 2.5, the compressive stress is almost completely relaxed by generation of misfit-dislocations during the growth and/or annealing of the second layer at the higher temperature, provided the thickness of the layers is sufficiently large. On cooling the layers to room temperature, stresses and strains are generated due to difference in the values of coefficients of thermal expansion α_{th} of the layer and the Si substrate. The values of α_{th} vary with temperature [52]. High temperature values of α_{th} of Si and the III-V semiconductors are given in Table H in the Appendix. Table H shows that values of α_{th} for the III-V semiconductors are larger than those for Si. Therefore the layers lattice matched to Si at a high temperature, experience a tensile thermal stress after they are cooled to room or lower temperatures. If the thickness of the layer is not sufficiently large, relaxation of stress at the growth temperature of the "second layer" is not complete and the compressive stress persists on cooling the layer to lower temperatures. The thickness of the layer at which transition from compression to tension takes place depends on the III-V semiconductor being grown and on the growth and annealing temperatures. For GaAs and InP it varies from 0.1 to 1 μm [52] and for GaP it is about 1 μm.

If the misfit strain is completely removed at the growth and/or annealing temperature T_g, and no plastic deformation takes place during cooling, the thermal misfit $f_{(m,th)}$ at room temperature (RT) is given by [52]

$$f_{(m,th)} = (\alpha_{th,\text{III-V}} - \alpha_{th,\text{Si}})(T_g - RT). \tag{4.4}$$

All equations derived for misfit strain can be used for thermal strain if $f_{(m,th)}$ for thermal mismatch is used instead of f_m for lattice mismatch. If plastic relaxation takes place up to a temperature T_f during cooling, T_g should be replaced by T_f in Eq. (4.4). The experimental values of thermal strain are lower than the values calculated using this equation [52].

The stress in the semiconductor layers can be determined using the observed stress induced changes in the PL and Raman spectra. The PL peak shifts and splits in to two components, designated as A and B, due to tetragonal strain see chapter (5). Similarly the stress induces a shift in the Raman frequency. The shifts (and the splitting of PL peak) can be used to determine the stress.

From room temperature X-ray measurements, σ_0 in GaAs/Si grown at $\sim$ 600°C is 1.2 kbar [286]. The shift in peak position gave a stress equal to 1.8 kbar. At low temperature the splitting between the A and the B peaks gave values of the order of 3 kbar for the stress. The value of stress measured at higher temperature is lower because of the smaller value of difference in the range of temperature through which the sample is cooled. Strain value measured by Sugo et al. [287] by the PL method is 1.3×10^{-3} for GaAs layers grown on Si at 700°C. The value measured by X-ray diffraction is 1.6×10^{-3} and by wafer curvature method, 1.7×10^{-3}. If it is assumed that the misfit strain was completely relaxed at the high growth/annealing temperature, this value should have been 2.2×10^{-3} [287].

Wuu et al. [288] measured the strain in InP/Si layers grown at 610°C using PL as well as X-ray methods. The transition from compressive to tensile strain

4.2. STRAIN IN III-V SEMICONDUCTOR LAYERS

Table 4.1: Experimental values of tensile strain $\epsilon \times 10^3$ at room temperature (RT) and low temperatures (LT) for GaAs, InP and GaP epilayers grown on Si determined by wafer curvature (curve), luminescence (L) and X-ray methods. The growth temperatures were between 600 and 700°C.

Method	GaAs	InP	GaP	T	Ref.
curve	1.7	0.4	0.7	RT	[287]
X-ray	1.6	0.4	0.9	RT	[287]
L	1.3	0.2	- -	RT	[287]
L	1.9	- -	- -	LT	[286]
L	1.4	- -	- -	RT	[286]
X-ray	1.2	- -	- -	RT	[286]
L	1.9	- -	- -	LT	[291]
L	- -	1.1	- -	LT	[288]
L	- -	1.1	- -	LT	[289]
L	- -	0.7	- -	LT	[290]
Eq. (4.4)	2.2	1.1	2.0	RT	[287]

took place at a thickness of 1 μm. Tensile strain measured by low temperature (2 to 77 K) PL spectra was about $\epsilon = 1.1 \times 10^{-3}$. X-ray measurements gave smaller values but the difference was not large. Values determined by Wehmann et al. [289] are also of the same magnitude. Grundmann et al. [290] measured the strain in the InP layers grown on Si at 640°C by the PL and by X-ray diffraction methods. The change from compressive to tensile strain took place at a thickness of about 0.3 μm and the strain saturated at about 0.7×10^{-3} at a thickness of > 1 μm. The values determined by Sugo et al. [287] in layers grown at similar temperatures are: $\epsilon = 0.4 \times 10^{-3}$ by the wafer curvature and X-ray methods and $\epsilon = 0.2 \times 10^{-3}$ by the PL method. The value of tensile strain in GaP layers grown on Si at 750°C was also measured by Sugo et al. [287]. The transition from compressive to tensile strain took place at 1 μm thickness. The values of strain were $\epsilon = 0.7 \times 10^{-3}$, and $\epsilon = 0.9 \times 10^{-3}$ determined by the wafer curvature and by the X-ray methods respectively. The experimental values of tensile strain in the III-V semiconductor layers grown on Si are given in Table 4.1. This table shows that the values of thermal strain measured by different authors are different. The values measured for the same sample but by different techniques are also different. Measured values are smaller than the values predicted by Eq. (4.4). The discrepancy can arise due to several reasons, e.g. sufficient number of dislocations might not have been generated to remove completely the mismatch strain at the growth/annealing temperature, or some of the thermal stress might be relieved by plastic relaxation during the cooling process [52]. The results depend on the method of measurement presumably because of different depth resolutions with different techniques. Difficulties can also arise because (i) there may be micro cracks present in the samples [291], and (ii) the number of dislocations may be different in different directions because of the difference in the nature of α and β dislocations [52]. To summarize, we can say that the measured strain in all III-V semiconductor layers grown on Si is tensile as expected but the scatter in the measured values is large and its magnitude is generally lower than that predicted by the thermal expansion theory.

Sacedón et al. [292] made Raman measurements of $In_xGa_{1-x}As$ layers of several thicknesses grown on (111)B GaAs. Raman frequency in unstrained GaAs layers grown on GaAs is 292 cm^{-1}. For a 175 nm thick layer and for $x = 0.22$, the result depended on the location of the laser spot on the layer. Measurements made in an area where the dislocation concentration (as determined by transmission electron microscopy) was very low showed the LO peak at 290 cm^{-1} i.e. shifted to lower frequencies by two wave numbers. At another location with large dislocation concentration, the LO peak was shifted to lower frequency by about 3 cm^{-1}. Sacedón et al. [292] estimated that the strain in the low dislocation density area was 0.017 and in the high dislocation density area, it was 0.012. This shows that the strain relaxation is highly non-uniform, there is small relaxation in some parts of the layer and up to 30% relaxation in the other parts.

4.3 Strain in II-VI layers

4.3.1 Relaxation of strain in low mismatched layers

4.3.1.1 Relaxation in CdZnSe and in the initial stages in ZnSe layers

Horsburgh et al. [275] investigated the mechanism of strain relaxation in ZnSe layers in the initial stages i.e. when the layer thickness is in the neighborhood of the critical thickness. The layers were grown on GaAs substrates by MBE. The strain was measured using synchrotron-based x-ray topography. The thicknesses of the layers were determined by x-ray rocking curve simulation and also by cross-sectional TEM. The mechanism by which dislocations are created in ZnSe was found to be different from that which is operative in InGaAs. In InGaAs the misfit dislocations are formed by the Matthews-Blakeslee mechanism [2]. The existing threading dislocations bend to form the misfit dislocations. Horsburgh et al. [275] found that this is not the dominating mechanism for ZnSe. There are other mechanisms which can generate threading and misfit dislocations. Stacking faults in ZnSe epilayers can dissociate and generate misfit dislocations. Frank partials bounding the stacking faults can also dissociate. This results in a Shockley partial and a threading dislocations. The threading dislocation is converted to a misfit dislocation by Matthews-Blakeslee mechanism.

Horsburgh et al. [275] studied the onset of misfit dislocations in ZnSe epilayers by increasing the thickness in the range 30 nm to 170 nm using the synchrotron-based x-ray topography. This technique can detect a single dislocation in a sample of macroscopic dimensions. No dislocation was detected in the layers with thicknesses below 95 nm. The first dislocation was seen at 100 nm thickness. The authors concluded that the critical thickness of ZnSe grown on GaAs is 97.5 nm. This value was quoted in chapter 3.

4.3. STRAIN IN II-VI LAYERS

We now discuss the strain relaxation in CdZnSe quantum wells. These quantum wells are widely used in ZnSe based blue-green laser diodes. Horsburgh *et al.* [275] determined thickness of ZnCdSe quantum wells beyond which the strain will to relax as a function of the Cd content. The thickness was determined by extrapolating the ZnSe data. Curves (1) and (2) in Fig. 4.2 show critical thickness for the CdZnSe wells. Curve (1) is based on the assumption that the critical

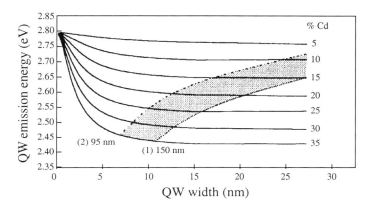

Figure 4.2: Variation of CdZnSe QW emission energy with well width (see text) [275].

thickness of ZnSe is 150 nm and Curve (2) on the assumption that the critical thickness is 95 nm. The QWs with thicknesses lying on the left of the shaded area are stable. If the thickness lies on the right of the shaded area, the wells will relax with the generation of dislocations. For good confinement the well width must lie in the shaded area [275]. Since critical thickness of ZnSe determined by Horsburgh *et al.* [275] is close to 95 nm, the wells with thicknesses lying in the shaded area will also be unstable.

4.3.1.2 Relaxation with increase of layer thickness

Extensive experimental work has been done on the relaxation of strain in the II-VI layers [125]. Recently Pinardi *et al.* [125] have compared these results with the existing theories. We discuss these results in this section.

Fontaine *et al.* [279] have studied the strain relaxation in thick uncapped CdTe epilayers grown on (111) $Cd_{0.96}Zn_{0.04}Te$ substrates. Lattice mismatch between CdTe and $Cd_{0.96}Zn_{0.04}Te$ is 0.23%. Measured values of stress as a function of thickness h are shown in Fig. 4.3. We have calculated [125] relaxation of strain in thick layers using the equilibrium theory[1] discussed in chapter 3. Curve 1 was calculated using the old theory based on Eq. (3.11). Curve 2 was obtained using Eq. (3.10) which takes into account the interactions properly [235]. The discrepancy between the theoretical curves and the experimental

---
[1]There was a numerical error in the calculations reported in [125]. Corrected values are given in Figs. 4.3 and 4.4. [125]

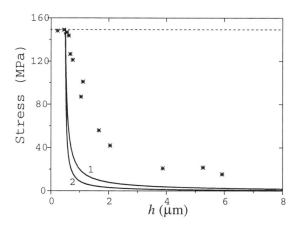

Figure 4.3: Plots of in-plane stress σ versus layer thickness h are shown for CdTe layers grown on a (111) $Cd_{0.96}Zn_{0.04}$Te substrate. Symbols are the experimental data taken from [279]. Curve 1 is calculated using Eq. (3.11) and curve 2 is calculated using Eq. (3.10) for the total energy of the layer, see text.

data is large. This result is similar to that obtained with GeSi and III-V semiconductors.

Lu et al. [293] have grown ZnSe layers on (InGa)P buffers deposited on GaAs substrates by gas source MBE. The buffer with thickness < 1 μm and containing 50 to 55% indium were pseudomorphic with GaAs substrate. ZnSe layers were grown on the pseudomorphic buffer layers. The lattice mismatch between ZnSe and the pseudomorphic buffer is 0.27%. At $T < 350°C$ the ZnSe layers grew in the 2D mode. X-ray diffraction studies showed that for thicker layers the strain relaxation occurred. The strain relaxation was nearly complete in a 1.9 μm ZnSe layer. More detailed studies of strain relaxation in ZnSe layers grown on (100) GaAs substrate (mismatch 0.27%) by MBE have been made by Petruzzello et al. [122]. Though thin layers were pseudomorphic and there was complete registry of the lattice across the interface, the surface of the layers was wavy. There were no misfit dislocations in layers with $h = 50$ nm and 87 nm. However stacking faults were observed even in these thin layers. For $h > 150$ nm, strain relaxation by the introduction of misfit dislocations was observed. The measured values of residual strain as a function of thickness h are shown in Fig. 4.4. Yao et al. [274] have measured lattice constant of ZnSe grown on (100) GaAs substrate. We [125] have determined the values of strain by using the observed values of the lattice constants and plotted them by open circles in Fig. 4.4. These values agree well with the experimental results of Petruzzello et al. [122]. The calculated curves are also shown. The calculated curves have been moved down by 0.4×10^{-3} to take into account the thermal strain (due to difference in the thermal coefficients of ZnSe and GaAs). The shift of 0.4×10^{-3} was quoted by Petruzzello et al. [122] based on their observations.

4.3. STRAIN IN II-VI LAYERS

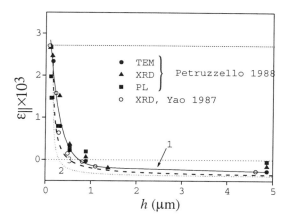

Figure 4.4: Plots of in-plane strain $\epsilon_\parallel$ in ZnSe layers grown on GaAs. Curve 1 is drawn through the experimental points shown by symbols (solid symbols are from [122] and open circles are from [274]) as an aid to eye; Curve 2 is calculated using Eq. (3.10) and curve 3 is calculated using Eq. (3.11) for the total energy, see text.

The calculated curves have been moved to the right by 0.11 μm to take into account the difference between the calculated critical thickness (40 nm) and the observed critical thickness (150 nm). The shift to the right implies that below the experimental critical thickness it is difficult for the misfit dislocations to nucleate. Once the critical thickness is reached and the dislocations set in the dislocations act as sources for generation of new dislocations. The agreement between the theory and the experiment is not so bad in this case. However the agreement is based on the shifts of the calculated curves we have used. We have used similar shifts for CdTe, InGaAs and GeSi layers but the large discrepancy between the theory and the experiment persisted. We note that ZnSe/GaAs (100) is the only known case where a reasonable agreement between theory and experiments exists. The discrepancy between the measured and calculated strain relaxation in the thick GeSi/Si and InGaAs/GaAs epilayers is very large [235].

4.3.1.3 Strain relaxation by electron irradiation

Chao et al. [32] investigated the relaxation of strain in a II-VI laser diode as a result of electron beam bombardment. The structure CdZnSe/ZnSSe/MgZnSSe of the diode was a separate confinement heterostructure. The thickness of the CdZnSe was 30 Å and Cd content was 20%. For damaging the sample a 30 kV electron beam with 160 nA current was used. CL spectra were mapped by scanning the sample at a lower beam current of 60 nA. A dark spot defect (DSD) at the spot where the beam irradiated the sample was seen. Dark line defects grew and spread along the $\langle 100 \rangle$ direction. The CL spectra were measured at

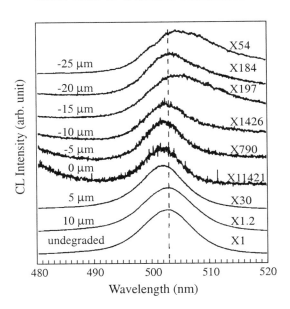

Figure 4.5: Room temperature CL spectra taken from various locations along the ⟨100⟩ direction. The multiplying factor 1421 indicates that the reduction in the luminescence due to degradation is by a factor 1421 [32].

several points by scanning the sample. The spectra taken at various point along the ⟨100⟩ direction is shown in is shown in Fig. 4.5. The spectrum from the DSD is marked 0 μm. As compared to the intensity from an unirradiated area the intensity at this point is decreased by a factor 1421. This spectrum is blue shifted due to the radiation damage. The curves above this spectrum gradually return to the unirradiated position i.e. they show increasing red shifts. The top curve is red shifted with respect to the lowest spectrum from the unirradiated area. The observed blue and red shifts were used to elucidate the mechanism of relaxation of strain and creation radiation damage. Three radiation induced changes can give rise to blue or red shifts of the CL spectra. Originally the CdZnSe quantum well is under a compressive strain due to lattice mismatch with the cladding layers. The lattice mismatch strain is estimated as 1.7%. The effect of strain on the bandgap is discussed in chapter 5. Using the theory given in section 5.1, the increase in the bandgap energy due to the compressive strain of 1.7% is estimated to be $\sim$ 30 meV [32]. The relaxation of strain due to radiation damage will reduce the bandgap and will cause a red shift of the CL spectrum. For a red shift of 5 meV, a 0.3% reduction of strain is required i.e. the strain should change from 1.7% to 1.4%. The blue shift is caused by the out diffusion of Cd and/or by intermixing at the interfaces. A change of Cd concentration from 20% to 19.6% will cause a blue shift of 1 nm. There is one more factor which can cause a blue shift. The capture cross section for the non-radiative recombination depends on the kinetic energy of the electrons. The

4.3. STRAIN IN II-VI LAYERS

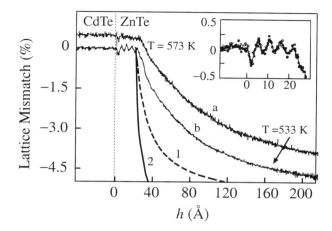

Figure 4.6: Mismatch due to relaxation in ZnTe layers grown on a CdTe relaxed buffer. Curves a and b: experimental results from [280], Curves 1 and 2: Strain relaxation calculated using Eqs. (3.11) and (3.10) for the total energy.

cross section is larger for the l0wer energy carriers. This results in a relatively large population of carriers with higher energy for radiative recombination. The CL spectrum blue shifts and becomes narrow. The peak position and halfwidth of the CL spectrum from different locations depends on all the three mechanisms. Chao et al. [32] found that it was difficult to determine the fraction of the shift at a location separately due to each of the three mechanisms. Direct determination of the structural and chemical changes in the damaged regions are required for a clear understanding of the processes that lead to the changes in the CL spectra and creation of the DLD defects.

4.3.2 Relaxation of strain in highly mismatched layers

4.3.2.1 Strain relaxation in the early stages of growth

The strain relaxation in the early stages of MBE growth of ZnTe on CdTe relaxed buffers (the mismatch is 5.8%) has been investigated by Kret et al. [280]. The buffers were deposited on (001) GaAs substrate. Two growth temperatures, 533 K and 573 K were used. The growth rate was 1.4 ML/s. The epilayer remained lattice matched to the buffer during the first 5 seconds. This gives an experimental critical thickness $h_c = 21$ Å. Fig. 4.6 shows the observed and calculated strain relaxation in the ZnTe layers for $h > h_c$. The calculated curves have been moved to the right by ~ 18.7 Å to take into account the difference between the calculated critical thickness (~ 2.3 Å) and the observed critical thickness (21 Å). There is a large discrepancy between the observed and the calculated curves. This large discrepancy is probably due to the fact that strain relaxation occurs by the production of islands in the 3D growth mode as discussed in chapter 3. The rate of relaxation (by production of islands)

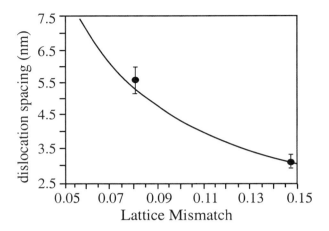

Figure 4.7: Equilibrium Lomer dislocation spacing (i.e. when there is zero strain) in epilayers as a function of lattice mismatch f_m. The solid line is the calculated spacing. The points with error bars are the experimentally determined average values for ZnTe ($f_m = 0.08$) and CdTe ($f_m = 0.146$) [299].

with increasing thickness is slower than the rate obtained by the introduction of dislocations.

Bauer et al. [149] have compared the strain relaxation in ZnTe epilayers grown on two different substrates. The layers were grown on GaAs and on ZnSe/GaAs substrates. In the second case the ZnTe buffer was pseudomorphic with GaAs. The mismatch of ZnTe with GaAs is 7.5%. As the thickness increased beyond the experimental critical thickness of 12 Å (see Table 3.1), the strain in the layers grown on both substrates began to relax. In the layers grown directly on GaAs substrate the rate of relaxation was larger and the residual strain became practically constant at –0.4 when the thickness was > 125 Å. The strain in the other sample relaxed slowly and saturated at –1.6. Residual strain was also determined by measuring average dislocation spacing by HRTEM and the corresponding values were –0.5 and –1.7.

4.3.2.2 Periodic distribution of dislocations

The lattice mismatch for CdTe/GaAs is 14.6% and for ZnTe/GaAs, 8%. The critical thickness for CdTe grown on GaAs is less than 1 mono layer. A few microns thick films of both systems were investigated by Schwartzman and Sinclair [299]. After a long anneal (100 h at 600°C), a periodic arrangements of Lomer edge dislocations was observed. The observed spacing between the dislocations in the annealed samples is shown in Fig. 4.7. The values calculated on the assumption that there was complete relaxation of strain ($-b_1/p = f_m$, see Eq. (3.3)) are also shown by solid curve. The experimental points lie on the calculated curve. Since edge dislocations can not glide on [110] planes, they

4.3. STRAIN IN II-VI LAYERS

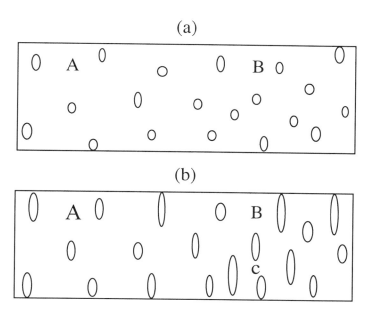

Figure 4.8: Homogeneous nucleation of loops in a highly mismatched layer. Fig. (a) shows the state when the loops are small in size and far apart from each other. Fig. (b) shows that when the loops have grown and some of them, as in region B, have come very close and interact with each other [300].

are not directly produced by nucleation. During the annealing process two 60° dislocations combine to produce a single 90° edge dislocation [300]. In most experiments the observed dislocation distribution in low mismatch (< 1.5%) systems is highly non-uniform. At high mismatch the periodic distribution of dislocations observed [299] in CdTe/GaAs and ZnTe/GaAs have been observed in other strained layers [300]. Kim et al. [301] have observed the periodic distribution in ZnTe epilayers grown on GaAs (7.5% mismatch). In GeSi/Si and InGaAs/GaAs layers a periodic distribution of dislocations is observed when the mismatch is more than 3% [235, 300][2]. At low mismatch the dislocation nucleation occurs only at defects where the energy of activation for nucleation is low [300]. The misfit dislocations are produced in clusters at these sites. If the mismatch is high, the activation energy for homogenous nucleation becomes small [235, 300]. The dislocation loops nucleate homogeneously and randomly throughout the volume of the film as shown in Fig. 4.8(a) [125]. In the beginning the loops are small in size and they are not close to each other. The growth is random and therefore the density and the radii of the loops are different at different points. After some time when the loops have grown larger in size, the situation changes as shown in Fig. 4.8(b). The loops in the region B are bigger and have come very close to each other. The interaction energy of the loops

---

[2]The behavior of III-Nitrides discussed in section 4.4 is different.

in this region increases. Firstly, it increases because the radii of the loops are bigger and secondly, because the loops have come close to each other. The effective misfit strain in region B becomes much smaller than in region A. Because of these factors, the critical radius for the loops in region B increases. Some of the smaller loops, like loop c in Fig. 4.8(b) diminish in size and disappear. The loops which are approximately at the equilibrium distance and are of equal size survive and form a periodic array of misfit dislocations.

4.3.2.3 Strain in magnetic superlattices

We now discuss the recent measurements of strain in magnetic superlattices using electron paramagnetic resonance (EPR or epr) as a probe [159]. Crystal field produces a fine structure in the electron paramagnetic resonance (EPR) spectra of paramagnetic ions present in the crystal. Atomic positions are rearranged and the crystal field is modified by strain. A change in the crystal field is reflected in the corresponding change in the crystal field splitting of the EPR lines. Furdyna et al. [159] have investigated the EPR spectra of ZnTe/MnTe, ZnTe/MnSe, and CdTe/MnTe superlattices deposited on (001) semi-insulating GaAs. The superlattice was symmetrically strained to a good approximation. The EPR measurements were carried out at X-band, 9.46 GHz at 4.2 K. The magnetic field was parallel to the [001] direction. The EPR spectrum of an isolated Mn^{++} ion consists of six hyper-fine lines produced by the interaction between the electron and nucleus spins. The lines are separated from each other by about 60 gauss. This is an intra-atomic splitting and is therefore insensitive to the crystal field or to the applied magnetic field. Crystal field splits the EPR line into five fine structure lines designated as p_1, p_2, p_3, p_4 and p_5 lines. Each fine structure p line consists of six hyperfine lines and thus the total spectrum consists of 30 lines and is very complex as shown in Fig. 4.9(a). The fine structure lines are not clearly resolved in this spectrum. The observed effect of strain on the EPR spectra of ZnTe layer in the ZnTe/MnTe superlattice is shown in Fig. 4.9(b). The spectrum in Fig. 4.9(b) is very different from that in Fig. 4.9(a) due to the effect of strain on the crystal field. The five fine structure groups are now spread over a range of more than 4000 gauss. Each group has six hyperfine lines. The six lines are resolved only in the central p_3 group, in the other groups they are significantly broadened presumably due to local fluctuations in the strain.

The spectra calculated by the method described below is shown in Fig. 4.9c. The approximate splitting of the lines due to crystal field and strain is described by the following equations [159]:

$$H\left(-\frac{5}{2} \to -\frac{3}{2}\right) = H_0 + 4D_0 + 2a_{fs}$$

$$H\left(-\frac{3}{2} \to -\frac{1}{2}\right) = H_0 + 2D_0 - \frac{5}{2}a_{fs}$$

$$H\left(-\frac{1}{2} \to \frac{1}{2}\right) = H_0 \quad (4.5)$$

4.3. STRAIN IN II-VI LAYERS

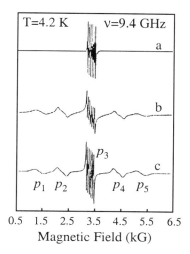

Figure 4.9: EPR spectra of very dilute Mn^{++} concentration in ZnTe for $H \parallel$ [001]: (a) observed in unstrained bulk ZnTe:Mn; (b) observed in a ZnTe/MnTe superlattice; and (c) calculated for the same superlattice using $D_0 = -503$ gauss and $\Delta D = 0.07 D_0$ to account for strain fluctuations, [159].

$$H\left(\frac{1}{2} \to \frac{3}{2}\right) = H_0 - 2D_0 + \frac{5}{2}a_{fs}$$

$$H\left(\frac{3}{2} \to \frac{5}{2}\right) = H_0 - 4D_0 - 2a_{fs}$$

where H is the magnetic field, $H_0 = h\nu/g\beta$, ν is the microwave frequency, β is the Bohr magneton, and the numbers in parentheses denote magnetic quantum number m_s for the initial and final state of each EPR transition. The symbol a_{fs} is the zero-field fine structure splitting parameter and D_0 is related to the strain by the following equation [159]:

$$D_0 = -\frac{3}{2}G_{11}\left(1 + \frac{2C_{12}}{C_{11}}\right)\frac{a_\parallel - a}{a}. \qquad (4.6)$$

Here G_{11} is a material parameter, it describes the splitting of the spin levels per unit strain. C_{11} and C_{12} are the elastic constants given in Table C in the appendix, $a_\parallel$ is the in-plane lattice constant of the strained layer and a is the lattice constant of the relaxed layer given in Table A. The value of G_{11} for Mn^{++} ion in ZnTe obtained from uniaxial stress experiments is 8.56 kG [159]. When hyperfine structure is included each of the five resonance fields in Eq. (4.5) becomes a sextet as seen in the experimental results given in Fig. 4.9(b). Fig. 4.9c shows the spectra calculated using Eqs. (4.5) and assuming $a_{fs} = 32$ gauss, $D_0 = -503$ gauss and $\Delta D_0 = 0.07 D_0$ was used to account for strain fluctuations. The calculated and observed spectra agree well. The procedure can be reversed, D_0 can be obtained by curve fitting and strain can be calculated using Eq. (4.6). The superlattice considered here is pseudomorphic. However strain in partially relaxed layers can also be measured by this method. This shows that in favorable cases EPR is a powerful tool for determining strain in the layers.

4.3.3 Strain oscillations

As discussed in section 3.3.5, if the mismatch between the epilayers and substrate is $> 2-3\%$ the growth of the epilayer is generally in the Stranski-Krastanow (SK) mode. The islands or 'patches' relieve the strain. The strain relaxes because the edges of the patches can expand or contract. This strain relaxation disappears as soon as the first layer becomes complete because at this moment there are no patches present at the surfaces. When the islands or patches grow on the first completed layer, the strain relaxes again. The process continues until misfit dislocations set in or a fresh layer nucleates before the previous layer is completed. This gives rise to strain oscillations in the early stages or highly mismatched layers. Such strain oscillations have been observed in ZnTe grown on CdTe [280, 294] and in InGaAs grown on GaAs [295]. Tatarenko *et al.* [294] observed strain oscillations in ZnTe layers grown by MBE at 280°C on (100) CdTe buffer layers. The buffer layers were grown on $Cd_{0.96}Zn_{0.04}Te$ substrates. Fig. 4.10(a) shows the oscillations in strain relaxation. Results of

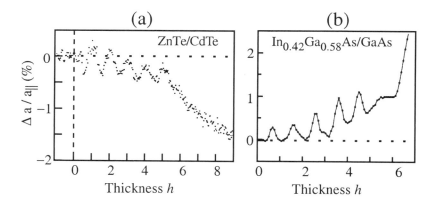

Figure 4.10: Observed strain oscillations in (a) ZnTe grown on CdTe buffer [294] and (b) $In_{0.42}Ga_{0.58}As$ grown on GaAs [295].

similar earlier experiments [295] on the growth of $In_{0.42}Ga_{0.58}As$ on GaAs are shown in Fig. 4.10(b). In both cases the average strain relaxation $\Delta a/a_\parallel$ (Δa is the change in in-plane lattice constant $a_\parallel$, it is zero if the layer is complete and pseudomorphic)[3] oscillates 5 to 6 times with an amplitude which varies from 0.3 to 0.4%. $|\Delta a/a_\parallel|$ is maximum when a monolayer is half complete and

---
[3]In Refs. [280, 295] the quantity $\Delta a/a_\parallel$ has been designated as lattice mismatch.

4.3. STRAIN IN II-VI LAYERS

is small when the layer is complete. After the completion of first layer $\Delta a/a_{\parallel}$ is zero. $\Delta a/a_{\parallel}$ does not go to zero after the completion of second monolayer. The magnitude of $\Delta a/a_{\parallel}$ becomes increasingly large after the completion of subsequent monolayers. This happens because a fresh layer nucleates before the previous layer is completed. Massies and Grandjean [295] have used a simplified version of the valence force field model to calculate the relaxation in patches. The patches were considered as linear chains of atoms. The substrate distortion was neglected. Massies and Grandjean [295] found that if a linear chain of 11 unit cells is considered, the calculated $\Delta a/a_{\parallel}$ within a patch is 0.6% and the average $\Delta a/a_{\parallel}$ is 0.3% which agrees approximately with the observed relaxation. The neglect of substrate stresses is not a good approximation [52, 141, 296]. Moreover the actual patches are not linear chains but are two dimensional [297]. Experiments show that the patches are not uniformly distributed but clustered [297]. In a cluster the average distance between the patches is much smaller than the average size of the patches. Under this situation, the elastic interaction (through the substrate) between the patches become large [141, 296].

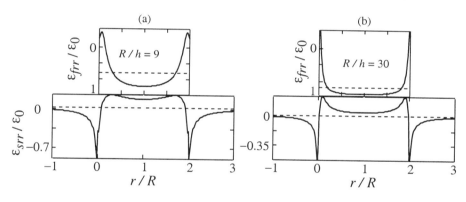

Figure 4.11: Calculated normalized strain $\epsilon_{frr}/\epsilon_0$ in circular patches and $\epsilon_{srr}/\epsilon_0$ in the substrate are plotted as a function of r/R for two samples, (a) $R/h = 9$ and (b) $R/h = 30$. The patch strain is calculated at the top surface and the substrate strain is calculated at a depth $= h/10$ measured from the interface. The variation of strain with height in the patch is not large. The value of $\epsilon_{frr}/\epsilon_0$ averaged over the area of the patch and of $\epsilon_{srr}/\epsilon_0$ averaged over an area of radius $4R$ in the substrate are shown by the dashed lines. ϵ_0 is the strain in a large area layer without edge induced relaxation. ϵ_0 is equal to 0.058 for ZnTe/CdTe and is equal to -0.03 for $In_{0.42}Ga_{0.58}As/GaAs$.

Pinardi et al. [125] used the continuum elasticity theory and made finite element calculations of the strain relaxation in the patches. Brandt et al. [298] have measured strain in a single InAs monolayer and three monolayers embedded in GaAs substrate. Their results show that the continuum theory of elasticity is not obeyed by the single layer but the theory becomes valid for the three monolayer case. For InGaAs layers grown on InGaAs wetting layers, the

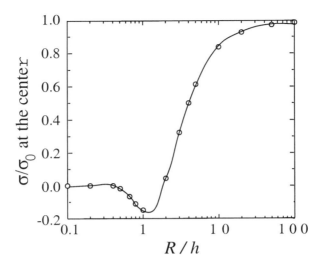

Figure 4.12: Normalized stress σ/σ_0 is plotted as a function of R/h for the circular patches.

assumption that the elasticity theory is valid seems reasonable. The strain in the patches of two different dimensions is shown in Fig. 4.11. Here R and h are the radius and thickness of the patches. The substrate stresses are included in the calculations but interaction between the patches is neglected. FE calculations show that if distance between the patches is more than $2R$ the interaction between them becomes negligible. The thickness of the substrate does not affect the result provided it is larger than $2R$. The strain distribution depends only on the ratio R/h and not on R and h separately. Fig. 4.11 shows that the stresses induced in the substrate are very large. The normalized stress σ/σ_0 (calculated by the FE method) at the center of the patches of several different dimensions is shown in Fig. 4.12. The patch in the first half finished layer has a thickness $h \sim 0.3$ nm and the radius $R \sim 2.2$ nm [295]. For these dimensions the ratio R/h becomes ~ 7. For this value of R/h the FE calculations give a value 1.5% for the average relaxation $\Delta a/a_\parallel$ in the patch. This is much larger than the relaxation of 0.6% observed in the patch. We mentioned earlier that STM pictures show that the patches are clustered [297]. The discrepancy is attributed to the neglect of non-uniform distribution and the effect of interactions.

Kret et al. [280] have also observed oscillations (with an amplitude up to 0.3%) in the lattice parameter of ZnTe grown on relaxed CdTe buffer. The oscillations are shown in the inset in Fig. 4.6. When the growth temperature was raised to 573 K, only one oscillation was observed. These results are different from those shown in Fig. 4.10(a). The maximum value of quantity $\Delta a/a_\parallel$ when a layer is half-finished is positive (except for the first half layer), a result not consistent with the fact that the ZnTe layer is under tension and on relaxation its lattice constant should decrease. The results of Kret et al. [280] are puzzling

4.4 Strain and defects in III-Nitride layers

4.4.1 Strain distribution and dislocations

Stress in the GaN and AlN layers has been measured by wafer curvature method and by the Raman measurements [302]-[305]. Yoon et al. [303] have studied the shift in the Raman E_2 mode as a function of the thickness of the GaN films grown on (0001) sapphire by GSMBE. The shift in the E_2 Raman mode showed that the residual stress changes from tensile to compressive as the epilayer thickness increases. Hiramatsu et al. [307] observed a compressive stress in the thin GaN epilayers and a rapid decrease in the stress of GaN as the thickness of the GaN layer increased to more than 4 μm. However Kozawa et al. [302] studied stresses in films of thickness up to 50 μm by the curvature method and observed only a small decrease in the compressive stress as the GaN films became thicker. A complex behavior of stress evolution in AlN films grown on Si substrate was observed by Meng et al. [229]. As mentioned earlier the layers were grown by the ultrahigh vacuum reactive sputter deposition. These authors also used the wafer curvature method to measure the stress. In the thin layers grown at room temperature the stress was compressive and small. The stress remained compressive but increased linearly as the thickness of the layers increased. The stress was 200 N/m in 1500 Å thick films. Some layers were grown at 800°C. In this case the stress was compressive and increased with thickness only up to 500 Å thickness. As the thickness increased further, the stress started decreasing. In the layers grown on (111) Si, the stress became practically zero in 1700 Å thick films.

Stress in GaAs, and InP epilayers grown on Si have been discussed in section 4.2. As discussed in this section, the layer can be regarded as "lattice matched" to the substrate at the growth temperature provided the layer thickness is sufficiently large. On cooling the layers to room temperature, stresses and strains are generated due to difference in the values of coefficients of thermal expansion of the layer and the substrate. If complete relaxation does not take place at the growth temperature, lattice mismatch strain persists. Some relaxation of thermal strain may take place during the cooling process. We therefore conclude that the evolution of stress in the conventional III-V semiconductors grown on Si substrates and in the III-Nitrides is a very complex process and can not be interpreted easily unless simultaneous measurements of micro-structure are also made.

We now discuss the stress distribution with distance z measured from the interface. The stress was measured by micro-Raman method in Refs. [304, 305]. The laser beam entered through the side of the sample. The stress does not vary in the x and y directions and therefore the observed Raman shift varies only with d z [304, 305]. The Raman shift and stress were maximum near the interface and decreased monotonically as the surface was approached. The shift measured by the Raman probe was corrected for the finite beam width and penetration depth of the laser beam [125]. The corrected observed Raman shift is shown by the solid lines in Fig. 4.13. Jain et al. [310] attributed the stress variation to the

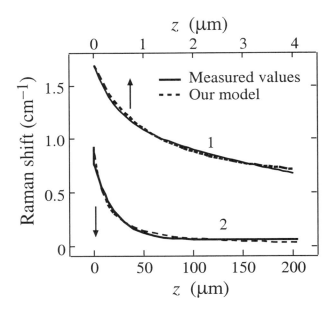

Figure 4.13: Dashed lines are the fits of Eq. (4.8) to the observed strain in GaN [304, 305] with $a_1 = -1$, $a_2 = 1.655$, $K = -0.365$, and $e^{-K_0} = 0.943$ for curve 1 and with $a_1 = -1$, $a_2 = -23.967$, $K = 0.004$, and $e^{-K_0} = -0.04$ for curve 2.

presence of large number of threading dislocations. The threading dislocations are not vertical, they have a horizontal component. The horizontal component relieves the strain. Jain et al. [310] gave a phenomenological model to interpret these experiments. In this model $d\epsilon/dz$ is given by a polynomial:

$$\frac{d\epsilon}{dz} = n\frac{b}{2}(a_1\epsilon(z) + a_2\epsilon^2(z)). \tag{4.7}$$

Here a_1 and a_2 are the adjustable parameters. The solution of Eq. (4.7) is:

$$Kz + K_0 = \frac{1}{a_1}\left(\ln\frac{\epsilon(z)}{a_1 + a_2\epsilon(z)}\right), \tag{4.8}$$

where $K = nb/2$ and K_0 is the boundary value at $z = 0$. Numerical fits of Eq. (4.8) with the (corrected) observed stress are shown by the dashed lines for GaN in Fig. 4.13. The agreement between the experiments and the model is very good.

4.4. STRAIN AND DEFECTS IN III-NITRIDE LAYERS

4.4.2 Structural defects

The point defects e.g. nitrogen vacancies, oxygen and other impurities, and donor-acceptor pairs are discussed in chapters 5 and 6. We discuss the structural defects in this section. The Nitride heterostructures contain characteristic one-dimensional (edge, mixed and screw dislocations) and two-dimensional (stacking faults and domain boundaries) defects. Relative concentration and nature of the defects depend on the growth conditions and structure of the sample. Measurements also show that the dislocation density in the LEDs is 10^8 cm^{-2} to $\sim 10^{12}$ cm^{-2} [195, 318]. Conventional III-V semiconductor light emitters will not be reliable if a dislocation density of $>\sim 10^5$ cm^{-2} is present in the device. The dislocations induce non-radiative recombination. This reduces optical output and generates heat. The rise in temperature causes multiplication of defects leading to failure of the lasers. The dislocations affect adversely morphology, and optical and electronic properties [319, 85, and references given there in]. In spite of the large dislocation density, the III-Nitrides LEDs work well. Many investigations have been made in an attempt to determine as to why the III-Nitride devices work in spite of the presence of the large concentration of dislocations. One possibility is that the dislocations in the III-Nitrides are benign i.e. they do not act as non-radiative recombination centers [318]. However recent work shows that dislocations act as pipes for the diffusion of metal such as Mg, increase the leakage current and reduce considerably the light output from the LEDs [213]. Experimental work also shows that the dislocations do act as non-radiative recombination centers [320, 313]. Sugahara *et al.* [313] took plan-view TEM and CL images from exactly the same spot of a sample. They found a one to one correspondence in the dark spots in the CL images and the dislocations seen in the TEM. The value of the diffusion length determined from CL mapping experiments is about 60 nm [85, 321]. The reason that the LEDs work is that the dislocation can not easily multiply and spread in to the active region because the activation energies for the nucleation and propagation of the defects are large. Also non-uniform distribution of In helps. There is considerable evidence that the distribution of indium in InGaN alloys is not uniform. A phase separation takes place and local regions of high In concentration are formed. These regions act as quantum dots. The carriers confined in these dots do not "see" the dislocations. This model has also been used to explain the high efficiency of light emission from the nitride semiconductors (see section 5.5.5.3).

4.4.3 Effect of defect clusters on the minority carriers in GaN

In GaN the dislocations are clustered in local regions of the epilayer leaving large volumes of the layer defect free [195]. Jain *et al* [322, 310] have constructed a model to interpret the effect of dislocations on minority carriers in GaN epilayers. The model is based on the concept that the trapping/recombination centers in a defect cluster (DC) can only cope with a limited number of recombination per unit time [322]. The ratio of carriers lost by recombination divided by the injected carriers decreases monotonically as the injected carrier density increases. For sufficiently large density of the injected carriers, the fraction lost by the recombination at the DC becomes small. The calculations based on this model show that there are two scenarios in which DCs will not be very effective in reducing the number of minority carriers by non-radiative recombination. In the first scenario, the capture cross sections are small and the dislocations become ineffective even when the injected carrier density is not very large. In this case only a very small volume near the DC looses carriers by recombination at the DC. In the second case the distance between the DCs is much larger as compared to the diffusion length. Large volumes of the layer do not 'see' the dislocations. Experimental work [313, 320] discussed above is consistent with the second scenario. For mathematical treatment of the model paper of Jain *et al.* [310] should be consulted.

4.4.4 Interfaces in III-Nitride epilayers

TEM lattice images of GaN and AlGaN show that the interfaces are atomically sharp i.e. the transition from the oxide to the nitride is abrupt and stable [195, 323]. The interfaces between two III-nitride epilayers are generally smooth and abrupt. The interface mixing and roughening observed in GeSi [3] and in III-V multilayers [52] are not so much of a problem in the III-Nitride multilayers. 2D gas of electrons or holes at the modulation doped interfaces in III-V nitride multilayers have been observed (see for example Ref. [324] and several papers in the MRS 97 symposium on III-V nitrides [87]).

Meng *et al.* [229] have studied the GaN/Si (111) interface by ion beam channelling. The thickness of the epilayers was 1000 Å. Strong dechanneling peaks generally associated with the presence of misfit dislocations were not observed. Localized strain contrast expected from the dislocations was also not observed in the TEM studies [229]. Meng *et al.* [229] suggested that the structure of the interface in the III-Nitride heterostructures is such that the whole interface accommodates the strain uniformly. Nano-pipes have also been observed in many studies in the III-Nitrides [314, 315]. The nano-pipes are N-terminated polarity inside the Ga terminated matrix having lower than matrix growth rate [316].

Annealing at high temperatures can have large changes in the defect structure and properties of the strained semiconductor layers and superlattices. Intermixing at the heterointerface takes place due to diffusion of atoms from one layer in to the other and vice-versa. Intermixing in the II-III and III-V semiconductors is well documented. We have discussed the intermixing in II-VI semiconductors in earlier chapters. In the II-VI semiconductors it is difficult to avoid the intermixing if the heterostructures are subjected to high temperatures for any length of time. The structure of the epilayer/buffer interface has a profound effect on the polarity and other properties of the epilayers [28]. For many applications sharp interfaces are required. However Chan *et al.* [27] have recently suggested that intermixing can be used to modify and tailor the bandstructure and optical properties of the quantum wells. The nitridation of the

4.4. STRAIN AND DEFECTS IN III-NITRIDE LAYERS

sapphire substrates is an important factor for obtaining flat unipolar films [317]. The interdiffusion in the Nitrides can occur only by the diffusion of Ga, Al or In because nitrogen is a common constituent in all the nitrides. Recently Chan et al. [27] and Sumiya et al. [28] have investigated intermixing at the interfaces and effect of annealing the buffer layers on the polarity of the GaN layer.

The effect of annealing on the PL of a $In_{0.2}Ga_{0.8}N$ MQW is shown in Fig. 4.14. The thickness of the wells was 2.5 nm and that of the GaN barriers, 7 nm.

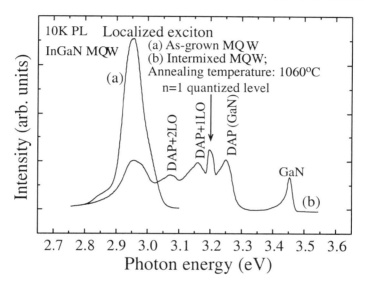

Figure 4.14: 10 K PL spectra of (a) the as-grown $In_{0.2}Ga_{0.8}N$ MQW and (b) the intermixed $In_{0.2}Ga_{0.8}N$ MQW annealed at 1060°C for 60 minutes [27].

The number of periods in the superlattice was 10. The superlattice was grown on a 2.3 μm GaN:Si layer on a sapphire substrate. The as grown MQW shows a strong localized peak at 2.950 eV. This peak shows intense blue emission. The spectrum changes drastically after thermal annealing of the MQW at 1060°C for one hour. The PL peak at 2.950 eV is suppressed considerably and blue shifted by 5 meV. Chan et al. [27] made theoretical calculations assuming that an interfacial layer of thickness from 0 to 4 Å is formed by the interdiffusion. For 4 Å thickness the blue shift predicted by the theory was 60 meV. This shows that if interface mixing can be described by an interfacial layer of a well defined thickness, the thickness should be less than 1 Å for this sample because the blue shift is only 5 meV. Several PL peaks are shown on the high energy side of the 2.955 eV peak in the annealed sample. These peaks were not visible in the spectrum of the as grown structure, presumably their intensity was very small. The 3.197 eV peak is attributed to the first quantized level of the exciton. The assignments of other features is shown in the figure. We can conclude from these experiments that high temperature annealing of InGaN/GaN quantum wells modifies their optical properties very significantly. The detailed processes that occur during the annealing and the changes in the defect structure are not well understood. However intermixing at the interface in the III-Nitrides seems to be much smaller than that obtained with the II-VI semiconductor heterostructures.

Sumiya et al. [28] studied the effect of high temperature annealing of the buffer layer on the surface morphology and polarity of the GaN layers. The two step method with MOCVD was used to grow the layers. The GaN layers were 1.2 μm thick and were grown on 160 nm GaN buffer and c-plane sapphire substrate. The substrate surface was cleaned in hydrogen gas at 1080°C for 10 min. The surface was then nitridated in ammonia at the same temperature for 5 min. The buffer layer was deposited at 600°C and then annealed for different durations at 1040°C. GaN layers with 20 nm buffer and without nitridation were also grown. The evolution of surface morphology of the GaN layers grown on 160 nm buffer layers with time of annealing of the buffers is shown in Fig. 4.15. For less than 10 min annealing of the buffer the surface of the GaN layer was smooth and featureless. Changes in the surface morphology are clearly visible in Fig. 4.15(a) after annealing the buffer for 30 min. As the time of annealing of the buffer layers increases, the size of the hexagonal columns and house shaped hexagonal structures (shown by the arrow in Fig. 4.15(c)) also increases. The maximum size of the hexagonal facets on 100 min annealing is about 60 μm. The annealing of the buffers has a large effect on the surface morphology of the main GaN layer even if the GaN layers are grown on identical buffers and under identical conditions.

The GaN films on c-pane sapphire grow along the c-direction and therefore they have a polarity as illustrated in Fig. 4.16. The polarity is important for understanding the effect of spontaneous electric field on the emission of light from the quantum wells. It was found [313] that GaN grown by MOCVD on low-temperature buffer layer (or without buffer) has N-terminated polarity with inclusions of Ga-terminated domains. An increase of growth temperature of buffer layers results in the formation of unipolar Ga-terminated films. More recently Sumiya et al. [28] identified the polarity of the buffer layers using the coaxial impact collision ion scattering spectroscopy (CAICISS). The polarity was determined by comparison of the CAICISS results of the buffer with those of ZnO. It was found that the as grown buffer layers have predominantly +c surface. This was common to both the thick and the thin buffer layers. As the annealing progresses, the fraction of -c surface increases and ultimately the surface becomes predominantly -c type of surface. Thus the surface of the buffer changes continuously from Ga terminated +c configuration to nitrogen terminated -c surface. Sumiya et al. [28] suggested that the surface of the as grown buffer layer has both +c and -c polarity due to imperfect nitridation of the surface. The layers which grow on the nitridated surface have -c nitrogen termination at the surface. On the other hand the layers which grow on portions of the (clean) sapphire substrate have +c Ga termination. The +c domains grow faster and also evaporate more easily. To interpret the experimental results, it was assumed that during the growth of the buffer at lower temperatures, the +c domains cover the -c domains. On subsequent annealing at high temperatures,

4.4. STRAIN AND DEFECTS IN III-NITRIDE LAYERS

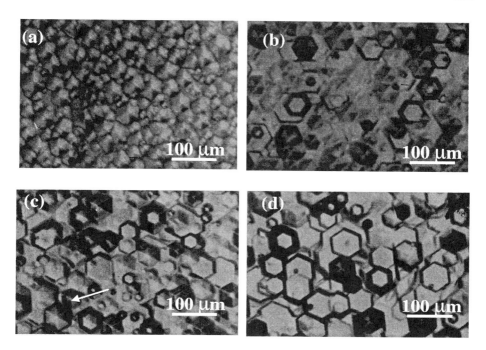

Figure 4.15: Surface morphology of GaN films deposited on the buffer layer whose polarity was controlled by the annealing time of (a) 30 min, (b) 40 min, (c) 60 min and (d) 100 min [28].

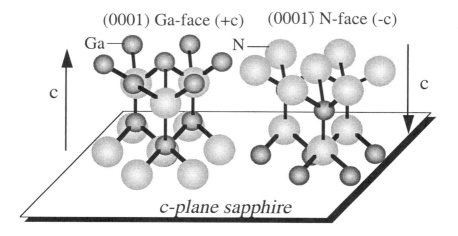

Figure 4.16: Schematic illustration of wurtzite GaN structure oriented along c-axis. Left side is (0001) Ga-face which is defined +c. Right side is (000$\bar{1}$) N-face which is defined -c. [28].

the +c domains evaporate leaving behind the -c domains. Presumably the polarity of the buffer surface propagates in to the GaN layer. This allows the control of polarity of the GaN layer by controlling the polarity of the buffer layer.

Chapter 5

Band structure and optical properties

5.1 Band structure

5.1.1 Zinc-blende semiconductors

Band structure of unstrained zinc blende semiconductors is shown schematically in Fig. 5.1(a). A semiconductor layer grown on a substrate of different material is generally strained. The strain is biaxial. We discuss the modification in the bandstructure caused by the strain in this section. At **k**=0, the band of

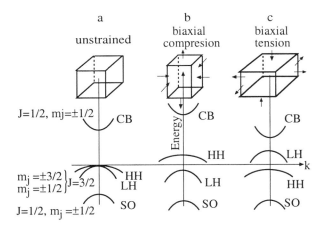

Figure 5.1: Schematic band structure (a) of an unstrained epilayer, (b) a layer under compressive biaxial strain, and (c) a layer under tensile biaxial strain is shown.

the unstrained crystal consists of a fourfold $P_{3/2}$ multiplet ($J = 3/2$, $m_J =$

$\pm 3/2, \pm 1/2$) and a twofold $P_{1/2}$ multiplet ($J = 1/2$, $m_J = \pm 1/2$). These bands are also described in an alternative notation: $|3/2, \pm 3/2\rangle$, $|3/2, \pm 1/2\rangle$ and $|1/2, \pm 1/2\rangle$ bands respectively. Holes in the $|3/2, \pm 1/2\rangle$ band have a smaller effective mass than the holes in the $|3/2, \pm 3/2\rangle$ band. The two hole bands are known as the light hole (LH) and the heavy hole (HH) bands respectively. The spin split-off band $|1/2, \pm 1/2\rangle$ is also shown in Fig. 5.1).

Biaxial strain has two parts, (i) a hydrostatic part which produces an isotropic volume change and (ii) a shear component which produces a tetragonal distortion and lowers the symmetry [52]. Effect of both components on the band structure of zinc-blende semiconductors is shown schematically in Fig. 5.1(b) and Fig. 5.1(c). Hydrostatic stress changes the distance between the conduction band and the center of gravity or the weighted average of the three uppermost valence bands (the heavy hole, light hole and spin-orbit split-off (SO) bands. The distance between the conduction band and the center of gravity of the valence bands increases under hydrostatic compression and decreases under hydrostatic tension. The shear component of the biaxial strain removes the degeneracy of the valence band multiplet. The two hole bands move in opposite directions away from the center of gravity of the two bands. Under compression the HH band moves up and determines the bandgap. Under tensile strain, the LH band moves up. It is closest to the conduction band and determines the bandgap.

For small strain, the eigen values of the orbital-strain Hamiltonian can be calculated and the shifts in bandgaps can be written as [325],

$$\Delta E_0 \left(\frac{3}{2} \pm \frac{3}{2}\right) = \Delta E_g^h = E_H + E_U, \tag{5.1}$$

$$\Delta E_0 \left(\frac{3}{2} \pm \frac{1}{2}\right) = \Delta E_g^l = E_H + \frac{1}{2}(\Lambda - E_U) - \frac{1}{2}(9E_U^2 + 2\Lambda E_U + \Lambda^2)^{1/2}, \tag{5.2}$$

$$\Delta E_0 \left(\frac{1}{2} \pm \frac{1}{2}\right) = E_H + \frac{1}{2}(\Lambda - E_U) + \frac{1}{2}(9E_U^2 + 2\Lambda E_U + \Lambda^2)^{1/2}. \tag{5.3}$$

Here Λ is the spin-orbit splitting parameter, $E_0(\frac{3}{2} \pm \frac{3}{2})$ and $E_0(\frac{3}{2} \pm \frac{1}{2})$ are the transition energies from the conduction band to the split components of the valence band i.e. they are the heavy hole bandgap E_g^h and light hole bandgap E_g^l respectively. $E_0\left(\frac{1}{2} \pm \frac{1}{2}\right)$ is the distance from the conduction band to the spin split-off band. The shift due to strain in these energies is denoted by $\Delta E_g^h = \Delta E_0(\frac{3}{2} \pm \frac{3}{2})$, $\Delta E_g^l = \Delta E_0(\frac{3}{2} \pm \frac{1}{2})$ and $\Delta E_0\left(\frac{1}{2} \pm \frac{1}{2}\right)$ respectively. The parameters E_H and E_U are the given by,

$$E_H = 2a\left[\frac{C_{11} - C_{12}}{C_{11}}\right]\epsilon \tag{5.4}$$

$$E_U = --b\left[\frac{C_{11} + 2C_{12}}{C_{11}}\right]\epsilon. \tag{5.5}$$

5.1. BAND STRUCTURE

Values of E_H and E_U depend on the hydrostatic deformation potential a, shear deformation potential b and elastic constants C_{11} and C_{12}. The values of deformation potentials and elastic constants are given in Table C for several II-VI semiconductors. Values for the ternary alloys are calculated using linear interpolation of the values of the constituent binary semiconductors.

In addition to band splitting, strain also changes the shape of the valence bands. Heavy hole mass is reduces under biaxial compressive strain. The reduction is very large in conventional III-V semiconductors.

We need values of strain to calculate the strain induced band shifts and band splitting. If the strain arises mainly due to lattice mismatch and thermal strain is negligible, it can be calculated using the lattice parameter of the substrate and the epilayer (see Tables given in Appendix A). If the misfit strain is completely removed at the growth and/or annealing temperature T_g, and no plastic deformation takes place during cooling, the thermal misfit $f_{(m,th)}$ at room temperature (RT) can by calculated as discussed in chapter 3. Using these values of strain, the bandgap shifts are calculated using Eqs. (5.4) and (5.5). Values of bandgaps of the strained layers can be obtained by adding the strain induced shifts to the bandgaps of the unstrained alloys. Values of bandgaps of unstrained semiconductors are given in the tables in Appendix A.

If the strain is compressive, bandgap increases due to hydrostatic strain. The increase is reduced by the valence band splitting and motion of the heavy hole band towards the conduction band. The heavy hole band is the lowest energy valence band in the layers under compressive strain. In the case of ZnSe the two effects nearly cancel each other and therefore the effect of strain on the HH band is small. In this case light hole band moves away from the conduction band. If the strain is tensile, bandgap decreases due to hydrostatic strain. This decrease is further enhanced by the motion of the light hole band towards the conduction band which now forms the band edge closest to the conduction band. Thus under tensile strain bandgap always decreases, decrease being much larger than the increase in the case of the compressive strain. It is important to remember that the shifts we have discussed in this section are only due to strain. If strain is changed by changing the composition of the layer (e.g. by changing x in the $ZnSe_{1-x}S_x$) alloy, the bandgap also changes independent of strain.

Photoluminescence (PL) or cathodoluminescence (CL) peak positions and shapes can be analyzed and shift in bandgap due to strain can be determined. This shift can be used to calculate the values of stress and strain using Eqs. (5.1) and (5.2). The peak positions depend on the binding energy of the excitons and of the excitons with the impurities. In practice the PL or CL in the unstrained and the strained layers are measured. The shift in the peak position in the two measurements is assumed to be equal to the change in the bandgap due to strain. The value of strain for which the observed and calculated shifts are equal is then easily calculated. If the transitions from the two hole bands are resolved, strain can be calculated by comparing the observed splitting with its calculated value given by the difference of Eqs. (5.1) and (5.2). If the strain is partially relaxed due to generation of misfit dislocations, it can still be determined by using Eqs. (5.1) and (5.2) and the luminescence measurements.

Using the values of elastic constants and deformation potentials (see Table C in Appendix A) in Eqs. (5.1) and (5.2), we obtain[1] for ZnSe layers grown on (100) GaAs substrate [326],

$$\Delta E_g^h = -0.19\epsilon \text{ eV}, \tag{5.6}$$

$$\Delta E_g^l = 5.06\epsilon \text{ eV}, \tag{5.7}$$

Effect of spin-orbit splitting has been neglected in these calculations. Similar equations for III-V semiconductors are given in Ref. [52, 141].

Suzuki and Uenoyama [327] have calculated the effect of strain on the zincblende GaN. They found that the effect of biaxial strain on the band structure of zincblende is similar to the effect on other III-V semiconductors. However reduction in the heavy-hole mass is not as large. The heavy hole masses in ZB GaN are much larger than in other III-V semiconductors. The masses remain quite heavy even under the biaxial strain and reduction in DOS is not so large as in other III-V semiconductors. Furthermore as compared to the other III-V semiconductors, the spin-orbit interaction is much weaker in zincblende GaN.

Spin-orbit parameters Δ_{SO} have also been calculated for the ZB III-Nitrides. The calculated values for the ZB III-Nitrides are given in Table 5.1. The main difference between the valence band structure of ZB and WZ crystals is the crystal field splitting in the WZ crystals. Because of the cubic symmetry crystal splitting is absent in the ZB structures.

Table 5.1: Band structure parameters for ZB AlN, GaN and InN.

Method	Δ_{SO}	Ref.
	ZB AlN	
Cal.	19, 20	[329, 327]
	ZB GaN	
Cal.	15, 20	[329, 327]
Expt.	17 ± 1	[327]
	ZB InN	
Cal.	6	[329]

5.1.2 Wurtzite (WZ) III-Nitrides

Several groups have discussed the band structure of WZ GaN [328, 285, 329, 330, 327]. Hexagonal wurtzite GaN, AlN and InN are direct bandgap semiconductors. Both the conduction band minimum and the valence band maximum are located at the center of the Brillouin zone (Γ-point, $k = 0$). The valence

---
[1]Numerical values given in Eqs. (5.6) and (5.7) are different from those given in [274] because of the uncertainty in the available values of elastic constants and deformation potentials.

5.1. BAND STRUCTURE

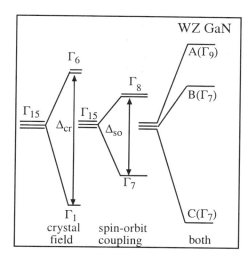

Figure 5.2: Effect of crystal field splitting and spin-orbit coupling on the valence band (near the Γ point) of WZ GaN [327]. At $k = 0$, the valence band is split by the combined action of crystal field and spin orbit coupling into $A(\Gamma_9)$, $B(\Gamma_7)$ and $C(\Gamma_7)$ states. See also figure Fig. 8.3 of [327].

band structure is modified by both spin-orbit splitting and crystal field splitting. The splittings are calculated using a quasi-cubic approximation [327]. A schematic diagram of the calculated splittings is shown in Fig. 5.2. The valence band splits into three bands designated as A, B, and C bands. The symmetry of the A-band is Γ_9 and that of the B and the C bands it is Γ_7. The calculated valence band structure near the Γ-point is shown in Fig. 5.3. The calculated and experimental values of ΔE_{AB} and ΔE_{BC} are given in Table 5.2. Numerical calculations show that $\Delta E_{AB} < \Delta E_{SO}$ and $\Delta E_{BC} \sim \Delta E_{CF}$.

According to Ref. [327], the calculated values of spin orbit splitting Δ_{SO} are 15.6 meV for WZ GaN and 20 meV for ZB GaN. The corresponding observed values are 11(+5,−2) meV and 17 ± 1 meV respectively for the two structures. The agreement between the theoretical and experimental values is reasonable. There is a considerable discrepancy in the values of crystal field splitting Δ_{CF} calculated by different groups. The calculated crystal field splitting Δ_{CF} is 72.9 meV in WZ GaN and −58.9 meV in WZ AlN according to Ref. [327]. The corresponding values calculated in Ref. [329] are −217 and 42 meV. It is generally believed that the crystal field splitting for AlN is negative (the order of Γ_6 and Γ_1 in Fig. 5.2 is reversed, for AlN Γ_1 is higher) and for GaN it is positive [329, 327]. The observed value of Δ_{CF} in WZ GaN is 21 to 22 meV. The calculated values of hole mass depend on k [328, 327]. However experimentally only average (Density of States) values have been determined.

As for the case of ZB semiconductors, the effect of strain on the band structure of the III-Nitrides is also large. The stress-strain relations for a hexagonal crystal (C_{6v} symmetry) are expressed by a 6×6 matrix. However if the crystal

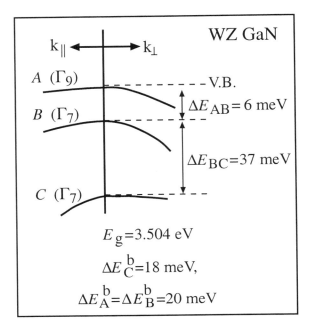

Figure 5.3: Calculated values of the band energy levels [328] near the Γ point. ΔE_A^b, ΔE_B^b, ΔE_C^b are the binding energies of the A, B and C excitons respectively.

Table 5.2: Values of ΔE_{AB} and ΔE_{BC} for WZ III-Nitrides. The values of temperature at which the measurements were made are given in rounded brackets.

Semicond.	ΔE_{AB}	ΔE_{BC}	Ref.
	Calculated		
AlN	211 meV	224 meV	[329]
GaN	7 meV	48 meV	[329]
GaN	6 meV	37 meV	[328]
InN	2 meV	43 meV	[329]
	Experimental		
GaN	6 meV (40 K)	–	[328]
GaN	8 meV (2 K)	25 meV	[331]
GaN	$\sim$ 6 meV (2 K)	–	[330]
GaN	8 meV (15 K)	25 meV	[332]
GaN	6 meV (2 K)	18 meV	[333]
GaN	9 meV (5 K)	37, 38 meV	[36]

is strained in the (0001) plane and is free in the [0001] direction, there are only three non-vanishing strain component [327]. Under the approximation of strain independent isotropic spit-orbit coupling, the effect of strain on the A, B, and C bandgaps (shown in Fig. 5.3) can be described by simple equations [330]. The band structure of the strained GaN [327] has been calculated in Ref. [327]. The uniaxial strain makes the effective hole masses different for k_x and k_y directions. For further details of the band-structure and effective masses of strained and unstrained WZ GaN the paper of Suzuki and Uenoyama [327] should be consulted.

5.2 Band offsets

5.2.1 General remarks

It is important to know the values of the band offsets for interpreting experiments and designing devices. The hydrostatic strain has a significant effect on the band offsets only if the pressure coefficients of the bandgaps of the two constituent epilayers are very different. The shear strain splitting causes a larger change in the band offsets of the valence band components. The band offsets at the heterovalent junction also depend on the atomic configuration at the interface [335]. The interfacial layer formed by intermixing affects strongly the band offsets [335]. The values of band offsets can be modified by depositing deliberately an ultra-thin layer of a suitable material at the interface [336]. In most cases the well known MMM (McCaldin-McGill-Mead) "common anion rule" [337, 338, and references given therein] is not valid. The MMM common anion rule states that the valence band offset between two semiconductors will depend only on the anion contained in each semiconductor. The validity of this rule and its applicability has been discussed in Refs. [337, 338]. According to this rule, the valence band offsets of CdSe/ZnTe, ZnSe/ZnTe and ZnSe/BeTe interfaces must be approximately the same. The values of valence band offsets are 0.98 eV for ZnSe/ZnTe, 0.64 eV for CdSe/ZnTe [337] and 1.50 to 1.56 eV for ZnSe/BeTe [336].

Band offsets of several heterojunction have been calculated and determined experimentally by many groups [325]-[361] In many cases experimental values of the band offsets of a given heterojunction determined by different authors vary considerably. Different techniques of measurements give different results. Part of the discrepancy is probably due to contamination and intermixing at the interface. Recent values of the band offsets of the III-V, II-VI and III-Nitride heterojunctions are discussed below. Review of the early work is given in Refs. [339, 340].

5.2.2 Band offsets: III-V heterostructures

Katnani and Margaritondo [341] determined the band offsets of III-V, II-VI and group IV semiconductors by analyzing the photoemission spectra. Schuermeyer

Table 5.3: Values of the valence band edges (in eV) of III-V binary compounds relative to AlAs. Values were compiled by Ichii et al. [343]. The abbreviation H is used for Harrison and S for Schuermeyer.

Material	Ichii [343]	Tersoff [339]	H [340]	Katnani [341]	S [342]
AlP	-0.48 ± 0.12	-0.22	-0.46	–	–
AlAs	0	0	0	0	0
AlSb	0.76 ± 0.08	0.60	0.90	–	0.90
GaP	0.29 ± 0.14	0.24	-0.43	-0.36	–
GaAs	0.48 ± 0.03	0.55	0.04	0.27	0.40
GaSb	1.16 ± 0.08	0.98	0.88	0.39	1.30
InP	0.20 ± 0.04	0.29	-0.07	-0.09	–
InAs	0.65 ± 0.06	0.55	0.36	0.27	0.80
InSb	1.08 ± 0.06	1.04	1.16	0.49	–

et al. [342] used x-ray photoemission spectroscopy (XPS) of several III-V heterostructures. Ichii et al. [343] have developed a method to predict empirically the valence band positions of III-V semiconductors and alloys. The values derived by them agree well with much of the existing experimental data as shown in Table 5.3.

Ichii et al. [343] also determined the valence band offsets of binary/ternary junctions of III-V compounds. These results are given in Table 5.4.

Table 5.4: Values of the valence band offsets (in eV) of III-V ternary compounds.

Material	Empirical	Experiment	Ref.
GaAs/AlAs	0.48 ± 0.03	0.55	[343]
InP/In$_{47}$Ga$_{53}$As	0.36 ± 0.05	0.346 ± 0.01	[343]
In$_{0.53}$Ga$_{0.47}$As/In$_{0.52}$Al$_{0.48}$As	0.23 ± 0.01	0.22 ± 0.05	[343]
GaAs/In$_{0.5}$Ga$_{0.5}$P	0.24 ± 0.08	0.24 ± 0.01	[343]

5.2.3 Band offsets: II-VI heterostructures

Most II-VI light emitting and other devices have been fabricated on (100) III-V substrates. Band discontinuity at the II-VI/III-V interface is of crucial importance in determining the carrier injection, contact resistance and overall efficiency of the device. Yang et al. [350] determined the band offset at the ZnSe/GaAs (100) interface by x-ray photo-electron spectroscopy (XPS). These [350] and other measurements [129, 335] showed that valence band offset is a large fraction of the bandgap difference between ZnSe and GaAs. The actual values of the valence band offsets depended on the value of the anion/cation ra-

5.2. BAND OFFSETS

tio during the nucleation of ZnSe on GaAs and varied between 0.6 and 1.2 eV. Nishi et al. [355] determined the band offset of ZnSe/GaAs heterojunction using a free electron laser and internal photoemission technique. The value of ΔE_c was found to be 0.113 eV. Measurements of band offsets of $Zn_{0.61}Cd_{0.39}Se/InP$ heterojunction have also been made [356]. The band offsets were determined using the $C-V$ technique. The value of the conduction band offset was –0.102 meV. Rubini et al. [335] determined the band offsets at the ZnSe/AlAs(001) interface. Valence band offset for Se rich interface was 0.2 eV and for Zn rich interface, it was o.9 eV. Values of band offsets determined by different workers are substantially different.

Values of band offsets of ZnSe/Si and ZnS/Si have also been determined. The values are $\Delta E_c = 1.7$ eV for ZnSe/Si(100) [362] and $\Delta E_v = 0.7$ eV for ZnS/Si(111) [363].

Using the model solid theory, Shahzad et al. [325] calculated band offsets at the interfaces of ZnS/ZnSe deposited on relaxed ZnSe layers. In this structure only ZnS barrier layers are strained. The band offset ΔE_v was 0.58 eV. Therefore the value of ΔE_c was 0.03 eV. Band offsets of thin layer $ZnSe/ZnS_xSe_{1-x}$ superlattice grown on relaxed ZnSe layers were also investigated. Calculations were made for several values of x. In this case also only the ZnS_xSe_{1-x} layers are strained. For $x = 0.25$, ΔE_v was 0.17 eV, and ΔE_c was 0.01 eV. If the superlattice is free-standing, ΔE_v is 0.21 eV, and ΔE_c was 0.01 eV. Shahzad et al. [325] have given numerical values of the band offsets for a large number of well and barrier thicknesses and compositions. Several other groups determined the band offsets of ZnSe/ZnS system in 1990s [364]. The values obtained by these subsequent workers have largely confirmed the results of Shahzad et al. [325]. The calculated values agreed with the experimental values within 0.05 eV. The band offsets of $ZnS_xSe_{1-x}/ZnSe$ heterojunctions have also been determined by Surkova et al. [360]. The values of x were in the range 0 to 0.5. Both layers were doped with $10^{17}/cm^3$ Ni ions. It is believed that the energy levels of transition metal ions are pinned to a certain universal reference level of a semiconductor. Therefore valence band offsets can be determined by measuring the energy levels of the Ni ions with reference to the valence band in the epilayers on the two sides of the heterojunction [360]. The bandgaps of the epilayers and energy levels of Ni ions were determined by reflection and absorption spectroscopy. A broad but strong absorption band in the range 1.7–2.2 eV was attributed to photoionization of Ni (2+ → 1+). The values of the band offsets of $ZnS_xSe_{1-x}/ZnSe$ heterojunctions obtained in this manner by Surkova et al. [360] were similar to the results of Shahzad et al. [325] discussed above. The value of ΔE_c was considerably smaller than the value of ΔE_v for all compositions of ZnS_xSe_{1-x}. However the absorption band due to Ni ions was broad and uncertainties in the experimental results were large.

If the quantum well in the $ZnSe/ZnS_xSe_{1-x}$ structures (for $x < 0.3$) is strained, the value of ΔE_c is further reduced and the conduction band-edge becomes almost flat across the junction [364]. The values of the conduction band are too small to cause significant confinement. If the strain in the well is large, the $ZnSe/ZnS_xSe_{1-x}$ structure changes from type I to type II. In the type II structure the electrons are confined in the barrier whereas the holes continue to be confined in the well. Hydrostatic pressure also changes the band offsets because the hydrostatic deformation potential for ZnS and ZnSe are quite different. The bandgap pressure coefficient is 72 meV/GPa for ZnSe and 64 meV/GPa for ZnS [364]. At some value of the applied hydrostatic pressure, change of the structure from type I to type II takes place [364, 365].

Miyajima et al. [353] determined the band offset of the ZnSe/ZnMgSSe heterojunction using PL and PL excitation spectroscopy. The composition of the ZnMgSSe alloy was such that the bandgap difference between ZnMgSSe and ZnSe was 220 meV. The value of ΔE_c was found to be $0.4 \Delta E_g$. In this case values of both ΔE_c and ΔE_v are comparable. Nishi et al. [355] suggested that for the ZnSe/ZnMgSSe heterojunctions, the ratio $\Delta E_c : \Delta E_v$ should be 0.45:0.55.

If one of the layers contains a Cd compound, the conduction band offset at the heterojunction is generally larger than the valence band offset [364, 351, 360]. The experimental values of ΔE_c at the ZnCdSe/ZnSe heterojunction range from 0.65 to 0.85 of the bandgap difference ΔE_g. Cingolani repeated the calculations using different values of the band offsets. The best results obtained by Cingolani [364] are $\Delta E_c = 0.8 \Delta E_g$. In CdZnS/ZnS also the conduction band offset is larger than the valence band offset. For a free standing $Cd_{0.3}Zn_{0.7}S/ZnS$ superlattice, the conduction band offset is 461 meV and the valence band offset is 88 meV. Menéndez et al. [366] measured resonance Raman scattering in strained pseudomorphic thin layer CdTe/ZnTe superlattices grown on $Cd_{0.1}Zn_{0.1}Te$ substrates. Menéndez et al. [366] showed that their results were consistent with the usually accepted result for these superlattices that valence band offset is small and most of the difference in the bandgap occurs as the conduction band offset for these heterojunctions also. Walecki et al. [349] have determined the band offsets of $Zn_{1-y}Cd_ySe/Zn_{1-x}Mn_xSe$ ($y = 0.15$, $x = 0.16$) heterojunctions grown on relaxed ZnSe buffer layers using magneto-optical measurements. Both $Zn_{1-y}Cd_ySe$ well layer and $Zn_{1-x}Mn_xSe$ barrier layers were pseudomorphic. The strain in the quantum well was estimated to be 0.3%. The excitonic bandgaps of the two layers were 2.612 eV and 2.850 eV respectively. The values of the band offsets were 30 meV for ΔE_v and 208 meV for ΔE_c. The band alignments in CdTe/ZnCdTe, ZnSe/ZnMgSe, CdTe/CdMnTe, and ZnSe/ZnMnSe are also similar. In all these cases the valence band offset is small and most of the bandgap difference appears as the conduction band offset [364, 367]. Band offsets at the interfaces of $Zn_{1-x}Mn_xSe/ZnSe$ multiple quantum wells have been determined by Klar et al. [357]. This work is discussed in detail in section 6.8.3. The value of the valence band offset was $(20 \pm 10)\%$ of the difference in the bandgap.

We now discuss the band offsets in type II heterojunctions (see Fig. 1.8). Yu et al. [337] have determined the band offsets of cubic CdSe/ZnTe heterojunction using x-ray photo-electron spectroscopy. Though normally stable phase of CdSe is wurtzite, thin films grown on ZnTe are of zinc blende cubic structure. The values of the band offsets obtained by Yu et al. [337] are $\Delta E_c = 1.22$ eV and $\Delta E_v = 0.64$ eV. The structure is type II, the valence band edge in CdSe is at higher energy. There is a large scatter in the values of ΔE_v obtained by

5.2. BAND OFFSETS

different workers [337]. Recently several authors have determined the band offsets at the BeTe/ZnSe interfaces [348, 336, 358]. The band offsets at the type II ZnSe/BeTe heterojunction have been measured by Platonov *et al.* [348] by PL spectroscopy. The values are 0.8 eV for the valence band offset and 2.2 eV for the conduction band offset. Since ZnSe has a large ionic binding and BeTe has a large covalent binding, the atomic structure and chemical composition of the ZnSe/BeTe interface has a large effect on the band offsets. Nagelstrasser *et al.* [358] investigated the band offsets of BeTe/ZnSe heterojunction for different interface terminations. The band offsets were determined by photoelectron spectroscopy. The valence band offsets was 1.26 eV for Zn-rich interface and 0.46 for Se-rich interface. These results show that band offsets can be modified by interface composition even for the isovalent heterojunctions. The results obtained by analyzing the PL data were different. For Zn rich interface, the valence band offset was 0.9 eV. The authors attributed the difference to an averaging effect because of the large size of the exciton, 10 nm, and short Deby length, 6nm, in the BeTe/ZnSe heterostructure. The valence band offset at the BeTe/GaAs heterojunction is small. There is a rather close valence band-edge match between GaAs and BeTe. Finally we give the values of band offsets of the p-ZnTe/p-ZnSe heterojunction [352]. The two layers forming the heterojunction investigated in Ref. [352] were the top layers of a laser structure. The thicknesses of the layers were 240 nm and 600 nm respectively. The bandgap difference of the two layers was 0.45 eV. The valence band offset was determined from the $C-V$ and $I-V$ measurements. Its value was between 0.7 and 0.8 eV. Since $\Delta E_v > \Delta E_g$, this heterojunction is also of type II.

5.2.4 Band offsets: III-Nitride heterostructures

Several groups have calculated and determined experimentally the band offsets of the III-Nitride heterostructures [228, 284, 285, 329]. The three nitrides form three heterojunctions: AlN/GaN, GaN/InN, and InN/GaN. The band offsets at the A/B heterojunction (A/B implies that epilayer A is grown on top of layer B) is different from the offsets in B/A heterojunction [285]. We therefore have 6 pairs of heterojunctions with 6 different band offsets for the three III-Nitrides. In all the six cases the band offsets are of type I. The measured [228, 284, 285] and calculated (using the LAPW method) [329] values of the offsets are given in Table 5.5. Calculated values for zincblende GaN are also included in the table. The substrate used in Refs. [284, 228] was 6H-SiC whereas in Ref. [285] both sapphire and 6H-SiC were used. Values of the band offsets depend on the strain in the two layers [228, 285]. The strain depends strongly on the experimental conditions. Therefore the scatter in the values is generally large. The value of the band offset for a given pair of layers A and B constituting the heterojunction is not unique. It depends on whether the pseudomorphic epilayer A is grown on the relaxed layer B or of the strained epilayer B grown on the relaxed layer A. The difference arises because strain gives rise to piezoelectric fields in the nitrides and induces band-bending. Since escape depth of the photo-electrons is small, the measured values correspond to a thin surface layer of the epilayer

Table 5.5: Valence band offsets (in eV) of WZ III-Nitride heterostructures[a]. The abbreviation used are: A for AlN, G for GaN and I for InN. A/B indicates top/bottom layers i.e. epilayer A grown on epilayer B.

Method	A/G	G/A	G/I	I/G	A/I	I/A	A/SiC
LAPW[b]	0.81	–	0.48	–	1.25	–	–
Exptl[c]	1.36	–	–	–	–	–	–
Exptl[d]	0.57	0.60	0.59	0.93	1.32	1.71	–
Exptl[e]	0.20	–	–	–	–	–	1.7

[a] For ΔE_V at the GaN/Al$_{0.2}$Ga$_{0.8}$N interface see [327].
[b] Ref. [329]. ΔE_V for ZB GaN is 0.84 eV.
[c] Ref. [284], Relaxed epilayer, see text.
[d] Ref. [285], Strained. A/B and B/A give different results.
[e] Ref. [228], Strained and thickness dependent. Values extrapolated to zero thickness are given.

the valence band offset is modified by the piezoelectric field. Since piezoelectric constants of the layers A and B are different, the valence band offset in the heterojunction A/B is different from its value in B/A heterojunction.

5.3 Optical properties of III-V semiconductors

Recently we have reviewed the optical properties of unstrained and strained conventional III-V semiconductors [52]. Discussion of quantum wires and quantum dots was also included in this review. We mention below some of our very recent work done after the publication of this review here.

The binding energy of the shallow acceptor states in GaAs/AlGaAs quantum wells (under external pressure) and InGaAs/AlGaAs strained quantum wells have been calculated by Zhao and Willander [368]. These calculations provide direct and strong evidence that the binding energy is affected considerably by both the confinement and strain. Strain distribution in an In$_{0.25}$Ga$_{0.75}$As quantum wire grown in V-grove has been calculated by Fu et al. [369]. The quantum wire was sandwiched between an AlGaAs buffer layer and a GaAs cap layer. The calculations showed that the strain in the quantum wire decreased linearly from the bottom buffer layer to the top GaAs layer. The change in PL energy distribution due to strain variation was also calculated and was found to be consistent with the experimental results..

In the last ten years study of fractional monolayaer heterostructures has become a subject of great interest [370, 371, and references given therein]. Zhao et al. [370] have studied MBE grown ZnTe islands sandwiched between CdTe layers. The CdTe layers formed wide ($\sim$ 78 nm) quantum well with CdMgTe or CdZnTe barrier layers. The concentrations of Mg was 25% and that of Zn was 8%. The structures were grown on CdZnTe substrate. Concentration of Zn

5.4. OPTICAL PROPERTIES OF II-VI SEMICONDUCTORS

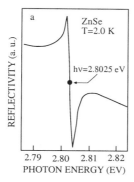
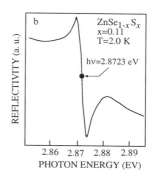

Figure 5.4: Reflection spectrum of a (a) ZnSe crystal and (b) ZnSe$_{0.89}$S$_{0.11}$ crystal in the exciton region at 2 K [372].

in the substrate varied from 0 to 5%. The distribution of the islands grown at lower temperatures was random. The distribution became uniform if the growth temperature was higher than 240

$$\Delta E = g_{eff} B \mu_B. \tag{5.8}$$

where μ_B is the Bohr magneton and B is the magnetic field. They found that the strain and confinement effects in the islands modify strongly the excitonic transitions due to the quantization of the excitonic center-of-mass motion. The modification occurs due to an increase in the hole g-factor. The electron g-factor remains unchanged. Self organization of CdSe islands sandwiched between ZnSe layers has been investigated by Sitnikova et al. [371]. CdSe layers were directly grown on GaAs-buffer/GaAs-substrate by MBE and MEE in the temperature range 280°C to 295°C. The structure of the sample was examined by TEM. The thickness varied between 0.15 and 2.8 monolayers. Self-organized islands were observed when the nominal thickness of the CdSe layer was 0.7 ML. The morphology of the fractional monolayer (i.e. the islands) was more homogeneous in the MEE grown samples.

5.4 Optical properties of II-VI semiconductors

5.4.1 Reflectance, absorption, and luminescence

The 2 K reflectivity spectra of ZnSe and ZnSe$_{0.89}$S$_{0.11}$ unstrained crystals measured by Mycielski et al. [372] are shown in Fig. 5.4. Free exciton feature at 2.8025 eV is clearly visible in the reflection spectrum of ZnSe in Fig. 5.4a. The corresponding feature in the reflection spectrum of ZnSe$_{0.89}$S$_{0.11}$ is at 2.8723 eV (see Fig. 5.4b). Measurements with different concentration of sulfur in the crystals showed that for $x \leq 0.15$ the FE energy in ZnSe$_{1-x}$S$_x$ crystals varies as

$$\text{FE energy (eV)} = 2.8025 + 0.600x \tag{5.9}$$

Strössner et al. [373] have measured the pressure dependence of the lowest direct absorption edge of ZnTe. The measurements were made at room temperature using a diamond anvil cell for pressures up to 11.9 GPa. The values of the bandgap determined from the observed edge are given by

$$E_g = 2.27 + 10.4 \times 10^{-2} P - 28 \times 10^{-4} P^2, \tag{5.10}$$

where E_g is in eV and the pressure P is in GPa. The effect of pressure on bandgap was computed using local empirical pseudo-potential and *ab initio* calculations. The experimental values agreed well with the calculated values. Direct absorption edge in Ge, ZnSe and many III-V semiconductors varies sub-linearly with pressure. However if the edge is plotted as a function of change in lattice constant $(-\Delta a/a)$, the plot becomes linear to a good approximation. Similar behavior is found for the ZnTe also.

PL and cathodoluminescence (CL) of unstrained ZnSe crystals have been studied by many authors. A detailed discussion of PL lines in bulk ZnSe, their nomenclature, observed peak positions, identification and references to earlier work are given in Refs. [118] and [362]. Typical 40 K CL spectra of two ZnSe crystals grown by the vapor transport method in He and in H_2 [374] is shown in Fig. 5.5. There are several lines in the excitonic region and there is a broad band

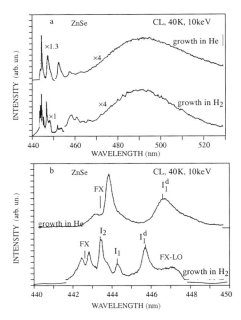

Figure 5.5: CL spectra of two ZnSe single crystals grown by the vapor transport method in He and in H_2, (a) shows the view of the whole spectra and (b) shows the expanded exciton region [374].

centered at 490 nm. The nature of the broad band is not clear. It is probably

5.4. OPTICAL PROPERTIES OF II-VI SEMICONDUCTORS

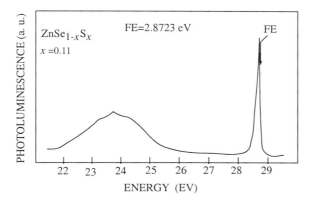

Figure 5.6: PL spectrum of a the ZnSe$_{0.89}$S$_{0.11}$ crystal at 4.2 K. The free exciton FE feature seen in the reflectivity (but not visible here) is indicated by the arrow. The sharp line at 2.864 eV is due to excitons bound to donors and the broad line at 2.38 eV is due to DA pairs [372].

related to structural defects, Zn vacancies and oxygen impurities. The expanded excitonic regions of the two crystals show some differences. In the crystal grown in He, all lines have larger widths and are shifted to longer wavelengths. Similar effect of the ambient gas has also been observed in ZnS crystals [374]. The effect is related to oxygen impurity. In the crystals grown in hydrogen ambient, oxygen form water molecules which condense in the colder part of the ampoule. If the crystals are annealed in Zn melt, I_1 line disappears, the intensity of the I_1^d line decreases and intensity of FX line increases. PL in the crystals grown by solid state regrowth method have also been studied [375, 376]. The peak positions in the spectra of the crystals grown by the two methods are in fair agreement. 4.2 K PL spectra of the top part (last to freeze) Cd$_{0.97}$Zn$_{0.03}$Te grown by the vertical gradient freezing method has been measured by Asahi et al. [112]. The conductivity of the crystals was p-type because the crystals were Te rich. The 1.4 eV peak generally observed in the low purity crystals was absent. As compared to CdTe, all peak positions in Cd$_{0.97}$Zn$_{0.03}$Te were at higher energies because of larger bandgap of ZnTe (see Table A in Appendix A). The shift of the peaks was about 5 meV for every 1% of Zn.

4.2 K PL spectrum of ZnSe$_{0.89}$S$_{0.11}$ crystal is shown in Fig. 5.6 [372]. The free exciton feature seen in reflection at 2.873 eV in Fig. 5.4b is not resolved in this figure. Its position is shown by the arrow. The sharp line at 2.864 eV is due to donor bound excitons. The broad band at 2.38 eV is due to DA pairs.

5.4.2 Effect of strain on the optical properties of II-VI semiconductors

Yao et al. [274] have made extensive measurements of reflection and PL spectra of strained ZnSe layers grown on (100) GaAs. The reflection spectra of nine samples with thicknesses in the range 0.07 to 22 μm are shown in Fig. 5.7. For $h = 0.07$ μm there are two reflection anomalies shown by features at 440.0 nm (2.817 eV) and 442.0 nm (2.8050 eV) respectively. The two features are attributed to the excitons associated with LH and the HH bands. As predicted by Eq. (5.1) the FE energy in the HH band (and therefore the HH bandgap) have increased from the bulk FE energy 2.8025 eV (see Fig. 5.4) to 2.8050 eV due to strain. The temperatures used in the measurement on bulk crystals is 2 K and on the epilayers is 4 K. At 2 K the FE energy in the epilayer will be somewhat larger. The higher energy position of the LH FE energy is also consistent with Eq. (5.2). As the thickness increases, both LH FE and HH FE move to lower energies, LH FE moves rapidly and the HH FE moves slowly. The separation between the two decreases. At 2 μm the two features merge into one. This behavior is consistent with the changes in strain and Eqs. (5.1) and (5.2). Compressive strain relaxes in thicker films and thermal tensile stress is produced due. The net compressive stress decreases rapidly and becomes practically zero at $h = 2$ μm. At larger thicknesses tensile stress dominates, the LH FE crosses the position of HH FE and continues to move to lower energies.

Yao et al. [274] made PL measurements for the same 9 samples for which reflectivity measurements are shown in Fig. 5.7. The intensity of PL increases with thicknesses in the range 0.15 to 0.8 μm because the density of defects in the surface layer decreases continuously [274]. The intensity saturates for larger thicknesses. For thickness less than 1 μm, excitonic line associated with the HH band dominates. For larger thicknesses dominant excitonic line is associated with the LH band. The excitonic lines shift to lower energy as the thickness increases because compressive strain decreases as explained earlier. The exciton line for $h = 2$ μm has the same position as in bulk crystal indicating that the two strains cancel each other at this thickness. At larger thicknesses tensile strain dominates because a tensile strain is generated during the cool down of the layer. The effect of strain on PL is consistent with the reflectivity measurements shown in Fig. 5.7 and with shift and splitting of the valence band shown in Fig. 5.1.

The PL spectra of nitrogen doped ZnSe was measured by Ohkawa et al. [326]. Thickness of the layers was large (2 μm) and they were fully relaxed at the growth temperature. Thermal tensile strain was created in the layers because the thermal expansion coefficients of the layers and the substrate are different. As discussed earlier under tensile strain separation between the HH band and the conduction band is larger than the separation between the LH band and the conduction band. A high energy PL peak and a low energy PL peak due to transitions to the HH and the LH bands were resolved. The relative intensities of the two lines depend on the population of the holes in the HH and the LH bands. As the temperature increases, holes are thermally activated from the LH into the HH band and relative intensity of the two lines changes.

Olego et al. [121, 264] have measured Raman spectra of ZnSe epitaxial layers grown on GaAs substrate. They used lasers of two different wavelengths, 413.1 nm and 488 nm. The absorption coefficient of 488 nm light in ZnSe is practically zero [264]. 413.1 nm light penetrates a distance of $\sim$ 50 nm in ZnSe [264]. The Raman spectra measured with the 180 nm thick layer is shown in Fig. 5.8. Since

5.4. OPTICAL PROPERTIES OF II-VI SEMICONDUCTORS

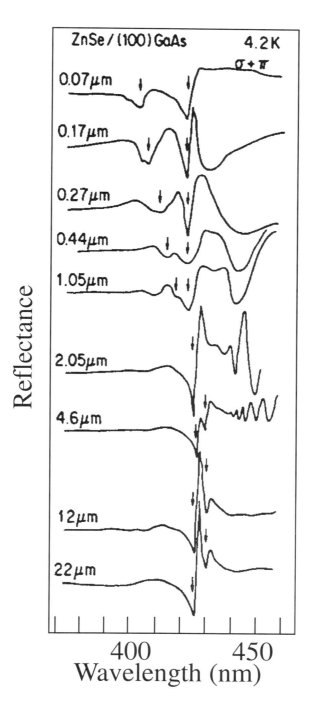

Figure 5.7: Reflection spectra ZnSe epilayers. [274].

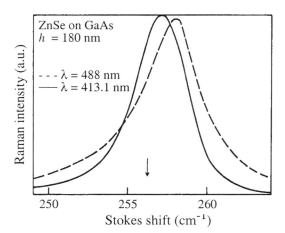

Figure 5.8: Raman shift of LO phonons of a ZnSe crystal grown on GaAs. Bulk ZnSe shift of 256.3 cm^{-1} is shown by the arrow [264].

lattice constant of ZnSe is bigger (mismatch between ZnSe and GaAs is 0.27%) thin ZnSe layers grown on GaAs are pseudomorphic and are under compressive stress. For the unstrained bulk ZnSe the LO Raman frequency is 256.3 cm^{-1}. The strain induced shifts in the Raman line measured with both the lights are shown in Fig. 5.8. The shift is ~ 1.14 cm^{-1} when 488 nm light is used. The shift measured with the 413.1 nm laser light is ~ 0.7 cm^{-1}. These shifts give 0.0027 and 0.0023 as the values of strain. As suggested by Olego et al. [264] the strain in the top part of the layer seems to be smaller than in the lower part. Olego et al. [264] assumed that the shift measured with 413.1 nm light was due to strain in the top 50 nm of the layer and that measured with the 488 nm light was the average over the whole thickness. Pinardi et al. [125] have discussed alternative interpretation of these experiments.

It is known that strain in partially relaxed layers can vary with depth of the epilayer. This happens because the threading dislocations are curved and their horizontal components relieve strain. The magnitude of this component varies with distance from the interface [310]. Following Olego et al. [264] Pinardi et al. [125] assumed that strain in the top part is 0.0023. It was assumed that the strain increases linearly with depth z. The strain distributions were calculated using different trial values of the strain at the interface. Raman shift was calculated using the method recently developed in Refs. [282, 308]. In this method the secular equation connecting the strain tensor and the Raman frequencies is solved to obtain the peak positions of the spectra at each point in the volume sampled by the laser light. Spectra for each of these peaks are calculated and superposed to synthesize the final spectrum [282]. The calculated values could be made to agree with the observed values if a value of 0.46% for the strain at the interface is assumed. This value is nearly twice as large as the lattice mismatch strain. Calculations were repeated with the assumption that

5.4. OPTICAL PROPERTIES OF II-VI SEMICONDUCTORS

strain increased exponentially with distance from the surface. These calculations also yielded a value of interface strain larger than the lattice mismatch. The strain near the interface can not be larger than the lattice mismatch. Therefore an alternative explanation must be found to interpret the difference between the 488 and 413.1 nm spectra.

The edge induced relaxation can cause lowering of strain in the top part provided the lateral dimensions of the layer are sufficiently small [296]. The calculations made by Pinardi et al. [377] showed that for a circular film of radius R and thickness h the ratio R/h should be less than 25 to obtain required decrease of strain in the top part. The value of R necessary to satisfy this condition seems to be too small for the layers used by Olego et al. [264]. The thickness 180 nm of the ZnSe layer is slightly more than the critical thickness. Due to some partial relaxation, the strain can decrease. Therefore the strain measured by the 413.1 nm light (0.23%) is probably the correct strain in the layer. Pinardi et al. [377] suggested that the Raman peak due to interfacial layer of Ga_2Se_3 observed at 258 cm^{-1} [378] overlaps with the strain induced shift and gives the composite line to 257.4 cm^{-1} observed with the 488 nm light. Since the intensity of 413.1 nm light decreases to < 2% at the interface, the contribution of the Ga_2Se_3 line is not contained in the spectrum of this light [125].

5.4.3 II-VI quantum wells and superlattices

5.4.3.1 Quantum confinement

It is well known that quantum confinement of electrons and holes increases the transition energies [325, 364]. The optical properties of quantum wells are determined by both, the strain and quantum confinement. Blue shift in the transition energy due to confinement depends on the thickness and the depth of the well. The depth is determined by the band-offsets at the interface between the well and barrier layers. Change in the transition energy is calculated by solving numerically the following equation [325]:

$$\cos(kl) = \cos(k_z l_z)\cosh(k_b l_b) + \frac{1}{2}\left[X - \frac{1}{X}\right]\sin(k_z l_z)\sinh(k_b l_b), \quad (5.11)$$

where l is the superlattice period and l_z and l_b are well and barrier widths. The expression for X is

$$X = \frac{k_b m_z}{k_z m_b}, \quad (5.12)$$

$$hk_z = (2m_z E)^{1/2}, \quad hk_b = [2m_b(V - E)]^{1/2}, \quad (5.13)$$

where m_z and m_b are the effective masses in the well and barrier layers and V is the potential barrier height. The barrier height V is equal to the band offset at the quantum well/barrier interface.

Table 5.6: Structure of ZnSe/ZnS$_{0.18}$Se$_{0.82}$ superlattices studied by Shahzad et al. [265]. d_1 and d_2 are the thicknesses of the individual layers, and h is the thickness of the whole superlattice.

Sample	d_1/d_2(Å/Å)	h (μm)	Buffer layer
1	40/43	0.10	ZnSe
2	41/43	0.30	ZnSe
3	37/37	0.56	ZnSe
4	46/49	1.0	ZnSe
5	55/60	4.6	ZnSe

5.4.3.2 ZnSe/ZnS$_x$Se$_{1-x}$ superlattices

Extensive work has been done on the optical properties of ZnSe/ZnS$_x$Se$_{1-x}$ quantum wells and superlattices [265, 325, 364]. Shahzad et al. [265] studied the effect of superlattice thickness on the PL in ZnSe/ZnS$_{0.18}$Se$_{0.82}$ superlattices. The superlattices were grown on a 1 μm thick ZnSe buffer layer deposited on GaAs substrate. Description of the 5 superlattices studied by Shahzad et al. [265] is given in Table 5.6. PL of the superlattices was measured at 5 K. A low energy peak at 2.796 eV due to donor-bound exciton was observed in all the sample due to transitions in the ZnSe buffer layer. A peak at higher energy observed in each sample was attributed to impurity-bound excitons in the ZnSe well. As total thickness h of a superlattice increased the position of this peak shifted to higher energies. Shahzad et al. [265] calculated the shift due to confinement alone, subtracted it from the observed shift and determined the shift due to strain. The shift due to strain increased as the total thickness of the superlattice increased. For $h = 0.1$ μm the observed shift was almost entirely due to confinement effect. The strain induced shift was negligible. If the total thickness' is small, the superlattice is pseudomorphic with the relaxed ZnSe layers. In this case strain in the ZnSe layers is expected to be negligible. For $h = 0.56$ μm, a strain induced shift of > 4 meV was found which corresponds to 0.3% strain in the ZnSe layers. As the thickness' increases to more than the critical thickness of the superlattice as a whole, effect of the substrate diminishes. The strain in the lattice corresponded more and more closely to the case in which the SLS + buffer layer are in a free standing configuration. In this configuration the ZnSe well layers become strained. The strain is compressive. For the thickest sample ($h = 4.6$ μm) the strain in the top part of the superlattice measured by Raman scattering was close to that expected from the free-standing superlattice.

In another publication Shahzad et al. [325] studied ZnSe/ZnS$_x$S$_{1-x}$ ($x < 0.30$) superlattices grown on 1 μm relaxed ZnSe buffer layers. The layer thicknesses, determined by TEM, ranged from 18 to 100 Å. These thickness are smaller than the critical thicknesses of individual layers. Altogether 8 samples, designated sample a to sample h were studied. The total thickness of sample

5.4. OPTICAL PROPERTIES OF II-VI SEMICONDUCTORS 135

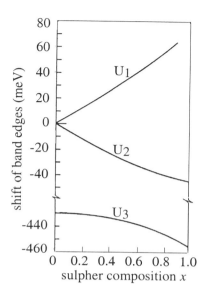

Figure 5.9: Splitting of the valence band of ZnSe well layers with equal thickness of ZnS_xSe_{1-x} barrier layers due to tetragonal strain [325]. The superlattice is assumed to be symmetrically strained. U_1 is the heavy hole band, U_2 is the light hole band and U_3 is the spin-orbit split-off band.

c was 1000 Å and that of sample e was ~ 4 μm. The total thickness of each of the other samples was ~ 1 μm. The sample e consisted of 801 50 Å thick alternating ZnSe and $ZnS_{0.19}S_{0.81}$ layers.

To interpret the PL experiments, Shahzad et al. [325] calculated the splitting of the valence band shown in Fig. 5.9. To calculate the bandgap the effect of hydrostatic strain on the band edges was taken into account. For calculating strain and confinement effects Shahzad et al. [325] used the effective mass values of electrons $0.17m_0$ for ZnSe and 0.27 for ZnS. The corresponding values for the heavy holes were $0.60m_0$ and $0.49m_0$ respectively. The 5 K values of the bandgap used in the calculations were 2.83 eV (436.0 nm) for ZnSe and 3.84 eV (322.8 nm) for ZnS. The 5-K PL spectra of three $ZnSe/ZnS_xZnSe_{1-x}$ superlattice samples is shown in Fig. 5.10. PL spectrum from a relaxed 4.9-μm thick ZnSe layer grown on (001) GaAs is also shown. In this spectrum, the peak at 2.8033 eV, labeled as $E_{Gx}^{n=1}$, is due to the recombination of free excitons. The feature I_{2o}^{Ga} is attributed to excitons bound to neutral Ga donors at the Zn sites. Two transition at 2.7913 eV and 2.7903 eV are also observed. The first transition is due to excitons bound to Li acceptors. Origin of the second transition is not clear. The bound excitons (BE) and free exciton (FE) features in the superlattices are shifted to higher energies due to strain and confinement effects. The FWHM of the PL peaks in the superlattices is larger as compared to the PL peaks in the thick ZnSe layer. In the case of the 4 μm thick sample e strain in the ZnSe wells calculated on the assumption that the superlattices were free standing agreed

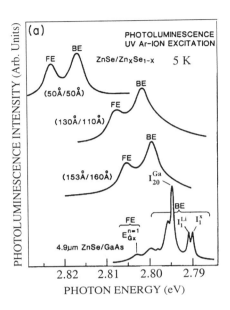

Figure 5.10: PL spectra of three ZnSe-ZnS$_x$Se$_{1-x}$ superlattices and of a relaxed 4.9-μm ZnSe layer [325].

with the measured strain. In other cases the strain calculated with the same assumption was smaller than that determined directly by Raman measurements. The three superlattices being discussed here (each about 1 μm thick) were not thick enough to acquire the free standing configuration. Excitation spectra of the superlattices were also measured. For the 130Å/110Å superlattice the peak due to transitions to the light hole band was resolved in the excitation spectra. The splitting between light hole and heavy hole bands was 8 meV. This value agreed with the calculated value of the strain induced splitting.

5.4.3.3 ZnSe/Zn$_{1-x}$Cd$_x$Se quantum wells

Babucke *et al.* [379] have measured the PL, reflection and transmission of Zn$_{1-x}$Cd$_x$Se/ZnSe quantum wells grown by MBE on 0.5 μm ZnSe buffer layer deposited on (001) GaAs substrates. In one sample, the number of Zn$_{0.83}$Cd$_{0.17}$Se wells was 10 and the widths of the wells and barriers were 4 and 6 nm respectively. In the other sample, the number of Zn$_{0.80}$Cd$_{0.20}$Se wells was 30, with the same widths (as in the first set) of the wells and the barriers. Measurements were also made under a vertically applied electric field (these measurements are discussed in section 6.6). PL measurements were made on samples with several values of x in the range 0 to 0.3. The widths of the wells were 2, 5 and 9 nm. PL measurements on a single Zn$_{1-x}$Cd$_x$Se thick bulk-like epilayer with same values of x were also made. For reflection measurements n-doped substrate was used.

5.4. OPTICAL PROPERTIES OF II-VI SEMICONDUCTORS

A transparent Schottky TiAu top contact and an ohmic AuGe back contact were used. For transmission measurements the substrate was removed by chemical etching and the transparent Schottky TiAu contact was deposited. The Schottky contact produces a depletion layer in the quantum well and an electric field exists even without the application of an external field. Assuming a free carrier concentration of 10^{16} cm^{-3} in the quantum well, the electric field at the applied voltage $U = 0$ V was estimated to be 9×10^4 V/cm. The field increased to 1.7×10^5 V/cm at a reverse bias $U = -5$ V and 3.7×10^5 V/cm at $U = -30$ V. (The effect of electric field on the optical properties of these samples is discussed in section 6.6.) An argon ion laser was used for PL measurements and halogen glow lamp for reflection and transmission measurements. The $Zn_{1-x}Cd_xSe$ quantum wells in these structures are under compressive in-plane strain which causes splitting of the heavy hole and the light hole bands. Heavy hole band is the lowest energy band. PL measurements at 5 K were used to determine the heavy hole bandgap of the $Zn_{1-x}Cd_xSe$ alloy (bulk like epilayers). The binding energy 19 meV of the exciton in ZnSe was taken into account. The difference in the binding energy of the exciton in ZnSe and CdSe (~ 15 meV) was ignored. The heavy hole bandgap determined in this manner is given by the following expression.

$$E_g(x) = 2.82 - 1.50x + 0.39x^2 \text{ (eV)}. \tag{5.14}$$

In the quantum wells the PL peak shifted to higher energy due to confinement effects. The confinement energies were calculated and compared with the experimental data treating the band-offsets as adjustable parameters. The best fit was obtained for the total combined band offsets (i.e. bandgap difference) $\Delta E_g = 1.35x - 0.35x^2$ eV. The conduction band offset was 70% of the total bandgap difference. The confinement induced change in the exciton binding energy was large. The binding energy was 27 meV for the 4 nm well with $x = 0.17$. The strong confinement effects are similar to those observed in III-V quantum wells. Large increase in the exciton binding energy shows that the quantum wells can be used for room temperature optoelectronic devices.

Cingolani and co-workers [364, and references given therein] have investigated optical absorption, luminescence and binding energies of excitons in ZnCdSe/ZnSe quantum wells. The samples were grown by MBE on thick ZnSe buffers which were deposited on a GaAs substrate. The layers were pseudomorphic and strained. For transmission measurements the samples were glued to a sapphire wafer and GaAs substrate was mechanically polished to a thickness of 80 μm. GaAs was selectively removed from 100 μm windows by wet etching. X-ray diffraction rocking curves showed distinct satellite peaks due to the superlattice periodicity. The peaks were broadened due to disorder. Comparison with the prediction of the dynamical theory showed that there was some intermixing at the interface.

Typical 10 K absorption and PL spectra of $Zn_{1-x}Cd_xSe$ MQWs are shown in Fig. 5.11. The strong absorption around 2.8 eV is due to ZnSe buffer and barrier layers. Distinct excitonic peaks are seen at lower energies. The features shift to higher energies with decreasing well thickness due to increase in the

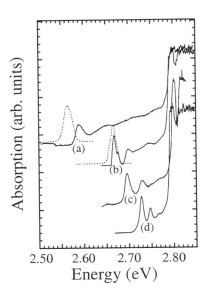

Figure 5.11: Absorption (continuous lines) and PL (dashed lines) spectra of $Zn_{1-x}Cd_xSe$ multiple quantum wells at 10 K. The compositions and well widths are (a) $x = 0.23$, $L_w = 3$ nm, (b) $x = 0.11$, $L_w = 7$ nm, (c) $x = 0.16$, $L_w = 3$ nm, and (d) $x = 0.11$, $L_w = 3$ nm [364].

quantum confinement. With increase in Cd content the spectra extend towards the green wavelength due to increase in the quantum well depth. Narrow wells ($L_w = 3$ nm) show only one absorption peak. The spectra become structured in wider wells. The PL spectra are the excitonic bands which show Stoke shifts from 2 to 20 meV. The shift increases with increase in Cd content and decrease in well width. Halfwidth of the PL peaks increases with Cd content. This shows that fluctuation of composition and localization of excitons increase with Cd concentration. Absorption measurements were also made at higher temperatures. In the wells with high Cd content ($x = 0.16$) excitonic absorption was observed up to room temperature. With lower Cd content ($x = 0.11$) the excitonic feature disappeared above 246 K. The exciton-phonon coupling increases with Cd content and compensates partly the reduction in the coupling due to confinement [364]. Several authors [364, and references given therein] investigated the photocurrent spectra in the ZnCdSe/ZnSe quantum wells grown of the *p-i-n* configuration. Results of these investigations show that photocurrent spectra closely reproduce the actual absorption spectra of the heterostructures.

The binding energy of the excitons in bulk ZnSe is about 20 meV. Calculated and experimental binding energies of the excitons in the ZnCdSe/ZnSe quantum wells is shown in Fig. 5.12. The values of the parameters used in these calculations are given in Table 5.7[2]. The binding energy of the excitons increases due

---

[2]Some values given in this table are different from those given in the Tables in appendix

5.4. OPTICAL PROPERTIES OF II-VI SEMICONDUCTORS

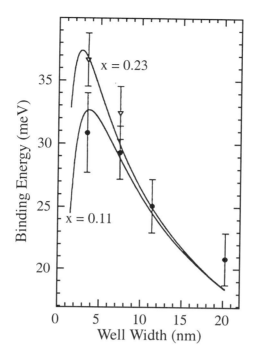

Figure 5.12: Symbols show the experimental exciton binding energies of $Zn_{1-x}Cd_xSe$ quantum wells. The continuous curves show the values obtained by variational calculations [364].

to quantum confinement. It increases from its bulk value of 20 meV to up to 40 meV. In the ZnCdSe/ZnSe quantum wells excitons become stable at room temperature. Therefore they play important role in ZnSe based LEDs and laser diodes. The enhancement of binding energy and reduction in exciton-phonon coupling depends on the composition, width and strain in the well. For small Cd content, ($x \leq 0.1$) the stabilization of the exciton due to confinement is negligible. For $x > 0.2$, a strong stabilization occurs. The integrated absorption of light is enhanced by a factor up to 3 in the multiple quantum wells [364].

The effect of well width on the interband transition energies of $Zn_{1-x}Cd_xSe$ quantum wells is shown in Fig. 5.13. The values of x are 0.11 Fig. 5.13(a) and 0.23 in Fig. 5.13(b). The curves represent the results of envelop function calculations in the effective mass approximation. The effect of strain was taken into account using equations derived in section 3.3.2. To obtain experimental values of the transition energies, binding energies of the excitons given in Fig. 5.12 were added to the HH and LH absorption features. The maximum discrepancy between theoretical and experimental values of the transition energies is $\sim$ 10 meV. A 1% fluctuation in the composition or monolayer fluctuations

Table 5.7: Electronic parameters of ZnSe, CdSe and ZnS [364].

Parameter	ZnSe	CdSe	ZnS
E_g (eV)	2.821	1.765	3.840
m_e/m_0	0.16	0.13	0.34
m_{hh}/m_0	0.6	0.45	1.76/0.61
m_{lh}/m_0	0.145	–	–
ε	8.8	9.3	8.32
a (Å)	5.6676	6.077	5.4060
C_{11} (N/m^2)	8.26×10^{10}	6.67×10^{10}	10.4×10^{10}
C_{12} (N/m^2)	4.98×10^{10}	4.63×10^{10}	6.50×10^{10}
α (eV)	–	–	-4.53
b (eV)	−1.2	0.8	−1.25
Δ (eV)	0.43	–	0.070

in the width can explain this discrepancy. Cingolani [364] also calculated and measured the transition energies of the quantum wells with $L_w = 3$ nm and values of x in the range 0.10 to 0.26. The transition energies decreased linearly with Cd content x. The agreement between the experimental and theoretical values was quite good.

Recently Toropov et al. [380] have measured PL of a CdSe/ZnSe quantum well with an embedded deeper ZnCdSe quantum well. They found that the localized and extended states of the excitons coexist. They interpreted their results in terms of an effective mobility edge. The states below this edge coupled with the phonon modes forbidden for free excitons. PL kinetics of these states corresponds to hopping relaxation within the tail of the localized states. The lifetime of the states above the effective mobility edge is small. They escape by tunneling at an extremely fast (on the ps scale) rate in to the deeper well.

5.4.3.4 CdTe/ZnTe superlattices

CdTe/ZnTe superlattices have been studied by several authors [278, 265, 366, 381, 382]. Miles et al. [381] studied several strained and unstrained CdTe/ZnTe superlattices grown on CdTe and Cd$_x$Zn$_{1-x}$Te relaxed buffer layers (see Table 5.8). The strain induces a shift of the bandgap of the superlattices grown on Cd$_x$Zn$_{1-x}$Te buffers of up to 350 meV. These shifts were not observed in superlattices grown on CdTe buffer layers. Luminescence intensities were high in all samples. TEM studies showed large density of dislocations near the interface but the density decreased as the distance from the interface increased. In sample 8 the density near the interface was $10^{10} - 10^{11}$ cm^{-2}. The density was reduced by 4 orders of magnitude near the top surface. Strain distribution was analyzed using Cu $K\alpha$ x-ray diffraction spectra measured in a $\theta/2\theta$ arrangement. The diffraction spectra were calculated both for the strained and unstrained superlattices using the kinematical theories. The observed spectra for sample 8 (see

5.4. OPTICAL PROPERTIES OF II-VI SEMICONDUCTORS

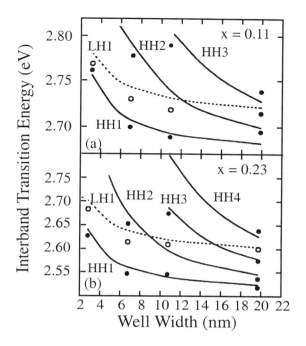

Figure 5.13: Interband heavy-hole (HH, filled circles) and light-hole (LH, open circles) transition energies of $Zn_{1-x}Cd_xSe$ quantum wells with $x = 0.11$ and $x = 0.23$ at 10 K. The continuous curves represent HH transitions with quantum number $n = 1, 2, 3, ...$, and the dashed curves represents the $n = 1$ LH transitions [364].

Table 5.8) agreed with the calculated spectra of the unstrained superlattice. In all other case the observed spectra were in-between the calculated spectra for the unstrained and the strained superlattices. These results show that most of the strain had relaxed in sample 8 but the residual strain in other samples was substantial.

Cibert et al. [278] fabricated single CdTe layers and CdTe quantum wells separated by ZnTe barriers on relaxed 1 to 2 µm ZnTe buffer layers (deposited on GaAs substrates) by MBE. RHEED oscillations were recorded during the growth. A few tens of oscillations were observed during the growth of ZnTe layers. When CdTe layers were grown, strong and clear oscillations were observed only during the first five monolayers. After the growth of the fifth monolayer, the intensity decreased abruptly. This shows that relaxation in CdTe layers sets in after the growth of the 5 monolayers. Therefore the critical thickness of CdTe on ZnTe is 5 monolayers or 17 Å [278]. PL of CdTe quantum wells with different thicknesses was measured. The PL was at ~ 2.35 eV for 1 ML and shifted to lower energies as the thickness increased. It was at ~ 1.98 eV from a quantum well of 5 ML thickness. As soon as the thickness increased to

142 CHAPTER 5. BAND STRUCTURE AND OPTICAL PROPERTIES

Table 5.8: Structure of CdTe/ZnTe superlattices grown on buffer/(100)GaAs substrates [381]. N is the number of periods in the superlattice and d_1 and d_2 are the thicknesses of the individual layers.

Sample	d_1/d_2(Å/Å)	N	Buffer layer
1	26/32	200	CdTe
2	31/23	200	CdTe
3	56/30	150	CdTe
4	27/30	200	ZnTe
5	24/30	200	$Cd_{0.46}Zn_{0.54}Te$
6	27/30	200	$Cd_{0.41}Zn_{0.59}Te$
7	35/32	200	$Cd_{0.46}Zn_{0.54}Te$
8	21/20	400	$Cd_{0.50}Zn_{0.50}Te$

Table 5.9: Observed Raman phonon frequencies (in cm^{-1}) in samples A and B described later in Table 5.10 and in bulk crystals [366].

Sample	LC_2	LZ_1	Calculated
A	184	205.5	–
B	180	–	–
Bulk CdTe	171	–	–
Bulk ZnTe	209qq	–	–
Bulk CdTe strained (A)	–	–	189
Bulk CdTe strained (B)	–	–	183
Bulk ZnTe strained	–	–	206

more than 5 MLs intensity of PL decreased drastically. Thus PL results support strongly the result obtained from RHEED oscillations that the critical thickness of CdTe on ZnTe is 5 MLs. TEM showed that if the thickness of the QWs was smaller than 5 MLs, no defects were present at the interfaces. In thicker QWs, stacking faults, partial dislocations, 60° dislocations and 90° Lomer dislocations were observed.

5.4.3.5 Raman studies of strained CdTe/ZnTe superlattices

The shift of the frequency of Raman phonons can be used to determine the strain in epilayers quantitatively [366]. Energies of the quantum well excitons can be obtained by measuring resonances in Raman cross section as a function of the laser wavelength. Optical phonons are used to probe the properties of the excitons. Menéndez et al. [366] have studied the LO vibrations in two samples of CdTe/ZnTe superlattices. The structure of the samples is described in Table 5.10. They observed a broad structure in the Raman spectra, labeled IF, and a series of narrow peaks, labeled LC_m and LZ_n from the CdTe and

5.4. OPTICAL PROPERTIES OF II-VI SEMICONDUCTORS

Table 5.10: Structure of CdTe/ZnTe superlattices studied by Menéndez et al. [366]. N is the number of periods in the superlattice and d_1 and d_2 are the thicknesses of the individual layers.

Sample	d_1/d_2(Å/Å)	N	Buffer layer
A	6/9, (21Å/27Å)	21	$Cd_{0.1}Zn_{0.9}Te/ZnTe$
B	15/23, (51Å/68Å)	15	$Cd_{0.1}Zn_{0.9}Te/ZnTe$

ZnTe layers respectively. Some of the observed frequencies are given in Table 5.9. It is seen from Table 5.9 that the frequency $\omega(LC_2)$ in both the samples is higher than ω(Bulk CdTe). $\omega(LZ_1)$ in sample A is lower than ω(Bulk ZnTe). The frequencies were calculated using a linear chain model [366, and references given there in]. Frequencies of strained bulk crystals were input parameters in the calculations. The input frequencies were adjusted until the calculated and the observed values of the frequencies $\omega(LC_2S)$ and $\omega(LZ_1)$ agreed. The adjusted frequencies were taken as the calculated frequencies of the bulk strained crystals also shown in Table 5.9. The strained induced shift is upward in CdTe, 18 cm^{-1} in sample A and 12 cm^{-1} in sample B. This is consistent with the fact that CdTe is under compressive strain and the strain is more in sample A than in sample B. It is downward in ZnTe, –3 cm^{-1} because these layers are under tensile strain. For a more quantitative discussion, phonon deformation potentials p and q or Grüneisen parameter γ and shear deformation parameter a_s are required. Relations between the deformation potentials and tetragonal strain are well known [366]. The expressions for γ and a_s in terms of p and q are given below.

$$\gamma = -\frac{p+2q}{6\omega_{LO}^2}, \tag{5.15}$$

and

$$a_s = \frac{p-q}{2\omega_{LO}^2}, \tag{5.16}$$

The measured value of Grüneisen parameter γ is 1.2 for ZnTe [366]. Since this parameter has not been measured for CdTe, the same value was used for CdTe also. The shear deformation parameter a_s is always smaller than 1 and increases with iconicity. A value $a_s = 0.6$, larger than the value 0.31 for Si was used. The calculated values of strain in this manner agreed with those calculated using the elasticity theory and known structure if it is assumed that the layers were lattice matched to the buffer layer in sample A and were free standing in sample B. Presumably sample B was partially relaxed by creation of misfit dislocations at the superlattice-buffer interface.

Olego et al. [264] have studied strain in ZnSe-ZnS$_{0.19}$Se$_{0.81}$ superlattices using the Raman method. Alternating layers of 5.5 nm ZnSe and 6 nm layers of ZnS$_{0.19}$Se$_{0.81}$ were grown on 1000 thick nm buffer. The buffer was grown on (100) GaAs. The total thickness of the superlattice was 4300 nm. Two laser wavelengths were used, 413.1 nm line of Kr$^+$-ion laser and 488 nm line of an

Ar$^+$-ion laser. With 488 nm line (which has small absorption coefficient and therefore samples regions close to the interface) the ZnSSe peak was red shifted and the ZnSe peak was blue shifted. The lattice constant of ZnSSe is 0.562 nm [264] and is smaller than the lattice constants of GaAs (0.5653 nm) ZnSe (0.5668 nm) (see Table A in Appendix A). Therefore ZnSSe is under tension and ZnSe is under compression. The red and the blue shifts are consistent with these stress conditions. We have seen in Fig. 5.8 that in ZnSe strained layers Raman shift was reduced when laser light was changed from 488 nm to 413.1 nm. In the case of the superlattice, ZnSe Raman shift increased considerably with 413.1 nm laser light. Olego et al. [264] suggested that the top part sampled by the short wavelength laser is considerably relaxed. In the case of the superlattice, on relaxation the top part becomes more like symmetrically strained superlattice (also known as free standing superlattice). It can be shown by simple calculations that in the symmetrically strained superlattice smaller lattice constant of ZnSSe layers increases the compression in the ZnSe layers and therefore the blue shift increases. On the other hand tension in the ZnSSe layers is reduced in the relaxed top part and Raman shift is expected to be smaller with the 413.1 nm light in agreement with the observation of Olego et al. [264].

5.4.3.6 Effect of hydrostatic pressure

Li et al. [383] have studied PL of ZnSe/Zn$_{1-x}$Cd$_x$Se strained superlattices under hydrostatic pressure. At 77 K PL spectra of ZnSe/Zn$_{0.74}$Cd$_{0.26}$Se superlattice (with no pressure) consisted of three lines denoted by P1, P2 and P3. The P2 line was at 2.569 eV and was the strongest. The P3 line was at 2.249 eV and P1 at 2.660 eV. At room temperature only the P2 line was observed. The P2 peak was assigned to the excitonic transition from the first conduction subband to the $n = 1$ heavy hole band. The P1 peak was assigned to the transition from the $n = 1$ conduction subband to the light hole band. The splitting of the heavy and light hole bands was estimated to be 91 meV for the superlattice. The origin of the P3 peak was not clear. Effect of hydrostatic pressure on P2 was studied up to 25 kbar at room temperature. The P2 line shifted sub-linearly to higher energies with pressure. The shape and line-width of the line did not change appreciably with pressure. The change of peak position $E(p)$ with pressure p can be described by the following equation:

$$E(p) = 2.345 + 0.067p - 0.013p^2 \quad (\text{in eV}), \tag{5.17}$$

where 2.345 eV is the room temperature position of P2 peak without any pressure. Similar sub-linear pressure dependence of PL is observed in bulk ZnSe and CdSe. The dependence has been attributed to the change of bulk modulus of these semiconductors with pressure. Li et al. [383] calculated the linear pressure coefficient of pressure to be 0.070 which is in good agreement with the experimental value of 0.067 eV.

The effect of hydrostatic pressure on short width CdSe/ZnSe single quantum wells has been studied by Hwang et al. [384]. The samples were grown on GaAs with 2-μm thick ZnSe buffer layer. CdSe layers were therefore highly strained.

The thickness of the CdSe layers varied from 1 to 4 monolayers. Pressure coefficient values of PL (Γ-Γ transition) decreased as the well width increased. For 3 and 4 monolayer wells the pressure coefficient decreased sub-linearly with pressure. Hwang et al. [384] attributed this behavior to the relaxation of strain by introduction of misfit dislocations and concluded that the critical thickness of CdSe on ZnSe substrate is less than 4 monolayers. However we have seen above that the sub-linear dependence is observed in bulk ZnSe and CdSe also and can be attributed to the change of the bulk modulus of these semiconductors with pressure [383].

Photoluminescence-excitation (PLE) measurements of $ZnSe/Zn_{1-x}Mg_xSe$ single quantum wells (SQW) have been made under hydrostatic pressure of up to 4 GPa [385]. The samples were grown by MBE on (001) and (110) GaAs substrates. The well transitions $1nH_{1s}$ and $1mL_{1s}$ with n=1,2,3,4,5 and m=1,2 were recorded. Excitonic signals from the barrier material and their pressure dependence were determined. A cross-over between the 11L and 13H as well as between the 12L and 14H transitions was observed above 1.2 GPa. The exciton binding energy was derived from the energy difference between the $11H_{1s}$ and the $11H_{2s}$ transitions.

5.5 Optical properties of III-Nitrides

5.5.1 Intrinsic luminescence

5.5.1.1 PL and splitting of the valence band

Typical near band-edge PL spectra of GaN at different temperature is shown in Fig. 5.14. Fig. 5.15 shows the PL spectra at different temperatures. At low temperatures and in the 3.5 to 3.2 eV range the spectra are dominated by the bound and free exciton peaks and their phonon replicas. The A, B and C excitons are clearly resolved. In the 3.15 to 3.20 eV range PL due to impurity related recombination is seen. The yellow band is seen around 2.2 eV. As the temperature increases from 5 K to 200 K, the A^0X and D^0X peaks shift continuously to lower energies due to reduction of the bandgap in all samples. The position of the free exciton band does not change up to 30 K [36] due to increase in the kinetic energy of the exciton which gives rise to polariton dispersion effect. The effect compensates the reduction in the bandgap. At higher temperatures the bound electrons can become free and this also changes the relative intensities of the different exciton bands. The intensity of the B excitons relative to the A excitons also increases with temperature. This shows that the B excitons are formed at the cost of A excitons at higher temperatures. The full width at half maximum (FWHM) of the exciton peaks up to 20 K and then increases linearly with temperature [36].

The splitting of the valence band can be determined from the low temperature PL peak positions due to excitonic recombination [330, 333]. Transitions from the $A(\Gamma_9)$ and $B(\Gamma_7)$ bands shown in Fig. 5.14 have been investigated by many groups [36, 162, 328, 330, 331]. Transitions from $C(\Gamma_7)$ band have been

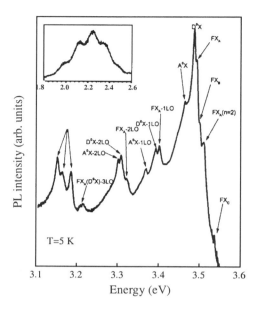

Figure 5.14: PL spectrum recorded at 5 K for undoped 2 μm thick GaN epilayer. The back-ground electron concentration was 8.4×10^{16} cm^{-3}. The inset shows the yellow band luminescence. The figure is taken from Ref. [36].

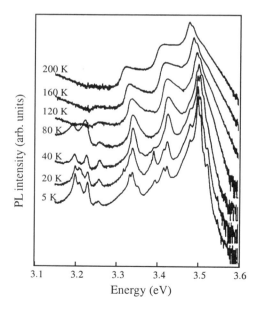

Figure 5.15: Temperature dependent PL of the same sample for which low temperature spectrum is given in Fig. 5.14. The figure is taken from Ref. [36].

5.5. OPTICAL PROPERTIES OF III-NITRIDES

observed in PL [36, 331] and in PR measurements [332]. We have used the observed PR and PL results to determine the splitting ΔE_{AB} and ΔE_{BC} of the valence band. We have already shown these values as well as the calculated values of the splittings in Table 5.2 in section 5.1.2.

5.5.1.2 Binding energies and lifetimes of the excitons

Binding energies of excitons ΔE_A^b, ΔE_B^b, and ΔE_C^b of the A, B and C free excitons have been calculated by Chen et al. [328]. The calculated values are shown in Fig. 5.3. The calculated binding energies are nearly equal for the three excitons. The energies of localization of the bound excitons have been determined from PL measurements by An em et al. [36]. The energies are 27 to 30 meV for acceptor bound excitons and 6 to 7 meV for donor bound excitons.

Exciton recombination dynamics and lifetimes of excitons have been reviewed by Monemar et al. [333]. Time resolved experiments of PL decay show that at 10 K the radiative lifetime of the free excitons is $\sim$ 35 ps and of the bound excitons, $\sim$ 55 ps [330]. Life time values of the excitons in III-V and II-VI semiconductors are similar [386, 387].

Several studies of erbium implanted GaN have also been made [391, and references given therein]. Extensive discussion of the optical properties of GaN containing Er is given in the recent reviews [13, 44].

5.5.2 Yellow and DAP bands and persistent photoconductivity

The spectra on the lower energy side of the exciton PL peaks is shown in the inset in Fig. 5.14. A yellow band with peak at 2.2 eV is generally observed in all GaN crystals [36, 330, 331]. The vibronic structure of the yellow band (resolved more clearly in the results of Götz et al. [331]) shows the strong electron-phonon coupling. Like the background donor centers, the nature of the centers which give rise to the yellow band is not known. There is some evidence that the layers which show strong yellow band also show large background donor concentration. It is possible that the centers responsible for both the yellow band and background n-type conductivity are the same. The PL lines observed by Götz et al. [331] and their assignments are given in Table 5.11. Götz et al. [331] attributed the CBA PL line at 3.294 eV to transitions of electrons originating at the conduction band edge and recombining with the holes at the acceptors. The DAP1 line (3.272 eV) is attributed to transitions between the Si donors and Mg acceptors. Analysis of these results gives the energy level of the Mg acceptors at 209 meV above the valence-band edge. The origin of the L_2 and L_3 lines is not known. Temperature effects on the yellow band are discussed in section 5.5.4.

The persistent photoconductivity (PPC) is present in WZ GaN epilayers grown by MOVPE or MBE [388]. We have already seen that origin of background n-type conductivity and yellow band is not known. Centers responsible for the persistent photoconductivity have also not been identified. Defects with

Table 5.11: PL line positions in the 2 K PL spectra of WZ GaN and their assignments. GaN was nominally undoped but contained unintentional background doping of 4×10^{17} cm^{-3} Si atoms. The background concentration of compensating acceptors was $< 1 \times 10^{16}$ cm^{-3} [331]. The abbreviations are, X: excitons. ND: neutral shallow donors, bt: bound to, DAP: donor-acceptor pair and CBA: transition from conduction band to acceptors.

PL line (eV)	FWHM (eV)	Nomenclature	Assignment
3.473	~ 2.2	BX	X bt ND
3.458	–	L$_1$	X bt acceptor
3.353	–	L$_2$	unknown
3.292	–	L$_3$	unknown
3.272	–	DAP1	D-A recombination
3.179	–	DAP2	D-A recombination
3.294 ± 0.004	broad	CBA	see text
2.2	broad	yellow band	see text

bistable character, nonstoichiometry induced potential fluctuations, and inclusions of cubic GaN have been considered as the possible centers [388]. Some correlation between the PPC and yellow band has been found. The yellow band and the PPC may originate at the same defect centers. No PPC is observed if the excitation is done with 1.38 eV light [388]. PPC increases rapidly as the energy of the incident photons increases. The threshold of the yellow band and the PPC are both at $h\nu \sim 1.6$ eV. For $h\nu = 1.77$ eV, the decay time was 3 hours. For $h\nu = 1.96$ eV, the decay time is more than a day. With white light, the PPC lasts several days. Reddy et al. [388] suggested a complex consisting of nitrogen antisite and a Ga interstitial, N$_{Ga}$-Ga$_I$ is the likely defect responsible for both. This defect is similar to As$_{Ga}$-Ga$_I$ defect which has been found in GaAs.

5.5.3 Effect of strain on III-Nitride layers

Strain in the epilayers arises due to lattice mismatch and thermal mismatch. In a large lattice mismatched system a thick epilayer is completely relaxed at the growth temperature by generation of misfit dislocations. The stress is mainly determined by the thermal mismatch during the cool down period. The films grown on GaN buffers have small compressive stress or no stress. The films grown on AlN buffers are under tensile stress. If there is no generation of misfit dislocations during the cool down period, the GaN layers grown on sapphire are under compressive strain and those grown on 6H SiC are under tensile strain. These are general trends, different results have been obtained in several cases. For example Edwards et al. [35] have observed compressive strain in the layers grown on 6H-SiC substrates. In fact the behavior of stress for different thickness of the GaN film was complex. The stress was compressive up to 9.7

5.5. OPTICAL PROPERTIES OF III-NITRIDES

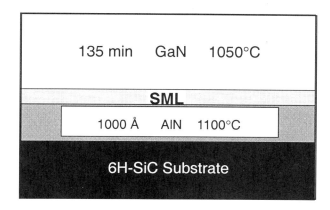

Figure 5.16: Structure, compositions, layer thicknesses, and growth parameters of SML samples studied by Edwards et al. [35].

μm thickness, tensile (4 kbar) up to 2 μm thickness and less tensile (1 kbar) for larger thicknesses. As the strain in the samples increases, the energy of all the three (A B and C) excitons shift upwards. However the shifts are not equal. The C exciton moves by up to 30 meV whereas the maximum shift in the A and B is 20 meV. It is expected that the excitons will move downwards in the GaN epilayers grown on 6H SiC. However the widths of the PL and PR spectra in the layers grown on SiC is large and such clear cut results have not been obtained [333]. Other experiments which show the effect of strain on optical properties have been discussed by Song and Shan [330].

Recently Edwards et al. [35] have studied a modified structure in which effect of strain can be tailored to suit specific requirements. The structure studied by these authors is shown in Fig. 5.16. By including a very thin strain mediating layer (SML) between the buffer layer and the main GaN film (see Fig. 5.16), stress behavior of the structure can be changed significantly. The SML does not degrade the optical or structural properties of the layers significantly. PL of the samples with and without SMLs was studied and the energy E_A of the A excitons was determined. The energy E_A and the inplane stress σ_{xx} are plotted as a function of GaN layer thickness in Fig. 5.17. Four samples with SML and several samples without SML were investigated. Sample 5, one of the samples without SML, has the largest tensile strain of 4.6 kbar and $E_A \sim 4.6$ eV. Sample 1 had a 37.5 nm thick SML. SEM results suggested that it was probably cracked. It has only a small stress of 0.26 kbar. The stress in samples 2, 3 and 4 has negative values, i.e. it is compressive. The values of SML thickness in these layers are 45.0, 75.0 and 37.5 nm respectively. Sample 4 was a $Al_{0.13}Ga_{0.87}N$ layer. It is seen that the energy E_A and stress σ_{xx} are considerably modified by the SML. There was some evidence that the SML can introduce strain anisotropy

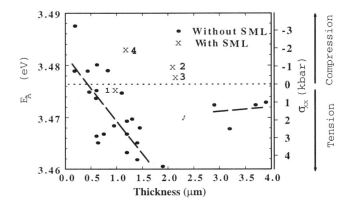

Figure 5.17: E_A and σ_{xx} versus thickness of GaN films on 6H-SiC. Filled symbols represent samples without strain mediating layers (SML) and points X with sample numbers represent samples with SMLs. The figure is taken from Ref. [35].

in the (0001) direction without cracking the film. Such strain anisotropy can be used to reduce the laser threshold values [35].

These experiments are important. They show that by using an SML the strain in the GaN layer tailored. This allows the valence band engineering and optimization of laser diodes for lowest thresholds [35].

5.5.4 Temperature dependence of the optical transitions

The temperature variation of the PL has been investigated by many groups [332, 333, and reference given therein]. Shan et al. [332] have used the Varshni empirical equation to describe the temperature dependence of the transition energies of the A excitons $[A(\Gamma_9)$-$\Gamma_7(CB)]$, B excitons $[B(\Gamma_7)$-$\Gamma_7(CB)]$, and C excitons $C(\Gamma_7)$-$\Gamma_7(CB)]$ (see Fig. 5.3). The Varshni empirical equation is given below,

$$E_0(T) = E_0(0) - \alpha T^2/(\beta + T), \tag{5.18}$$

where $E_0(0)$ is the transition energy at 0 K, and α and β are the temperature coefficients. the values of $E_0(0)$, α and β are given in Table 5.12. The directly observed 15 K values of E_0 for the three excitons are 3.485 eV (A exciton), 3.493 eV (B exciton) and 3.518 eV (C exciton) respectively. Similar values have been obtained recently by An et al. [36] at 5 K. As we discussed earlier, the transition energies depend on the strain and strain in the epilayer changes with temperature. The effect of strain and temperature have not been separated out in the above experiments. The temperature dependence of the exciton transition energies determined by different workers do not always agree [333].

An et al. [36] have investigated the effect of temperature on the yellow band. The peak position blue-shifts on raising the temperature from 5 K to 40 K.

5.5. OPTICAL PROPERTIES OF III-NITRIDES

Table 5.12: Values of parameters $E_0(0)$, α and β obtained by fitting experimental data to Eq. (5.18) [332].

Exciton	$E_0(0)$ (eV)	α (eV/K)	β (K)
A	3.486	8.32×10^{-4}	835.6
B	3.494	10.9×10^{-4}	1194.7
C	3.520	2.92×10^{-3}	3698.9

From 40 K to 200 K a red shift is observed. At higher temperatures the shift becomes blue again. The intensity of the yellow band decreases somewhat up to 40 K, starts increasing and obtains a maximum at 150 K and then decreases again. The reason for this complex behavior is not understood.

5.5.5 Optical properties of InGaN alloys

5.5.5.1 Bandgap and PL

At low temperatures the bandgap of $In_xGa_{1-x}N$ alloys can be varied in the range ~ 1.9 eV to 3.5 eV by changing the alloy composition. The 10 K bandgap of the alloy is given by the following equation [392],

$$E_g(x) = 3.5 - 2.63x + 1.02x^2 \text{ eV}. \quad (5.19)$$

At 295 K the band gap decreases by about 0.06 eV for all compositions. The alloy epilayers are suitable as active layers in the $GaN/In_xGa_{1-x}N/GaN$ double heterostructures. Nakamura and Mukai [393] have grown high quality $In_xGa_{1-x}N$ films by the two-flow MOVPE method at a growth temperature in the range 780 to 800 °C. X-ray diffraction measurements showed that the quality of the films was good. PL of the layers containing In mole fraction in the range 10 to 30% was measured and PL peak wavelength was plotted as a function of In concentration. The plot was a straight line. Many other groups have studied the structural and optical quality of the InGaN films at low and room temperatures [39, and references given therein]. Wagner et al. [39] have studied the InGaN pseudomorphic films with In concentration up to 15%. The films were grown by MOCVD. Composition and strain in the films were determined by SIMS and high resolution x-ray diffraction techniques. The room temperature photoreflection (PR) spectra of the alloy layers showed sharp resonance at 3.43 eV originating at the GaN buffer layer. In addition broadened resonances due to the alloy layer were seen. Bandgap values were derives by fitting the calculated curves to the observed spectra. The effect of strain was included in the calculations. The bandgap energy determined in this manner is plotted in Fig. 5.18 as a function of In concentration. Results for both strained and relaxed layers are given. Results for the relaxed alloy were obtained by subtracting strain induced shifts from the observed bandgaps. The strain induced shift was taken to be $\Delta E = 1.02x$. PL experiments were also performed and the bandgap derived

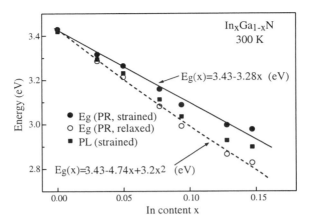

Figure 5.18: Composition dependence of the room-temperature bandgap energy of strained $In_xG_{1-x}N$ layers on GaN as obtained by PR spectroscopy (filled circles), and after correction for the strain induced band gap shift (open circles, see text). (InGa)N alloy composition was deter-mined by SIMS. The full and dashed curves indicate linear and quadratic fits to bandgap for strained $In_xG_{1-x}N$ and for relaxed $In_xG_{1-x}N$, respectively. For comparison, room-temperature PL peak positions are also shown (filled squares). The figure is taken from Ref. [39].

from the PL measurements are also included in the figure. Room temperature bandgap can be represented by the following equation:

$$E_g(x) = 3.43 - 3.28x \quad \text{eV} \tag{5.20}$$

for the strained alloy and,

$$E_g(x) = 3.43 - 7.4x + 3.2x^2 \quad \text{eV} \tag{5.21}$$

for the relaxed unstrained alloy. Wagner et al. [39] also made spectroscopic ellipsometry (SE) measurements and determined the energy at which the peaks in the real part $\langle \varepsilon_1 \rangle$ of pseudo-dielectric function are observed. The peaks were attributed to overlapping A and B excitonic recombination. These peak positions also give the excitonic bandgap energy [39]. These peak positions are plotted as a function of bandgap $E_g(x)$ of the alloys in Fig. 5.19. There is a good correlation between the PL, PR and SE measurements.

5.5.5.2 Bowing parameter of InGaN

Early absorption measurements of InGaN alloys by Matsuoka et al. [395] suggested a large value of the bowing parameter. Recently McCluskey et al. [396] have studied experimentally and theoretically the bowing parameter of $In_xGa_{1-x}N$ alloys with low values of x. These authors measured absorbance for several values of x in the range $0 \le x \le 0.12$. Values of bandgap E_g were

5.5. OPTICAL PROPERTIES OF III-NITRIDES

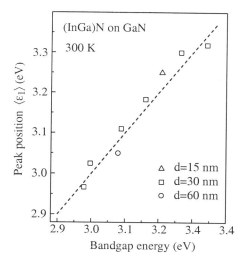

Figure 5.19: Energy position of the (InGa)N related peak in the $\langle \varepsilon_1 \rangle$ spectrum versus (InGa)N band gap energy derived from the PR measurements. Data are shown for different (InGa)N layer thicknesses L given in the figure. The dashed line indicates a one-to-one correspondence. The figure is taken from Ref. [39].

determined from the absorbance versus photon energy plots. The strain induced shifts of the bandgap were calculated and subtracted from the bandgaps of the strained layers. After subtracting the strain induced shift the bandgap of the relaxed layers can be described by: [396],

$$E_{g,\text{In}_x\text{Ga}_{1-x}\text{N}}(\text{relaxed}) = 3.42 - 4.95x \text{ eV.} \quad (5.22)$$

The bowing parameter for the relaxed alloy layer was determined by comparing these values with those calculated using a Vegard type of equation. The value of the bowing parameter was found to be 3.8 eV for $x = 0.1$. Recently Akasaki and Amano [81] have studied the PL of the strained InGaN alloys. Values of x in the range 0.05 and 0.2 were used. The average value of the bowing parameter was 3.2 for $0 < x \leq 0.2$ (see Fig. 5(b) of Ref. [81]). We have discussed the PR and SE measurements [39] of the alloys earlier. The bowing parameter obtained from these measurements is also 3.2 The values obtained by previous workers were lower, only 1 eV at $x \sim 0.1$ [396, and references given therein]. Though there is some discrepancy between the results of different measurements, it is clear that the bowing parameter for these alloys is quite large.

5.5.5.3 Nonuniform distribution of indium

Chichibu et al. [321] measured the temperature dependence (in the range 10 to 300 K) of the luminescence spectra of GaN and undoped and Si-doped $\text{In}_x\text{Ga}_{1-x}\text{N}$ 3D alloys. The temperature dependence of the emission spectra in InGaN was much weaker than in GaN. Chichibu et al. [321] suggested that the PL centers in the alloy are localized in local high concentration regions of In. Subsequently structure analysis of $In_{0.20}Ga_{0.80}N$(3 nm)/$In_{0.05}Ga_{0.95}N$(6 nm) superlattice was performed by using TEM and energy dispersive X-ray microanalysis [397]. A contrast in the light and shaded regions of the well was observed which was attributed to fluctuation in the In composition. It was suggested that the In rich regions act as quantum dots and give rise to efficient radiative recombination. Recently Behr et al. [398] have studied the Raman scattering from GaN and $In_{0.11}Ga_{0.89}N$ epilayers at different temperatures in the range 300-900 K. From the analysis of their results they also concluded that the In concentration in the layers must be non-uniform. The output power of the LED containing a small amount of indium in the active layer was about 10 times greater than that of the diode containing no In. The In concentration was so small that the wavelength of the emitted light changed from 368 nm only to 371 nm. This can be explained if one assumes that local high concentrations act as quantum dots and provide an efficient mechanism of light emission [85].

Some experiments contradict this point of view, Crowell et al. [399] obtained information about the inhomogeneity by PL measurements and by near-field optical spectroscopy. Near field scanning microscopy with a 100 nm diameter fiber was used to study the spatial distribution of the PL. The distance between the tip and the surface was kept constant at a value less than the wavelength of light used. At RT the intensity of the deep level band increased by a factor about 10 when ever the tip was over a pit at the surface. The 2.95 eV emission was observed from the entire area. A variation of PL intensity which could be attributed to the efficient emission of light from the quantum dots formed by local high concentration of In was not observed. The authors concluded that they did not find any evidence of PL from strongly localized carriers.

5.5.6 AlGaN alloys: Optical absorption, bandgap and bowing parameter

Optical properties of the $Al_xGa_{1-x}N$ alloy were measured in late 1970s and 1980s by several groups [401, and references given there in]. The early results were conflicting. The bandgap plots of the $Al_xGa_{1-x}N$ alloy as a function of Al molar fraction x were reported by different authors to bend upward, downward and also remain linear. The first reliable optical measurements of the alloy epilayers measurements were made by Koide et al. [401] and values of the bandgap were derived. The bandgap could be represented by the following equation,

$$E_{g,Al_xGa_{1-x}N} = xE_{g,AlN} + (1-x)E_{g,GaN} - bx(1-x). \tag{5.23}$$

A value 1.0±0.3 was found for the bowing parameter b. Angerer et al. [402] have repeated these experiments with the plasma-induced MBE grown $Al_xGa_{1-x}N$ epilayers. They covered the whole composition range of the alloy from $x = 0$ to $x = 1$. They determined the lattice constants and showed that the Vegard's law is valid for the lattice constants to a good approximation. The bandgap

5.5. OPTICAL PROPERTIES OF III-NITRIDES

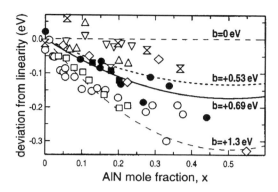

Figure 5.20: Measurements of the deviation of the bandgap from linearity. AlGaN layers were grown on sapphire by MOCVD using low temperature buffer. The filled circles and filled squares are the data of Ref. [37]. Open symbols are the data taken from published literature. The references to the open symbols are given in Ref. [37] from where this figure is taken.

values were determined using the results of absorption measurements. A value 1.3 eV of the bowing parameter was obtained. Akasaki and Amano [81] have determined a value 0.25 of the bowing parameter of $Al_xGa_{1-x}N$ layers from the PL measurements for $x < 0.25$.

Recently Lee et al. [482] have reviewed the values of the bowing parameter in AlGaN alloys and have given a very comprehensive list of the papers in which the bowing parameter has been investigated. Their results show that the bowing parameter depends strongly on the method used for growing the layers and on the experimental conditions. They found that only three groups of workers reported positive deviation of the E_g, X plots from linearity. These results are classified as Group A. These results were obtained in the early days and the material used were probably not of high quality. The other 13 groups reported negative deviation of the plots from linearity. The bowing parameter values reported in these 13 papers varied from 0.53 to about 3. The results of these 13 papers were further divided in Groups B and C. In group B the layers were grown without the buffer layers or without optimizing the growth conditions. In Group C low temperature buffer layers were used on sapphire substrates or highly optimized conditions were used for growth on SiC substrates. The values of bowing parameter in both groups were positive but the values in Group C were smaller. The measurements which fall in Group C are shown in Fig. 5.20. Lee et al. [482] also made carefully fresh measurements. Their results are included in Fig. 5.20. This figure shows that in carefully prepared films the bowing parameter should lie between 0.5 and 1.3. Even in carefully prepared films the buffer layer thickness, stoichiometry and growth rates can affect the results.

156 CHAPTER 5. BAND STRUCTURE AND OPTICAL PROPERTIES

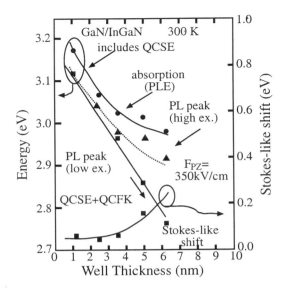

Figure 5.21: High and low excitation PL and apparent bandgap positions and apparent Stokes-like shift in InGaN wells as a function of L. Apparent bandgap is the energy where PLE intensity drops to half of its maximum value. The figure is taken from Ref. [34].

5.5.7 III-Nitride quantum wells and superlattices

The earliest studies of GaN quantum wells with AlGaN barriers were made by Khan et al. [75] and by Itoh et al. [76]. Khan et al. [75] investigated three GaN/AlGaN quantum well structures. 0.2 μm $Al_{0.14}Ga_{0.86}N$ layers were used as barriers. The thickness of the GaN well layers were 10, 15 and 30 nm respectively. The GaN PL peaks shifted to higher energies by 38.7 meV for the 30 nm well and by 43.6 meV for the 15 nm well. In the 10 nm well the PL peak was not resolved. The calculations of quantum confinement effect gave a shift of 2.3 meV for the 30 nm well and 8.50 meV for the 15 nm well. These values are smaller than the observed values. The PL spectra of the GaN wells was also measured by Itoh et al. [76]. Discrepancy between the calculated confinement energy and observed shift was observed in these investigations also. In the doped semiconductors band-tails and other heavy doping effects can change the effective bandgaps [403]. The strain present in the GaN quantum well layers due to lattice mismatch and thermal mismatch can also modify the bandgaps. These effects were not calculated.

Recently Chichibu et al. [34] have investigated the effect of well width and In concentration on the optical properties of the InGaN quantum wells. PL and PLE (PL excitation) measurements were made. The results of these measurements are shown in Fig. 5.21. PLE peaks and PL peaks for both low and high excitation are plotted as a function of well-width. Calculated results are

5.5. OPTICAL PROPERTIES OF III-NITRIDES

also shown. Both quantum confined Stark effect and Franz-Keldysh effect were included in the calculations. Analysis of these results show that the PL and PLE peaks in the quantum well depend on a number of factors. There exists an electric field across the quantum well plane. The field arises due to spontaneous and piezoelectric effects. In the case of doped *p-n* junctions the electric field is also created by the junction. The electric field causes a red shift. The absorption and PLE spectra are affected by quantum confined Stark effect and Franz-Keldysh effect. These effects become very large when the potential drop across the well due to the electric field becomes larger than the valence band offset. If the quantum well is doped, the free carriers cause screening and the electric field is reduced. Calculation of shift due to quantum confinement is not sufficient to interpret the PL results from the III-Nitride quantum wells. The experimental results can agree with the theoretical values if quantum confined Stark effect and Franz-Keldysh effect along with other effects (strain and band-filling effects) are taken in to account in the theory. The relative importance of these different effects depend on the doping concentration and structure of the sample.

Chapter 6

Electrical and magnetic properties

6.1 Electrical properties of II-VI semiconductors

6.1.1 n-type doping

Considerable work has been done on doping ZnSe, ZnTe, MgSe, MgTe and ZnMgSSe alloy [334, 405]. ZnSe, ZnS, CdSe, and CdS can be doped n-type more easily than p-type. Theoretical work shows that Al and Ga form stable deep centers in most II-VI semiconductors, similar to DX centers in AlGaAs. Chlorine is an effective n-type dopant in ZnSe, and wider bandgap ZnMgSSe alloy [405, and references given there in] and BeMgZnSe alloy [406]–[408]. Chlorine doped ZnSe epilayers have been grown by conventional MBE as well as by GSMBE. In the recent growth of Cl doped ZnSe by GSMBE, Ho and Kolodziejski [405] used an effusion source of solid anhydrous $ZnCl_2$. The effusion cell temperature was varied from 140°C to 320°C. The Cl concentration obtained at the highest effusion cell temperature was $\sim 10^{21}$ cm^{-3}. The growth rate remained constant at about 0.6 μm/hr up to 280°C but decreased abruptly at 300°C. The reason for the abrupt decrease is not understood. The PL of the Cl doped samples was dominated by donor bound excitons for Cl concentrations $\leq 4 \times 10^{18}$ cm^{-3}. No deep level PL was observed. The intensity of the donor bound excitonic PL increased with Cl concentration up to 1×10^{17} cm^{-3} and then started decreasing. For Cl concentration of 8×10^{19} cm^{-3}, the donor bound excitonic PL disappeared and a broad band due to deep levels was observed. The broad band extended from 1.8 to 2.4 eV.

For n-doping an electron concentration of up to 3×10^{18} cm^{-3} can be easily obtained in ZnMgSSe alloy by addition of $ZnCl_2$ [409]. The optical quality of the epilayers does not degrade. BeMgZnSe can be degenerately doped n-type with Cl up to a bandgap of 3.2 eV [406, 407].

6.1.2 p-type doping

A review of theoretical and experimental work done on doping of ZnSe-based wide-bandgap semiconductors has been published by Morkoç et al. [53], Lischka [410], Van de Walle [334] and Han and Gunshor [129]. As grown ZnSe is generally n-type. Doping ZnSe p-type with large concentration of holes was not achieved until recently. Early attempts of p-doping of ZnSe were made using P, As, O, Na and Li [129]. The p-type conductivity could be obtained only with Li. Control of doping profile was difficult because of the large diffusion length of Li. The maximum concentration of holes obtained by Li doping was $< 10^{17}$ cm^{-3}. Evidence that N is a shallow acceptor in ZnSe was obtained in 1980s (see the reviews [53, 129, 410, 411]). However large concentrations of nitrogen could not be incorporated in ZnSe by using the conventional techniques. First efficient p-type doping of ZnSe was accomplished by using radio frequency plasma sources of nitrogen by Park et al. [130] in 1990 and independently by Ohkawa et al. [131, 132] in 1991. The maximum hole concentration of a few-times 10^{17} cm^{-3} was achieved. In the beginning the nature of the species responsible for doping (i.e. whether they were nitrogen ions, excited molecules or atomic nitrogen) was not clear. Later work in which the plasma emission spectrum was analyzed and correlated with the doping concentration showed that atomic species of nitrogen are the dopant species. The success of p-doping by this method made it possible to fabricate the first ZnSe-based injection lasers [133, 134] in 1991. Nitrogen doped ZnSe epilayers have also been grown by GSMBE [405]. However hole concentration in the epilayers doped heavily with nitrogen remains small due to hydrogen passivation and due to large ionization energy (120 meV) of nitrogen acceptors.

Fujita et al. [150, 189] have reviewed the recent experimental work on the nitrogen doping of ZnSe and have compared the results obtained by different workers. They have also discussed the nitrogen doping of ZnSe during MOVPE growth [189]. Concentration of holes as a function of nitrogen concentration in ZnSe is shown in Fig. 6.1. The maximum hole concentration is 1×10^{18} cm^{-3} (see also [53]). Nitrogen doping has also been achieved by electron-cyclotron plasma and by using thermally cracked nitrogen oxide, NO [129]. Lugauer et al. [412] doped ZnSe with N using a nitrogen plasma diluted with Ar (Ar:N=10) and obtained a hole concentration of 5×10^{17} cm^{-3}. Experimental work discussed by Van de Walle [334] shows that as the nitrogen concentration increases, initially the hole concentration also increases. As nitrogen concentration increases further, hole concentration saturates as shown in Fig. 6.1. At still higher nitrogen concentration, a precipitous decrease in the hole concentration is observed. Theoretical work shows that as the concentration of nitrogen increases, the concentration of the compensating defects Zn_i increases faster than the concentration of nitrogen. This can explain the observed saturation of the hole concentration but not the precipitous decrease. Van de Walle [334] suggested that the decrease is probably due to degradation of crystal quality above a certain dopant concentration. PL experiments support this suggestion [405].

Unlike ZnSe, ZnS, CdSe, and CdS, generally as grown ZnTe is p-type [129].

6.1. ELECTRICAL PROPERTIES OF II-VI SEMICONDUCTORS

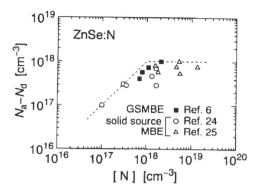

Figure 6.1: Observed net acceptor concentration $N_d - N_a$ as a function of nitrogen concentration $[N]$ in ZnSe layers grown by GSMBE and MBE. The solid symbols are for GSMBE, and open circles are for MBE growth. The figure is taken from Ref. [150].

It is difficult to dope ZnTe n-type. The concentration of holes is 1×10^{16} cm^{-3} in Te-rich and 1×10^{15} cm^{-3} in Zn-rich ZnTe [413]. Therefore the resistance of the as grown material is quite high. The p-doping in ZnTe can be obtained using P, Sb and N acceptors. Using the RF plasma technique, a hole concentration of 1×10^{19} cm^{-3} can be obtained in nitrogen doped ZnTe [367]. The temperature dependence of the carrier concentration is weak. The growth temperature also does not have a large affect on the carrier concentration and mobility. These results suggest that at high concentrations of nitrogen an impurity band is formed in ZnTe. The quality of the epilayers does not deteriorate at these high concentrations of nitrogen. Like ZnTe, BeTe can also be doped to high hole concentrations [406]–[408].

Wienecke et al. [413] have achieved p-type conductivity in CdTe and ZnTe by implanting ^{107}Cd. ^{107}Cd transmutes into Ag. ^{115}Cd transmuting into In was also implanted. The maximum fluence was up to 10^{12} cm^{-2}. A hole concentration up to 10^{17} cm^{-3} was obtained in Ag doped materials. Lugauer et al. [412] doped CdTe with As by using an Ar/As plasma cell during its MBE growth. A hole concentration of 2×10^{16} cm^{-3} was obtained. They also doped CdTe to a hole concentration of 5×10^{17} cm^{-3} by using the Ar diluted nitrogen (Ar:N=10) plasma during its MBE growth. Higher nitrogen doping levels have been obtained in the MBE growth. Both theory and experiments show that under identical experimental conditions the concentration of nitrogen in ZnTe is an order of magnitude larger than in ZnSe. CdTe has also been doped by implanted As and P ions. Thermal annealing and pulsed laser annealing of P implanted CdTe produces a hole concentration up to 10^{17} cm^{-3} in CdTe [414]. A high hole concentration $= 5 \times 10^{17}$ cm^{-3} has been produced by pulsed beam electron annealing of P implanted CdTe [414]. Valdna et al. [415] have fabricated CdTe samples by isothermal sintering CdTe with 2 wt% CdCl$_2$ flux on

Al$_2$O$_3$ substrates in evacuated quartz tubes. The samples were cooled under Te pressure. Stable p-type resistivity lower than 100 Ωcm was obtained. Shallow acceptor complexes (Te$_i^{2-}$-Cl$_{Te}^+$)$^-$ and/or (V$_{Cd}^{2-}$-Cl$_{Te}^+$)$^-$ were responsible for the p-conductivity. The p-type doping of BeSe, MgSe, and BeTe with Li and Na has also been studied [334, and references given there in]. The solubility of the dopants in these semiconductors is lower than in ZnSe except for BeTe.

ZnMgSSe is an important alloy in II-VI laser technology. Its bandgap is tunable up to ~ 4.5 eV. It forms type I heterostructure with ZnCdSe and can be lattice matched to GaAs [409]. Using ZnMgSSe cladding layers and ZnSe based MQW as active layer blue light emitting laser diodes have been demonstrated [409]. If Mg concentration is not large ($E_g \leq 2.95$ eV) ZnMgSSe can be doped to a hole concentration of $\sim 1\times 10^{17}$/cm^3 using RF or ECR source of nitrogen [409]. The maximum hole concentration that can be achieved decreases rapidly as the concentration of Mg increases. Ishibashi [409] has shown that the difficulties caused by low p-doping of the wide bandgap semiconductor can be over come to some extent by using a (ZnMgSSe)$_n$(ZnSe)$_m$ superlattice. The band structure of the superlattice is such that for the same effective bandgap as that of a ZnMgSSe alloy, the superlattice can be doped to 10 times larger concentration of holes. By using the superlattice (with $n = 10 \sim 15$ and $m - 4 \sim 6$) cladding layers it is possible to design and fabricate a CW laser diode operating at $450 \sim 460$ nm instead of the 463 nm laser diode obtained with the ZnMgSSe alloy cladding layers [409].

Using plasma activated nitrogen, BeTe can be doped to a concentration of 1×10^{20} cm^{-3} [407]. Incorporation of Te in ZnMgSeTe and ZnSeTe increases the maximum p-doping. However since lattice parameter of ZnTe is large, the lattice mismatch between these alloys and ZnSe increases as Te concentration increases. This causes difficulties in using cladding layers of these alloys. Lattice matched p-doped BeZnSeTe should provide better cladding layers. High p-doping in BeZnSeTe alloy necessary for device fabrication can be accomplished only for bandgaps smaller than 2.9 eV. For higher bandgaps, compensating defects increase faster than the hole concentration. This result is similar to that obtained with ZnMgSSe.

6.2 Electrical properties of n-type GaN

6.2.1 Undoped GaN

There is always a large background donor concentration in the as grown GaN layers. The identity of these donor centers is not known. In the early days the electron concentration in the as grown samples used to be very large. Now samples with carrier concentration $\sim 10^{16}$ cm^{-3} are easily grown. Several authors [99, 276, 416] have investigated theoretically the nature of the background donors. Nitrogen vacancies, Ga interstitials and residual impurities (Si, Ge, O) have been considered as the possible background donors. Perlin *et al.* [416] made first principle calculations and found that the energy of formation of nitro-

6.2. ELECTRICAL PROPERTIES OF N-TYPE GAN

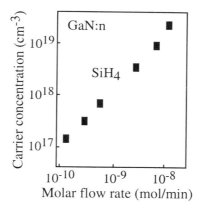

Figure 6.2: Incorporation of Si in GaN in MOVPE growth of GaN using silane as the precursor [417].

gen vacancy is lower than that of Ga interstitials. This suggests that Nitrogen vacancies rather than interstitial Ga atoms are responsible for the back-ground n-type conductivity. Popovici et al. [99] deposited GaN films using different flow rates of ammonia. SIMS and Hall measurements showed that as the nitrogen flow rate increased the background donor concentration decreased. This result also supports the view that nitrogen vacancies are responsible for the n-type conduction. Careful measurements of the impurities and electron concentrations were also made. The electron concentration did not correlated with the impurity concentration. These experiments appeared to rule out the possibility that residual oxygen and Si are responsible for the background conductivity. However Neugebauer and Van de Walle [276] found that the energy of formation of nitrogen vacancies is 4 eV. This energy is so large that at room temperature the vacancies will not be present in sufficiently high concentrations. Abernathy [194] found that GaN epilayers containing larger concentration of nitrogen than Ga showed a large background electron concentration [194]. These two results argue against the nitrogen vacancies as being the donors. Recent experiments of Götz et al. [331] (discussed in the next subsection) suggest that in the MOVPE grown GaN the background n-type conductivity is due to the residual concentration of Si atoms. A consensus regarding the nature of the background donors has not yet emerged.

6.2.2 Si and Ge doped GaN: Carrier concentration

Both Ge and Si substitute Ga sites in GaN and form shallow donors. Nakamura et al. [417] determined the carrier concentration in the MOVPE grown GaN layers (with sapphire substrate and GaN buffer) as a function of the flow rates of germane and silane. The results for Si doped layers are shown in Fig. 6.2. It can be seen from this figure that incorporation efficiency of Si in GaN is

Table 6.1: Values of optical and electrical activation energy E_A of Si donors in GaN.

Method	E_A (meV)	Ref.
Hall effect	30	[418]
Optical	22	[331]
Electrical	12 to 17, 32	[331]

very high. The surface of the layers was mirror like, no cracks or etchpits were observed. However the FWHM of the (0002) X-ray diffraction peak increased with increasing Si concentration. In the case of Ge doping, the quality of the epilayers containing carrier concentrations $> 1 \times 10^{19}$ cm^{-3} degraded. Etch-pits were observed at the surface in the highly Ge doped samples.

The early measurements of the activation energy of Si in GaN were made by Tanaka et al. [418] using Hall effect measurements. The authors found a value 30 meV for the activation energy. More recently Götz et al. [331] have studied extensively the Hall effect in the Si doped MOVPE grown GaN layers. One nominally undoped sample (sample 1) and four samples doped with different concentrations of Si (samples 2 to 5) were studied. Hall measurements were made at different temperatures in the range 70 to 500 K. PL measurements made at 2 K in sample 1 have already been discussed in chapter 5. The carrier concentration was determined by assuming that the Hall constant R_H for GaN is unity. From the analysis of their experimental results Götz et al. [331] concluded that there were two donor levels in the Si doped GaN epilayers. The values of activation energy for one donor level was the range 12 to 17 meV for and for the other level 32 to 37 meV. The concentration of the donors with higher energy of activation was small. The electrical properties of GaN are determined mainly by the donors with the lower activation energy. The activation energy determined from PL measurements was 22 meV. The measured activation energies are given in Table 6.1. Generally accepted values from electrical measurements are less than 20 meV. The optical activation energy is higher than the electrical activation energy.

Recently Albrecht et al. [40] have simulated Hall effect in n-type GaN using the Monte-Carlo technique. The authors calculated the Hall factor R_H (ratio of Hall mobility and drift mobility), carrier concentration and mobility in the samples containing three different doping concentrations (see Fig. 6.3). Calculations were made at different temperatures in the 70-400 K range. R_H depends weakly on doping concentration. R_H has a value ~ 1.6 at 100 K. As the temperature increases R_H decreases continuously to about 1.2 at 400 K. The values at room temperature are between 1.2 and 1.3 depending upon the impurity concentration. Fig. 6.3 shows that there is a large fraction of carriers frozen at liquid temperature. For the highest carrier concentration used in the calculations the carrier concentration starts decreasing from the very beginning as the temperature decreases. The decrease in concentration becomes very rapid below 150 K

6.2. ELECTRICAL PROPERTIES OF N-TYPE GAN

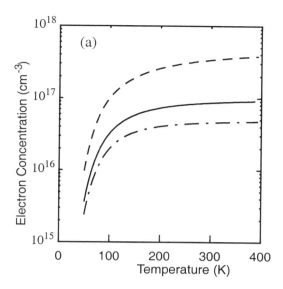

Figure 6.3: Electron concentration vs. temperature for GaN with donor doping densities of 5×10^{16} cm^{-3} (dashed-dotted line), 1×10^{17} cm^{-3} (solid line) and 5×10^{17} cm^{-3} (dashed line) [40].

in all cases. This carrier freeze out gives rise to a large concentration of neutral impurities. If dislocations and other defects are not present the reduction in mobility due to neutral impurity scattering should become important at low temperatures.

6.2.3 Electron mobility in n-type GaN

6.2.3.1 Measurements of electron mobility in n-doped GaN

We first discuss the effect of layer thickness on the mobility. In undoped n-type GaN layers both electron Hall mobility and electron concentration depend strongly on the thickness d_b of the buffer layers [219]. The mobility increases sharply as d_b increases from 100 Å to 200 Å and then starts decreasing slowly for larger values of the thickness. For 200 Å thick layer the mobility reaches a value of 600 cm^2/Vs at room temperature and 1500 cm^2/Vs at 77 K. The electron concentration also depends on d_b. It increases monotonically as d_b increases beyond 20 nm. It also increases at lower thicknesses. Thus It has a minimum at $d_b = 20$ nm. Look and Molnar [419] measured the Hall effect in the temperature range 10 to 400 K in GaN layers grown on ZnO pre-treated sapphire substrates by HVPE. The average concentration of electrons was $\sim 10^{17}$ cm^{-3} at room temperature. The carrier concentration decreased at higher temperatures. It became minimum at $T \sim 100$ K and then started increasing again at lower temperatures. It became practically constant at $\sim 2 \times 10^{17}$ cm^{-3} below 30 K.

The results could be interpreted on the assumption that a thin degenerate n-type layer at the GaN/sapphire interface and a layer with a lower conductivity above it exist. In a more recent paper Götz et al. [420] studied extensively the effect of varying the value of layer thickness d on the Hall mobility and electron concentration. The mobility increased and the carrier concentration decreased as the thickness d of the epilayers increased. A detailed modeling of the Hall effect measurements showed that the quantitative behavior depends on the treatment of the substrate surface on which the layers are deposited. The observed mobility in the layers grown on GaCl pre-treated substrate agreed with a two layer model used earlier by Look and Molnar [419]. The results obtained with the ZnO pre-treated substrate did not agree with this model. In this case the results agreed with a model which assumed that there was a continuous reduction of electron concentration and increase of mobility with the thickness of the epilayers.

Many other groups [55, 222, 331, 417] have measured the electron mobility in Si and Ge doped GaN layers. In another publication Götz et al. [331] measured Hall mobilities of electrons in Si doped GaN layers at different temperatures. At high temperatures the mobility varied as $\sim T^{-1.5}$. As the temperature decreased, the mobility obtained a maximum value and then started decreasing again. The reasons of the decrease of the mobility with decreasing temperature in the low temperature range were not clear at the time measurements were made. Recent work of Weimann et al. [23] (discussed later in this section) seems to suggest that this decrease is due to dislocation scattering. The electron concentration dependence of the mobility was also complex. In three samples the mobility decreased with increasing electron concentration. In one sample which had a higher electron concentration the mobility was the highest. It was 501 cm^2/Vs at 300 K and 764 cm^2/Vs at 160 K. Moustakas et al. [222] also found that when the electron concentration in their samples decreased to less than 10^{16} cm^{-3}, the mobility started decreasing. Moustakas et al. [222] suggested that if the concentration of background deep level defects is more than the electron concentration, the material is fully compensated and the conduction is dominated by the hopping transport. Nakamura et al. [421] have also measured the effect of temperature on the electron mobility. Starting from the value of 900 cm^2/Vs at room temperature, it increases to $\sim$ 3000 cm^2/Vs at 70 K and then decreases at lower temperatures. Above 70 K it varied as T^{-1}.

Weimann et al. [23] made measurements of the mobility as a function of electron concentration in epilayers containing different concentrations of dislocations. The epilayers were grown by ECR MBE on sapphire substrates. The substrate surface was nitridated by the nitrogen plasma before the films were grown. 30 nm GaN buffer layers were grown at 550°C before the main GaN layers were grown at temperatures in the range 700 to 800°C. The time for which nitridation was done and the structure of the buffer layer affected strongly the dislocation concentration in the main GaN epilayer. The thickness of the samples was in the range 0.5 μm to 2 μm. Si doping concentration was varied by changing the temperature of the Si source cell. Samples prepared with two different substrate surface treatment were investigated in detail. The carrier

6.2. ELECTRICAL PROPERTIES OF N-TYPE GAN

Table 6.2: Carrier concentration n (cm^{-3}), dislocation density ρ (cm^{-2}) and Hall mobility (cm^2/V s in two GaN samples studied by Weimann et al. [23].

Sample	n	ρ	mobility
A	3.0×10^{17}	8×10^9	306
B	1.57×10^{18}	2×10^{10}	208

concentrations, dislocation densities and measured Hall mobilities in the two samples are given in Table 6.2 The mobility obtained with the two samples (and other samples with same dislocations but different carrier concentration) are shown in Fig. 6.4(a) by triangles and squares. Total transverse mobility using the Conwell-Weisskopf formula for ionized-impurity scattering, a constant mobility of 450 cm^2/Vs due to phonon scattering and including the effect of dislocation scattering was calculated. The dislocations trap the electrons and the dislocation lines become negatively charged Coulombic scatterers. In samples having low carrier density, the effect of dislocation scattering dominates. In general the effect of dislocation scattering becomes significant for dislocation densities more than 10^9 cm^{-2}. If dislocation density is negligible, the mobility is limited by lattice scattering if other defects are not present. The scattering by the ionized donors and lattice phonons was included in the calculations. The lines in Fig. 6.4(a) show the calculated transverse mobility for different dislocation densities. The measured data for the two layers containing different concentrations of the dislocations agree well with the theory.

Nakamura et al. [417] measured the mobilities for electron concentration in the range $\sim 10^{16}$ to $> 10^{19}$ cm^{-3}. The mobility values measured by Nakamura et al. [417] and by other groups are plotted in Fig. 6.4(b). For carrier concentration $< 6 \times 10^{16}$ cm^{-3}, undoped layers were used. The mobility values do not depend upon the nature of the dopant, they were the same for Si and Ge dopants. As expected the mobility decreases with the increase of carrier concentration [417]. As mentioned earlier, a high room temperature mobility of 501 cm^2/Vs at an electron concentration of 2 to 3×10^{17} cm^{-3} was achieved by Götz et al. [331]. This value is shown by the diamond symbol in Fig. 6.4(b). The highest reported room temperature value of electron mobility is 900 cm^2/Vs [55, 55]. The values are shown by horizontal filled rectangle [421] at an electron concentration $\sim 3 \times 10^{16}$ cm^{-3} and by the open rectangle [55] at a higher electron concentration $= 5 \times 10^{16}$ cm^{-3}. A log-log plot of the mobility versus electron concentration is a good straight line [222]. This line extrapolated to lower values of electron concentrations predicts a mobility of 10^4 cm^2/Vs at an electron concentration of 10^{14} cm^{-3}, which is comparable to the mobility values in GaAs. The highest published room temperature value of electron mobility in GaN is 1700 cm^2/Vs (not shown in the figure) in an AlGaN heterostructure [57]. Filled and open circles show the mobility values in MBE grown layers. These values are lower.

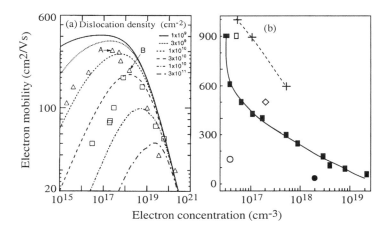

Figure 6.4: (a) The lines show the calculated transverse mobility for different dislocation densities in GaN. The measured data for the two epilayers with different dislocation densities are indicated by triangles and squares [23]. (b) Measured room temperature Hall mobility of electrons in GaN. Filled vertical rectangles: [417], filled circle [224] and open circle [162]: MBE grown undoped films, and diamond: [331]. The highest experimental mobility values are shown by the open rectangle [55] and by the horizontal filled rectangle [421]. Plus lines connected by the dashed line are the Monte-Carlo simulated values [40].

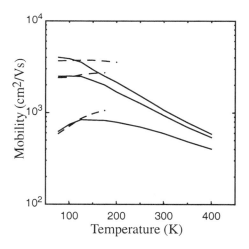

Figure 6.5: Monte Carlo values of electron mobility in GaN are plotted as a function of temperature. Top curve is for electron concentration $n = 5 \times 10^{16}$ cm^{-3}, middle curve is for $n = 10^{17}$ cm^{-3} and bottom curve is for $n = 5 \times 10^{17}$ cm^{-3} [40].

6.2. ELECTRICAL PROPERTIES OF N-TYPE GAN

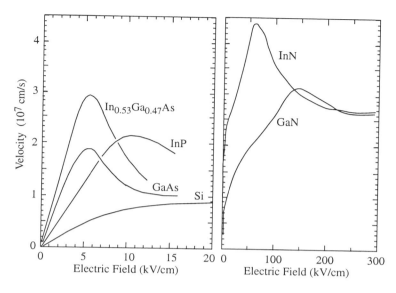

Figure 6.6: Electron drift velocity as a function of electric field for Si, GaAs, InP, InGaN and GaN. The data for constructing this figure have been taken from Ref. [24] for Si, GaAs and InP and from Ref. [25] for GaN and InN.

6.2.3.2 Monte Carlo simulations of electron mobility and velocity

An analytical multivalley conduction band model based on empirical pseudopotential band structure calculations is available for WZ GaN [58, and references given therein]. Using the analytical model Albrecht *et al.* [58, 40] calculated the electron mobilities μ_e by Monte Carlo simulations. The calculations were made for different concentrations of the ionized impurities and at different temperatures. The calculated values of the mobility are shown in Fig. 6.5. The mobility increases at lower temperatures up to about 150 K. It starts decreasing as the temperature decreases below 150 K. This decreases is attributed to neutral impurity scattering because of the significant carrier freeze out at low temperatures. Dislocation scattering was not included in these calculations. The values predicted by these simulations are higher than the experimental values discussed earlier (see Fig. 6.4). Theoretical values of the mobilities calculated by other workers have similar high values [59]. As the technology of growth improves, the experimental mobility values should increase and approach the calculated values. The values in cubic GaN are expected to be 40% higher [95]. However the predicted high mobilities and high field velocities can be obtained only in dislocation free crystals. As already discussed recent work shows that the effect of dislocation scattering on the mobility is very large.

Monte Carlo simulations of electron velocity versus electric field have been made for many semiconductors [25, 55, 60]. The results of these simulations are shown in Fig. 6.6. The peak velocity in GaN is close to 3×10^7 cm/s and the saturation velocity is 2.5×10^7 cm/s [25]. (The saturation velocity in GaN quoted in Ref. [55] is 1.5×10^7 cm/s.) The peak velocity for InN is $> 4 \times 10^7$ cm/s but the saturation velocity in InN is the same as in GaN. Peak and saturation velocities in Si and GaAs are much smaller. Transient Monte Carlo calculations show that even though conduction band effective mass in GaN is $0.22m_0$ as compared to $0.067m_0$ in GaAs, electrons in GaN are expected to show much larger overshoot effects, though fields required for the effect will be considerably larger [95]. The overshoot effects in InN are expected to be larger than in GaN.

6.3 Electrical properties of p-type III-Nitrides

6.3.1 Resistivity, hole concentration and mobility

Until ten years ago efficient p-type doping of GaN was not achieved. Amano et al. [72, and references given therein] showed that Mg-doped GaN can be converted into p-type GaN by low-energy electron beam irradiation (LEEBI). A higher activation ratio of the Mg atoms was obtained by Nakamura et al. [422] with GaN layers deposited on GaN buffer layers. Nakamura et al. [74, 423] also showed that the effect of LEEBI is in fact to heat the layers. The heating causes dissociation of hydrogen atoms which form complexes with Mg. The process of dissociation activates the Mg dopant atoms. Thermal annealing in vacuum or in nitrogen atmosphere also activates the Mg acceptors without the LEEBI treatment [74, 423]. High temperature annealing of the MOVPE grown Mg doped layers increases the conductivity by up to 6 orders of magnitude. A hole concentration $> 10^{18}$ cm^{-3} was obtained at room temperature by the thermal treatment. Nakamura et al. [74, 423] also found that the effect of thermal treatment is reversible. By heating the activated layers in hydrogen atmosphere, the Mg atoms are de-activated. Subsequently Götz et al. [424] also activated the Mg impurity by thermal annealing and confirmed that on thermal annealing the resistivity decreases by more than 6 orders of magnitude.

Recently Fujita dt al. [38] have studied more extensively the effect of annealing the GaN:Mg layers in nitrogen and hydrogen ambients (see Fig. 6.7). First the layers were annealed in N_2 atmosphere for 20 min at different temperatures shown in Fig. 6.7(a) as a parameter. The concentration of the annealed layers was measured at temperatures in the range 100 to 300 K. The concentration increases as the annealing temperature increases up to 900° but starts decreasing on annealing the layers at temperatures higher than the growth temperatures. The increase in the carrier concentration is due to breaking of Mg-H complexes and activation of Mg. The decrease at higher temperatures of annealing is due to the generation of compensating defects. The carrier concentrations also decrease in the initial stages as the temperature of measurements decreases. The concentration plots show minima and increase again at lower temperatures as shown in Fig. 6.7(a). This increase is probably related to the formation of a defect conduction band. In the presence of such a band the one electron model used in the analysis is not valid.

6.3. ELECTRICAL PROPERTIES OF P-TYPE III-NITRIDES

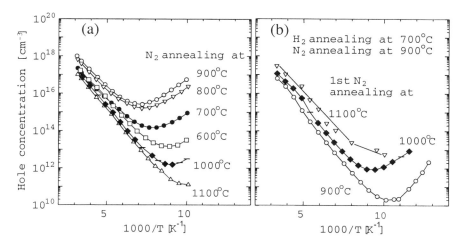

Figure 6.7: (a) Temperature dependence of hole concentration in GaN:Mg samples subjected to different dehydrogenation (N2 annealing) temperature. (b) samples which have been subjected to dehydrogenation, hydrogenation, and re-dehydrogenation, successively. The 1st dehydrogenation temperatures are shown as a parameter and the successive hydrogenation and re-dehydrogenation were done at 700°C and 900°C, respectively. The figure is taken from Ref. [38].

Fig. 6.7(b) shows the effect of repeated hydrogenation and dehydrogenation on the carrier concentration. This figure shows that the actual processes which control the activation of Mg are very complex and depend strongly on the experimental conditions.

Miyachi et al. [427] fabricated GaN p-n diode using Mg dopant. To activate Mg, different diodes were annealed under different biasing conditions. Under forward bias of 8 V, the resistance of the Mg doped dropped to a low value on annealing at 300 to 350°C. These experiments show that Mg can be activated at lower temperatures in the presence of minority electrons.

We now discuss the hole mobilities in GaN. Brandt et al. [428] studied ZB GaN containing both Be and oxygen dopants. The layers were grown on (113) plane of GaAs by MBE. In a sample containing 10^{20} cm^{-3} Be atoms, the free hole concentration was 5×10^{18} cm^{-3}, and the mobility was 70 cm^2/Vs. In another sample, with 1×10^{18} cm^{-3} holes, a mobility of 150 cm^2/Vs was observed. These high values of hole mobilities are attributed to the formation of Be-O pairs which do not scatter the holes efficiently. We have compiled values of the hole mobility in GaN obtained by different groups and plotted them in Fig. 6.8. The measured hole mobility in LEEBI GaN:Mg samples [422] are shown by filled circles. The value shown by open circle is for a MOVPE grown layer by Akutsu et al. [425]. Lin et al. [426] deposited ZB GaN epilayers on vicinal (100) GaAs substrates by ECR plasma-enhanced MBE. Mg was also injected into the ECR source using N$_2$ carrier gas. Hole concentrations between 8×10^{16} cm^{-3}

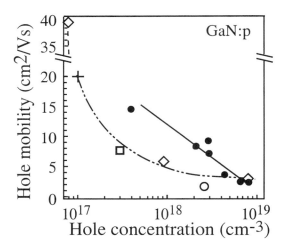

Figure 6.8: Hole mobility in Mg doped GaN as a function of hole concentration. Filled circles are from Ref. [422] for LEEBI treated GaN, open circle is from Ref. [425], and open square is from Ref. [225] for a MBE grown layer. Open diamonds show the results for ZB cubic MBE grown GaN [426]. The plus sign shows the recent value obtained by Fujita et al. [38]. Solid and dot-dashed lines are drawn as an aid to the eye.

and 8×10^{18} cm^{-3} were obtained. The mobility at a few typical concentrations used by Lin et al. [426] are shown by open diamonds in Fig. 6.8. The highest mobility is 39 cm^2/Vs. This high value is obtained because layer with very low hole concentrations could be fabricated. At higher hole concentrations, the mobilities are lower and comparable to the mobilities obtained by Nakamura et al. [422].

Fujita dt al. [38] have studied the effects of annealing the GaN:Mg layers in nitrogen and hydrogen ambients on the hole mobility (see Fig. 6.9). In all cases the mobility shows a maximum around 200 K. The hydrogenation and dehydrogenation have substantial effect on the mobility. This is not unexpected. The annealing changes the number of free carriers and ionized impurities. The scattering of the carriers is affected and the mobility changes.

6.3.2 Activation energy of Mg acceptors in GaN

Earliest determination of the activation energy of Mg was made by Akasaki et al. [429] using the temperature dependence of the PL intensity due to donor-acceptor recombination. The GaN layers were Mg-doped and LEEBI treated. Values in the range 155 to 165 meV for the ionization energy of Mg were obtained. Moustakas and Molnar [222] measured and analyzed conductivity and Hall effect measurements. Mg concentrations in the range 10^{17} to 10^{19} cm^{-3} were4 used. Corresponding values of hole concentrations were in the range 10^{13}

6.3. ELECTRICAL PROPERTIES OF P-TYPE III-NITRIDES

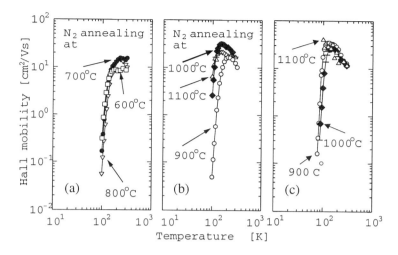

Figure 6.9: (a) Temperature dependence of hole Hall mobility of GaN:Mg samples subjected to different dehydrogenation (N annealing) temperatures. (c) Temperature dependence of hole mobility of GaN:Mg samples subjected to hydrogenation and re-hydrogenation successively. The first dehydrogenation temperatures are shown as a parameter and the successive hydrogenation and re-dehydrogenation were done at 700°C and 900°C respectively [38].

to 10^{18} cm^{-3}. The slope of the high temperature (conductivity) versus $1/T$ plot was analyzed to extract the energy of ionization of the Mg atoms. Assuming that $N_{Mg} \gg p$ (p is concentration of holes) the slope yielded an activation energy of 150 meV. Götz et al. [424] determined the hole concentration by Hall measurement and plotted the concentration as a function of $1/T$. Analysis of the plots yielded values of the activation energy 182 to 200 meV. Activation energy of Mg in GaN determined by Tanaka et al. [418] by Hall effect measurements is 150 meV.

Johnson et al. [430] have determined the ionization energy of Mg acceptors in GaN using PL measurements. They obtained an optical ionization energy of Mg acceptors as 221 meV. The value determined from the 2 K PL measurements by Götz et al. [331] is 209 meV (see section 5.5.2). Both values determined by the PL method are considerably larger than the energy of ~ 160 meV determined from the temperature variation of conductivity and Hall effect measurements. The difference in the values could arise due to the large difference in the temperatures at which the electrical and the optical measurements were made. Even at the same temperature the thermal ionization energies can be lower than the optical ionization energies [431, and references given therein].

The values of the activation energy determined by different groups are shown in Table 6.3. The activation energies are so high that only a small fraction, < 5% of Mg atoms can be ionized at room temperature. In most experiments the observed fraction of activated Mg atoms is less than 1%. The maximum

Table 6.3: Energies E_A required to activate Mg acceptors in WZ GaN.

Method	E_A (meV)	Ref.
PL[a]	155 to 165	Akasaki 1991 [429]
Hall effect	150	Moustakas 1993 [222]
Hall effect	150	Tanaka 1994 [418]
Hall effect	170 to 182	Götz 1996 [424]
PL	209	Götz 1996 [331]

[a] Determined from temperature variation of the intensity of D-A recombination PL in GaN:Mg.

hole concentration is $\sim 2 \times 10^{18}$ cm^{-3} in MOCVD GaN:Mg layers after LEEBI or thermal annealing. In the MBE layers a concentration of $\sim 2 \times 10^{19}$ cm^{-3} was obtained without any posts growth treatment (quoted in [81]). The Mg concentration is higher by at least a factor of 20 than the free hole concentration. It seems that un-ionized Mg atoms scatter holes and reduce the hole mobility.

In view of the difficulties with Mg as a p-type dopant, other dopants are being tried to obtain high p-type conductivity [13, 99, 194]. Carbon, cadmium and calcium are the other potential acceptors. Abernathy [194] and Popovici [99] has reviewed the work on III-Nitrides doped with other acceptors. Zn gives rise to 4 levels above the valence band at 0.57 eV, 0.88 eV, 1.2 eV and 1.72 eV [62].

6.4 Electrical properties AlN, InN and alloys

6.4.1 AlN and AlGaN

In this section we discuss the electrical properties of III-Nitride alloys. As grown AlN has a large concentration of C, O and H impurities [194]. However unlike GaN and InN, carrier concentration and mobility in AlN are very low, AlN is semi-insulating. The carrier concentration is $< 10^{16}$ cm^{-3} and mobility is less than 2 cm^2/Vs.

The cladding regions of a III-Nitride based laser diodes consist of $Al_xGa_{1-x}N$ layers. It is therefore important to be able to dope $Al_xGa_{1-x}N$ layers both n- and p-type. Stampfl and van de Walle [432] have discussed the experimental work on doping these alloys and given a list of references which describe the earlier work. As Al mole fraction x increases from 0.4 to 0.6, the concentration of electrons and mobility decrease by up to 5 orders of magnitude and the epilayers become semi-insulating. Doping experiments have been performed on $Al_xGa_{1-x}N$ alloys with x varying from 0 to 0.75. For $x > 0.4$, addition of Si was not effective in decreasing the resistance of the $Al_xGa_{1-x}N$ layers.

Activation energy of Si was determined by Tanaka et al. [418] by Hall effect measurements in both GaN and $Al_{0.1}Ga_{0.9}N$. The activation energy of Si was about 30 meV in GaN and appeared to be smaller in $Al_{0.1}Ga_{0.9}N$. Recently Korakakis et al. [231] have made extensive measurements of electrical conductivity and Hall mobility in AlGaN layers grown by ECR microwave plasma. Conductivity of n-type Al_xGa_{1-x} epilayers was measured for AlN mole fraction up to 60% [231]. The epilayers were grown on sapphire as well as on SiC substrates. Both undoped and Si-doped layers were investigated. The conductivity did not depend on the substrate used. It decreased rapidly with the increase of Al fraction x in both the doped and the undoped samples [231, and references given therein]. Hall effect measurements showed that both the carrier concentration and mobility of the carriers decrease rapidly as the Al fraction increases. The suppression of the mobility can be due to alloy scattering. The resistivity of a GaN and two Al_xGa_{1-x} samples (for $x = 0.08$ and 0.18 respectively) was also measured as a function of temperature. The conductivity was thermally activated at low temperatures. The activation energy determined from the temperature variation of the carrier density was 17 meV for the GaN samples and 54 meV for the AlGaN samples. The activation energy for the donors increases (i.e. the donor states become deeper) as the Al fraction increases. This explains the suppression of carrier density with the increase of Al concentration.

The situation for p-type doping is worse. Attempts to dope $Al_xGa_{1-x}N$ layers p-type with Mg have also been made. Activation energy of Mg in GaN and AlGaN was determined by Tanaka et al. [418] by Hall effect measurements. It is about 180 meV in $Al_{0.075}Ga_{0.925}N$, higher than the value 150 meV in GaN measured by the same authors. It has not been possible to dope the layers p-type with Mg if $x > 0.13$ (see references in [432]). However Akasaki and Amano have mentioned in a recent publication [81] that a hole concentration of $\sim 2 \times 10^{17}$ cm^{-3} can be obtained in MOCVD $Al_{0.2}Ga_{0.8}N$ layers.

Stampfl and van de Walle [432] have made first principle calculations to examine the causes which make it difficult to dope the $Al_xGa_{1-x}N$ layers. They have assumed that the centers responsible for the background n-type conductivity are the residual oxygen and/or Si impurities. Their calculations show that as Al concentration increases to more than 40%, oxygen atoms change from the shallow donor configuration to deep DX centers. The effect is similar to that obtained with GaN under hydrostatic pressure. Both experiments and calculations show that at a hydrostatic pressure of about 20 GPa, the carriers in oxygen-doped n-type GaN freeze out and the layers become highly resistive. In both cases i.e. GaN under hydrostatic pressure and in $Al_xGa_{1-x}N$ layers the bandgap widens and the shallow donor oxygen atoms are converted into deep DX centers. This interpretation is not valid for Si doped GaN. Si donors do not form deep DX states. The calculation of Stampfl and van de Walle [432] show that formation energy of the triply charged Al vacancies, V_{Al}^{3-} is low for $x > 0.4$, particularly in the highly n doped $Al_xGa_{1-x}N$ layers. The authors suggested that these defects compensate the n-type doping and reduce the conductivity when dopants are Si atoms.

Stampfl and van de Walle [432] have also studied the p-doped $Al_xGa_{1-x}N$ layers. Some acceptors form AX deep level centers, analogous to the DX centers formed by the donors. However Mg acceptors do not form the deep level centers in $Al_xGa_{1-x}N$ layers. Calculations of Stampfl and van de Walle [432] show that the efficiency of p-type doping in the $Al_xGa_{1-x}N$ layers decreases on two account. The formation energy of the triply charged nitrogen vacancies V_N^{3+} is low. These defects compensate the Mg acceptors and reduce the hole concentration. Stampfl and van de Walle [432] have shown that ionization energy of Mg acceptors increases as x in the $Al_xGa_{1-x}N$ layers increases. This effect also suppresses the efficiency of p doping. The formation energy of interstitial Al_I^{3+} is large in WZ crystals and is unlikely to be the center responsible for reducing the efficiency of the p doping [432].

InAlN has not been studied extensively [194]. Carrier concentration in InAlN epilayers grown in hydrogen ambient have been measured [194]. If the In mole fraction is less than 30%, the epilayers show AlN like behavior. Both carrier concentration and mobility are very small. As the In concentration increases, both the carrier concentration and the mobility increase approximately linearly with In concentration up to l00% In concentration.

6.4.2 InN

Abernathy and co-workers [194] have made Hall effect measurements in InN films grown by MBE at different temperatures. As grown InN films are always n-type with a high concentration ($> 10^{20}$ cm^{-3}) of electrons [194]. The carrier concentration is low in the layers grown at higher temperatures. At high temperatures the nitrogen atoms are more likely to evaporate and concentration of nitrogen vacancies should increase [194]. This result shows that the nitrogen vacancies are not responsible for the background doping. Similar results have also been reported by Ren [433] for MOMBE grown InN at growth temperatures in the range 400 to 530°C. The highest mobility was 120 cm^2/Vs at 525°C and the highest carrier concentration was 6×10^{20} cm^{-3} at 400°C.

6.5 Schottky barriers and ohmic contacts

6.5.1 Contacts on II-VI heterostructures

An ohmic contact is defined as a metal-semiconductor contact that has a low resistance. The resistance should be so low that the voltage drop across the contact is negligible as compared to the drop across other parts of the semiconductor structure. The contact must be stable at temperatures through which the device will be heated during processing or to which it will be heated during its usage. High quality, low resistance and stable ohmic contacts are necessary for reducing parasitic resistance and improve over-all performance of the electronic and optoelectronic semiconductor devices. The carriers across the contact are transported by one or more of the three mechanisms: (1) The thermionic emission, (2) Field emission or tunneling, and (3) Thermionic field emission. If the Schottky barrier at the contact is thick, tunneling can not take place and thermionic emission dominates. The contact is rectifying. On the other hand if

6.5. SCHOTTKY BARRIERS AND OHMIC CONTACTS

the barrier is thin, tunneling takes place and the contact is ohmic. The thickness of the barrier is related to the width of the depletion layer in the semiconductor. The depletion width is inversely proportional to the free carrier concentration in the semiconductor near the junction. If the barrier is wide at the bottom and thin near the top, the carrier need be excited thermally only up to the thin region and they then tunnel through. It requires lesser energy than and occurs at lower temperatures then the thermionic emission.

The valence band edge in ZnSe is at the Γ point and is very deep. The hole affinity is ~ 6.7 eV. Since a metal with sufficiently large work function is not available it is very difficult to make ohmic contacts to p-ZnSe. Most recent work is concentrated to improving the ohmic contacts to p-type ZnSe. Prior to 1992, Pt and Au metal contacts were used but there is a rather large Schottky barrier with these metals [129]. The energy barrier between Au contact and ZnSe is ~ 1.6 eV and therefore contact resistance is very high [354]. After the demonstration of the first ZnSe-based laser diodes [133, 134] the need to develop low resistance ohmic contacts became urgent. Because the contact resistance in these laser structures was very high, they operated at high voltages and the threshold current was also high. Ohmic contacts are also necessary to make $I - V$ and Hall effect measurements. Holes could be injected by using the reverse biased Schottky contact but the high voltages (threshold voltage was > 25 V) at which the laser operated resulted in the breakdown of the Schottky contacts [129]. Efforts to inject holes by using a p-GaAs layer on p-ZnSe were also not successful because of the large (~ 1 eV) valence band offset between GaAs and ZnSe.

The valence band offset and the contact resistance of Au metal contact can be reduced considerably by using intermediate InAlP layer. Iwata et al. [354] have grown good quality $In_{0.5}Al_{0.5}P$ layers on ZnSe by MBE. They could obtain a high hole concentration of 2×10^{18} cm^{-3} in the $In_{0.5}Al_{0.5}P$ layers grown at a low temperature of 350 °C. Be has a low activity in the alloy $In_{0.5}Al_{0.5}P$ and hence a high temperature (500 °C) of Be cell could be used. Though Be concentration was high. However Be precipitation or segregation was not observed. Surface morphology was good. These results show that good ohmic contacts to p-ZnSe can be formed by using $In_{0.5}Al_{0.5}P$ layer between Au and ZnSe.

ZnTe can be doped p-type with a high concentration of holes (see section 6.1.2). Au forms a good ohmic contact to p-ZnTe. Efforts were made to make ohmic contact to p-ZnSe by using Au/p-ZnTe/p-ZnSe. The energy band diagram of the p-ZnTe/p-ZnSe heterojunction is shown in Fig. 6.10(a). In the structure shown in Fig. 6.10(a) the valence band offset between p-ZnTe and p-ZnSe is ~ 1 eV and forms a barrier for hole injection. Therefore early attempts to form ohmic contacts with p-ZnTe inter-layer were not very successful [129]. In later work the energy spike in the valence band was removed by using a graded p-type ZnTeSe layer as shown in Fig. 6.10(b). Though the graded inter-layer contacts were ohmic and had low resistance, it was difficult to make reproducible contacts using MBE [129, 409]. The main difficulty was in controlling the quality of the epilayers and concentration of Te due to competition between Se and Te species for incorporation in the graded epilayer and due to clustering of Te. Han and

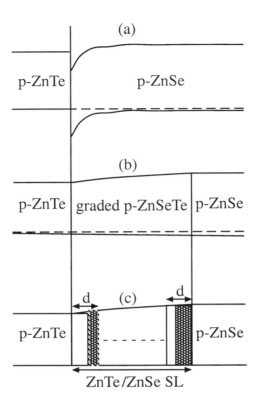

Figure 6.10: (a) Schematic energy band diagram of the p-ZnTe/p-ZnSe heterojunction and (b) p-ZnTe/graded p-ZnSeTe/p-ZnSe heterostructure. (c) Schematic structure of the pseudo-graded Zn(SeTe) superlattice (SL) contact layer (see text) [129].

Gunshor [129] replaced the graded layer by a ZnTe/ZnSe superlattice (SL) as shown in Fig. 6.10(c). The superlattice consisted of 17 periods. The thickness of each period, consisting of p-ZnTe and p-ZnSe bilayer, was kept constant at 20 Å. However thickness of individual p-ZnTe and p-ZnSe layers was varied continuously. The thickness of ZnSe layer was 18 Å and that of ZnTe was 2 Å in the period in contact with ZnSe. The thickness of the ZnSe layer was 17 Å and that of ZnTe was 3 Å in the second period. The thickness of ZnSe layer decreased and that of ZnTe increased continuously in going from ZnSe towards ZnTe. $I-V$ measurements showed that the contact using the SL scheme was ohmic at both 300 and 77 K. The $I-V$ curves showed some non-linearity at high current densities (500 A/cm^2).

It is not so difficult to make ohmic contacts to p-CdTe. Good ohmic contacts to p-CdTe can be made using p-HgTe or copper-doped ZnTe [434]. A graphite paste doped with either Au, Cu or Hg has also been used for making contacts to p-CdTe. On annealing, these layers form p^+ Au$_2$Te, Cu$_2$Te, Hg$_x$Cd$_{1-x}$Te

6.5. SCHOTTKY BARRIERS AND OHMIC CONTACTS

respectively. However because of the high diffusion coefficient of metals in these layers, these contacts are liable to degrade with usage [434]. To avoid this problem Ni/Al contacts have been used. Low resistance Ni–P contacts to p-CdTe have been fabricated by Miles *et al.* [434]. The contact fabrication involved autocatalytic deposition of Ni–P alloy coatings onto a Te rich surface of p-CdTe followed by a thermal anneal. Various annealing temperatures were used in the range 20°C to 300°C. Contacts with low resistance (0.08–0.1 Ωcm^2) were obtained by annealing at 250°C. During the thermal annealing P diffuses into CdTe and produces a heavily p-doped surface layer which facilitates quantum mechanical tunneling and formation of the low contact resistance. Cu–Au and Hg doped graphite paste contacts were also tried. In these cases, the contact resistance was larger, ~ 0.5 Ωcm^2.

It is possible to grow pseudomorphic graded p-doped BeTe-ZnSe layers between GaAs and device structure for making low resistance ohmic contacts to p-ZnSe based alloys [367, 407]. We mentioned earlier that there is a close valence band-edge match between GaAs and BeTe. It is therefore possible to fabricate devices on p-GaAs substrate in a "p-contact down" configuration [435]. This provides another degree of freedom in the design of devices. Since bandgap of BeTeZnSe is larger, these contacts have additional advantage that they do not attenuate the laser light.

6.5.2 Contacts on III-Nitrides

6.5.2.1 Schottky barriers

Good Schottky contacts are essential for realizing high quality Metal Semiconductor Field Effect Transistors (MESFETs). Good Schottky contacts should have large barrier heights and large breakdown voltages so that reverse leakage current is small and higher levels of drain current can be controlled. Schottky diode rectifiers are also used when fast switching is required. In the II-VI semiconductors most studies of contacts are confined to ohmic contacts. Since GaN has a potential for high temperature electronic devices, several authors have investigated the properties of Schottky contacts made with different metals on GaN [55, 433, 436, 437][1]. In the case of Si and conventional III-V semiconductors, the Schottky barrier height is not sensitive to the work function of the metal with which the contact is made. In these semiconductors, the surface state density is high and the Fermi level is pinned at the surface. The barrier height is determined mainly by the properties of the semiconductors. For the III-Nitrides Schottky barriers are sensitive to the work function of the metal. According to Schottky model, the barrier height is equal to the difference between the work function W of the metal and the work function (electron affinity) χ of the semiconductor. The measured work function for n-type GaN is 4.1 eV [433]. A metal with a higher work function should form a Schottky contact and a metal with a lower work function should form an ohmic contact with GaN.

---

[1] Additional references and values of Schottky barriers are given in the recent review of Pearton *et al.* [13]

Table 6.4: Work function W of metals used to make contacts with GaN.

metal	W (eV)	Ref.
Au	5.1	[438]
Au	4.82	[433]
Ni	5.15	[438]
Pd	5.12	[438]
Cr	4.5	[438]
Al	4.08	[433]
Ca	3.1	[329]
Cu	4.6	[329]
Ti	4.3	[55]
Sc	3.50	[439]

Table 6.5: Values of Schottky barrier $q\phi_b$ (in eV) determined by $I-V$, $C-V$ and Photoemission (PE) methods by Yu et al. [440].

Metal	Nitride	$I-V$	$C-V$	PE
Ni	GaN	0.95	0.96	0.91
Ti	GaN	0.65	0.68	–
Ni	$Al_{0.15}Ga_{0.85}N$	1.25	1.26	1.28
Ti	$Al_{0.15}Ga_{0.85}N$	0.84	1.10	–

Experiments show that this indeed is the case [433]. The values of work functions of metals are given in Table 6.4. The work function of Al is 4.08 eV and it does form an ohmic contact [433] on GaN. Au with a work function of 5.1 eV makes a Schottky contact.

Recently Yu et al. [440] have measured the Schottky barriers of Ni and Ti contacts on GaN and $Al_{0.15}Ga_{0.85}N$ epilayers grown by MOVPE on n-type 6H-SiC substrates. A 0.1 μm thick AlN buffer layer was deposited on the SiC substrate before the GaN and the $Al_{0.15}Ga_{0.85}N$ epilayers were grown. The thickness of the $Al_{0.15}Ga_{0.85}N$ layer was 1 μm and the electron concentration due to background doping was $\sim 1 \times 10^{17}$ cm^{-3}. The background electron concentration in GaN layers was presumably higher. Al/Ti (71 nm/30 nm) with Ti layer at the bottom are used for ohmic contacts. The Schottky contact dots consisted of 100 nm Ni or 100 nm Ti and were of 125 μm diameter. The Schottky barrier heights $q\phi_b$ were extracted from the experimental current–voltage ($I-V$) and capacitance–voltage ($C-V$) plots. In the case of Ni contacts, internal photoemission measurements were also made and Schottky barrier heights were derived from these measurements also. The results of these measurements are summarized in Table 6.5. For Ni and Ti contacts on GaN the results obtained by different techniques agree with each other. For Ti contacts

6.5. SCHOTTKY BARRIERS AND OHMIC CONTACTS

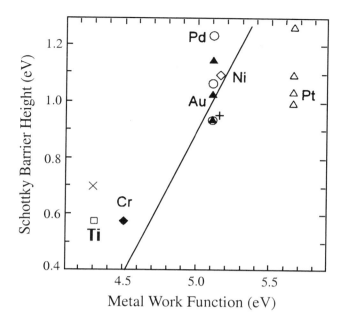

Figure 6.11: Schottky barrier height versus metal work function in metal-GaN heterojunctions. The Schottky model $\phi = W - \chi$ is shown by the straight line for reference. The data shown by × for Ti and by + for Ni are taken from Table 6.5 and other data are taken from [55].

on AlGaN the values obtained from $I - V$ measurements is lower by 0.26 eV as compared to the value obtained by the $C - V$ method. The reason of this discrepancy is not known. The heights of the Schottky barriers measured with different metal contacts on GaN are shown in Fig. 6.11. Table 6.5 and Fig. 6.11 show that the experimental data do not agree with the Schottky model very well. However metals with higher work functions do have larger Schottky barriers as predicted by the Schottky model. The work function of Ti is the lowest and it has the smallest barrier. Ti produces the lowest barriers (0.1 to 0.6 eV). Fig. 6.11 shows that the scatter in the values of the barrier height is rather large. The values for the same metal obtained by different groups or by the same group with different techniques differ significantly. The scatter in the values of the Schottky barriers can arise due to surface roughness and surface cleanliness. Local stoichiometry near the contact is also important [93]. In view of the large scatter, the agreement with the Schottky model may be regarded as reasonable.

As mentioned earlier, Al has a work function lower than that of Ti. In this case the barrier is close to zero. This value can not be shown in the figure as it is beyond the scale used in the figure. PtSi contacts on GaN have also been studied. PtSi makes a Schottky contact with barrier height ~ 0.8 eV. Pt contact is stable up to 400°C for one hour and PtSi contact is stable up to

500°C for the same duration [93]. Au/GaN contacts have shown high breakdown voltage. However the reverse leakage current is high and value of the effective Richardson's constant A^* are small. Until recently the maximum experimental value of A^* was 6.61 A/(cm^{2}K^2) [437]. The small value is attributed either to tunneling of electrons through the tunneling barrier or to decrease of the effective contact area. Recently Ishikawa et al. [437] have fabricated Pd/GaN Schottky diodes with improved performance. The reverse breakdown voltage was about 100 V, the reverse leakage current was 3.26×10^{-6} A/cm^2. The barrier height was 1.53 to 1.55 eV. The value of A^* was 23.2 A/(cm^{2}K^2) in good agreement with the theoretical value of 26 A/(cm^{2}K^2). The barrier height of the Pd/GaN measured by Ishikawa et al. [437] is larger than the earlier values. The difference is probably due to doping concentration but also due to the surface treatment. Ishikawa et al. [437] treated the surface with $H_3PO_4+H_2SO_4$.

6.5.2.2 Ohmic contacts

Murakami et al. [441] have discussed the techniques and methods that are used to prepare ohmic contacts on GaAs and wide bandgap materials. Before depositing the contact metals, the semiconductor is heavily doped p- or n-type. The doping must be $> 10^{17}$ cm^{-3}. The high doping decreases the width of the depletion layer and facilitates transport of the carriers by tunneling. The surface of the semiconductor is cleaned and etched to remove the native oxide and other defects. The metal is chosen with a value of the work function which makes the barrier height as small as possible. After the contact metals are deposited, the semiconductor is annealed at suitable high temperature. During the annealing process, an intermediate layer at the interface between the contact metal and the semiconductors is formed. If the intermediate layer is heavily doped and has low barrier height, the contact is ohmic with low resistance, otherwise it is a Schottky contact with high resistance. The interfacial layer between most metals and n-GaN is the nitride of the metal. Nitrogen is extracted from the surface layers of GaN and the surface layer becomes heavily n-doped.

Making good ohmic contacts to wide-bandgap semiconductors is a difficult problem, particularly if the semiconductors can not be doped with sufficient concentration of free carriers e.g. AlN or p type GaN [433]. Since nominally undoped and n doped GaN and InN have large concentrations of electrons, making ohmic contacts with these materials is less difficult. Considerable progress has been made and contacts with very low resistivity have been obtained since then. Most work has been done Al and Ti/Al contacts. Contacts with single- and multi-layers of other metals have also been studied. A review of early work on ohmic contacts on GaN has been published by Ren [433].

As deposited Al makes ohmic contacts on n-type GaN. However the contacts degrade after they are heated to 575°C [433]. Ti also produces ohmic contacts on n-GaN. Ti contacts require a higher temperature (900°C) annealing [433]. Luther et al. [442] studied the effect of ambient gas on the properties of Al contacts on n-type GaN. They found that low resistivity Al contact can be obtained by annealing at temperatures in the range 400 to 600°C, provided the anneal-

6.5. SCHOTTKY BARRIERS AND OHMIC CONTACTS

Table 6.6: Best published values of specific resistivity, ρ_c, of ohmic contacts on n- and p-type GaN. T is the annealing temperature and t is the time for which the samples were annealed.

Contact	p or n type	ρ_c (Ωcm^2)	T,t (°C,s)	Ref.
Al	n type	8.0×10^{-6}	400^a,1200	[442]
Ti/Al	n type	8.0×10^{-6}	900,30	[436]
Ti/Al	n type	5.0×10^{-6}	600^b,15	[442]
Ti/Al/Ni/Au	n type	8.9×10^{-8}	900,30	[443]
Pd/Al	n type	1.2×10^{-5}	650,30	[444]
Cr/Au	p type	$\leq 4.3 \times 10^{-1}$	900,15	[438]
Ni/Au	p type	$\geq \times 10^{-2}$	–	[93]
W	n type	8.0×10^{-5}	1000,60	[445]
WSi	n type	$\leq 10^{-5}$	1050	[93]

^{a}In Ar/4%H$_2$ forming gas.
^{b}In Ar gas.

ing is done in the forming gas. They annealed the Al contact at 600°C in the Ar/4%H$_2$ gas for 20 min and obtained specific resistivity ρ_c of 8×10^{-6} Ωcm^2. Al contacts annealed at 600°C for one hour degraded only slightly. The severe degradation of the Al contacts, which was observed on high temperature annealing in the earlier work [433], was not observed in these experiments. The degradation observed earlier was probably due to oxidation of Al which is suppressed in the forming gas ambient. We have compiled the values of specific resistivities ρ_c of the contacts on GaN reported by different groups in Table 6.6. Low resistivity contacts have been obtained on n-type GaN with Ti/Al by several group. The lowest values obtained by two groups are shown in Table 6.6. After annealing the Ti/Al layer at 900°C, a resistivity of 8×10^{-6} Ωcm^2 was reported by Lin et al. [433, 436]. Luther et al. [442] showed that low resistivity Ti/Al contact can be obtained by annealing at lower temperatures, in the range 400 to 600°C, provided the annealing is done in the Ar gas. Ti/Al contact annealed in Ar gas at 600°C for 15 s had a contact resistivity of 5×10^{-6} Ωcm^2. This is the lowest resistivity of the Ti/Al contacts reported so far. Luther et al. [442] performed depth profiling by the XPS experiments. These experiments showed that during the annealing process, mixing of Ti and Al occurs and an Al-Ti intermetallic phase is formed. The intermetallic phase has a low work function, similar to that of Al. Ti reduces the native Ga oxide. The contact between GaN and the intermetallic phase forms the good ohmic contact. Experiment with different thickness of Al and Ti were made. It was found that a thicker Ti layer gives better stability while a thinner Ti layer gives lower contact resistance. The optimum layer thicknesses which gave good stability as well as low resistivity was (50 nm)/(100 nm) for the Ti/Al contact bilayer.

Ti/Al/Ni/Au multilayer contacts [93, 443, and references given therein] are also ohmic with low resistivity. The multilayer contact seems to give a wider processing window [93, 443]. TiN has been observed at the interface between the metal multilayer (containing one Ti layer) and GaN [442, 443]. Presumably nitrogen diffuses out during the annealing process producing TiN interfacial layer. Simultaneously GaN becomes more heavily doped n-type with nitrogen vacancies near the interface. This facilitates transport of electrons by tunneling and formation of low resistance ohmic contact. The lowest published value of specific resistivity of the contact on n-GaN is 8.9×10^{-8} Ωcm^2 for the Ti/Al/Ni/Au multilayer annealed at 900°C for 30 s [443]. The interface was good but not completely abrupt, it contained reaction products involving all the species. The reaction product probably involved formation of TiN by out-diffusion of N as already discussed.

Ping et al. [444] have fabricated Pd/Al contacts on n-GaN with fairly low value of the specific resistivity. As deposited contacts were rectifying. The lowest value of the specific resistivity was 1.2×10^{-5} Ωcm^2 and was obtained after annealing the Pd/Al bilayer at 650°C for 30 s. These authors also fabricated Ti/Al contacts. The resistivities of the Ti/Al contacts was lower but they required higher temperature annealing. Low temperature processing of the contacts is advantageous because of the thermal constraints in the over-all processing of the devices. W contacts on n-GaN have also been studied [445]. W contacts on n-type GaN after annealing in the temperature range 600-1000°C were ohmic. The integrity of the W contacts did not degrade even after annealing at these high temperatures. β-W_2N and W-N phases were observed at the W/GaN interface after annealing at temperatures in the range 600 to 1000°C. This resulted in heavy n-type doping at the interface and helped in the formation of ohmic contacts. After RTA at 1000°C a low contact resistivity of 8.0×10^{-5} Ωcm^2 was achieved [93, 445]. WSi contacts also require high temperature anneal. Cole et al. [446, 447] have investigated in detail the integrity of the interface between WSi and n-type GaN. No elemental diffusion or interfacial reaction was observed on annealing at 600 and 700°C. On annealing at temperatures $>$ 800 °C, N is released from GaN and reacts with W to form β-W_2N at the interface. A roughening of the interface was observed as a result of the formation of the β-W_2N layer. Out-diffusion of Ga was not observed. The β-W_2N layer acts as a barrier to the out-diffusion of Ga. WSi contacts have low resistivity as shown in Table 6.6 [93].

Contacts on p-type GaN have not been studied so extensively and low resistance ohmic contacts on p GaN have not yet been fabricated. The maximum hole concentration that has been reported in Mg-doped p-type grown GaN is about 8×10^{18} cm^{-3}. In most practical cases the concentration is lower. At these concentrations, the barrier thickness is not sufficiently small for tunneling to occur. Moreover the valence band edge is $\chi + E_g = 7.5$ eV below the vacuum level. To have small barrier height we need a metal with a work function equal to this value. The work functions of the metals were given in Table 6.4. Metals with such high values of the work function are not available. The barrier heights are therefore very large. The most commonly used contact to p-type GaN is Ni/Au. Trexler et al. [438] have examined the characteristics of three different metal contacts to the p-GaN, Ni/Au, Pd/Au, and Cr/Au. The metals were chosen because they have large work function values. The contacts were electron beam evaporated thin films on MOVPE grown Mg doped GaN containing a hole concentration of 9.8×10^{16} cm^{-3}. A 50 nm film of Ni, Pd or Cr was first evaporated. A 100 nm capping layer of Au was then deposited. The Ni/Au gold contacts were rectifying as deposited and did not become ohmic even after the heat treatment. These results are different from earlier work in which near linear $I - V$ curves observed [438]. The behavior Pd/Au contact was similar. The contacts just after deposition as well as after heat treatment were rectifying [438]. As deposited and annealed at 600°C Cr/Au contacts were also rectifying. However after 15 s RTA at 900°C the $I - V$ curves became linear. The estimated specific resistivity of the contact was $\leq 4.3 \times 10^{-1}$ Ωcm^2. There was evidence of formation of Au:Ga phase and perhaps CrN was formed at the interface after the RTA. The specific contact resistivity of NiAu contact on p-type GaN quoted by Pearton et al. [93] is $\geq 10^{-2}$ Ωcm^2. Efforts to find better multilayer contacts have not been successful so far [93]. A possible solution might be to use heavily p-doped GaInN layer on top of GaN. A p doping of $\sim 10^{17}$ cm^{-3} has been achieved [93] in InGaN with 15% In.

6.6 Effect of applied electric field

6.6.1 Franz-Keldysh and Quantum Confined Stark Effects

Numerous studies of the effect of electric field on the optical properties of III-V semiconductor layers, quantum wells and superlattices have been made [453]–[455]. Franz-Keldysh electric field effect on the optical properties is well known [448] and [450]. It was found experimentally that the absorption edge of the semiconductor layers varies with field E as $1/E^n$, the exponent n has values between 1 and 2. Exciton energy was found to decrease with applied electric field in copper and lead halides. In these materials the exciton Bohr radius is of the same order of magnitude as the unit cell and the excitons are of Frenkel type. Properties of III-Nitrides under applied electric field have not been investigated extensively [449, 450]. Binet et al. [449] and Duboz et al. [450] measured the effect of external electric field on exciton energies in GaN. The excitons in GaN are of Wannier type. The work of Binet et al. [449] and Duboz et al. [450] is discussed later in this section. Work of others [450, 451, 452] on the studies of III-Nitrides under external electric field on GaN is discussed in chapter 7 in the section on modulators. Effect of electric field on II-VI semiconductor quantum wells have been investigated by many groups, e.g. CdTe/Cd$_{1-x}$Mn$_x$Te quantum wells [456, 457], Zn$_{1-x}$Cd$_x$Se/ZnSe quantum wells [379, 458], CdZnSSe/ZnSSe quantum wells [459] and Type-II CdS/ZnSe strained layer superlattices [460, 461]. Such studies are important for developing waveguide modulators, optical switching and signal processing devices [454]. Experimental studies also show that when an electric field is applied perpendicular to the quantum well layer, the exciton optical absorption edge remains resolved but shifts to lower energies.

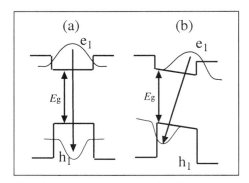

Figure 6.12: (a) Schematic potential profile of a quantum well, and (b) the profile in the presence of an electric field.

This electroabsorptive effect in quantum wells is known as the Quantum Confined Stark Effect (QCSE). In the theoretical work on QCSE, the Hamiltonian of the electron and hole includes, in addition to the electrostatic potentials and electron-hole interaction energy, the confinement potential. The eigen values of the Hamiltonian have been calculated for the lowest states by a variety of approximate methods. These calculations show large shifts of the exciton peaks to lower energies as observed in the experiments. The electric field can also quench the PL. In the presence of the electric field, electrons and hole are pulled to opposite sides of the well. The potential profile of a quantum well and electron and hole wavefunctions are shown in Fig. 6.12. This figure shows that the electrons and holes are confined in the triangular potential wells in the two opposite walls of the well. It is seen from Fig. 6.12 that spatial separation of the wavefunctions of electrons and holes increases as the thickness L_w increases. This results in a decrease in the binding energy. The binding energy decreases approximately as $1/L_w$. The decrease in the overlap of the wavefunctions also decreases the oscillator strength. The oscillator strength decreases as $1/L_w^2$. In modulation doped structures, the shape of the potential well in Fig. 6.12 also affects the transfer of charge carriers from the doped barriers into the wells. The electrons from the barriers which are raised in energy are transferred more efficiently into the well. They then move to the opposite corners. However if the charge carrier density is high, they screen the field and change the potential profile. If the screening is efficient a blue shift of the PL takes place. The figure also shows that the transition energy shifts to the red as the thickness L_w of the well increases. The potential energy of the system is reduced because the charges are now closer to the electrodes. This reduction is partly compensated by the increase in the kinetic energy because now the electrons and holes are confined more tightly near the walls of the well [453]. When these effects are taken into account, the calculated values of the exciton transition energies agree well with the observed values.

6.6. EFFECT OF APPLIED ELECTRIC FIELD

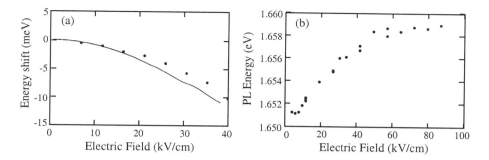

Figure 6.13: (a) Shift of the PL peak of sample T284 with applied electric field. The dots are experimental points while the curve is calculated using the transmission resonance method. (b) The PL peak position of sample T302 versus electric field. The luminescent was excited at 5.35 K with 25 μW in the first case and 45 μW in the second case with 514.5 nm light from an argon ion laser [456].

6.6.2 Experimental results: II-VI quantum wells

6.6.2.1 CdTe/Cd$_{1-x}$Mn$_x$Te quantum wells

Hartwit *et al.* [456] have investigated the effect of an applied perpendicular electric field on the PL of CdTe/Cd$_{1-x}$Mn$_x$Te quantum wells. Two samples were fabricated with different thicknesses and different composition of the barrier layers. Sample 1, designated as sample T284 consisted of 20 periods (i.e. 21 barriers) and sample 2, designated as sample T302 consisted of 60 periods. The thickness of CdTe quantum well was 11.8 nm in sample T284 and 4.23 nm in sample T302. The corresponding thicknesses of the barrier layers were 25 nm and 2.35 nm respectively. The barrier layers in the two samples consisted of Cd$_{0.82}$Mn$_{0.18}$Te and Cd$_{0.87}$Mn$_{0.13}$Te. The (111) oriented samples were grown on 296.5 nm CdTe buffer layers which were deposited on (100) conducting GaAs substrates. A third sample, similar to sample T302 was grown with higher concentration of Mn (Cd$_{0.77}$Mn$_{0.23}$Te) in the barrier layers. Semitransparent gold Schottky contacts were used. The $I-V$ plots gave a value 0.71 eV for the Schottky barrier height. The peak position in sample T284 (larger quantum well width and wide barriers) was at 1.591 eV. The peak position in sample T302 was at higher energy as shown in Fig. 6.13(b). The peaks were attributed to E_1-heavy hole radiative recombination. The shifts of the PL peak with applied electric field observed in the two samples are shown in Figs. 6.13(a) and 6.13(b). Fig. 6.13(a) shows that the PL peak shifts continuously to lower energies as the electric field increases. This red shift is due to the QCSE discussed earlier. The shift was calculated using the resonance method. The values of the material parameters used are: with $E_{\text{gap}} = 1.595 + 1.592x$, $m_e = 0.09 m_0$, $m_{hh} = 0.4 m_0$ and assuming ratio $\Delta E_c / \Delta E_v = 10 : 1$ ($\Delta E_c = 0.261 eV$, $\Delta E_v = 0.0261$ eV). The agreement between the calculated and observed values shown in Fig. 6.13(a) is quite good. Fig. 6.13(b) shows that the behavior of sample T302 is quite different. Now the shift due to applied electric field is to higher energies. This sample has very thin barriers so that the quantum wells interact and form a superlattice. In this case a miniband is formed due to overlap of the wave functions of charge carriers in different quantum wells. The applied electric field reduces the coupling between the wells [456]. The wave functions get more and more localized in the wells (known as Wannier-Stark localization). In the limit the energy becomes more like that of a single quantum well. Energy of the coupled quantum wells and width of the mini-band were calculated using the Kronig-Penney model. Based on these calculations, the saturation of blue shift is predicted to be at 25 meV. The observed saturation is only 8 meV. Two causes can make the shift smaller. The coupling may be limited only to the nearest neighboring wells and calculation of the mini-band energies may not be accurate. The binding energy of the exciton changes with the applied electric field. This will also affect the blue shift. The PL of the this superlattice was also measured at lower electric fields in the range 5 to 18 kV/cm. At these fields the peak shifts to lower energies with applied electric field. The peak is probably due to recombination of charge carriers in the two neighboring wells, the so called $n = -1$ transition of the Stark ladder [456]. Higher order transitions were not observed. In GaAs transitions for $n = -6$ and $n = +3$ have been observed [456]. This suggests that coherence length in II-VI semiconductors is much smaller than in the III-V semiconductors. The PL of the third sample with $x = 0.23$ should be interesting because CdMnTe of this composition undergoes a magnetic phase transition at 3 K from paramagnetic state to a spin glass state. The PL of this sample was measured in the range 10 K to 1.9 K. The PL did not show any change in the neighborhood of 3 K. It is possible that the phase transition in the thin 2D layers is affected by the reduced dimensions.

6.6.2.2 $Zn_{1-x}Cd_xSe/ZnSe$ quantum wells

Recently QCSE in ZnSe/ZnCdSe multiple quantum wells has been investigated by Thompson et al. [457]. The sample was a p-i-n structure. The i region consisted of the ZnSe/ZnCdSe multiple quantum wells surrounded on the two sides by n- and p-doped ZnSe. The structure was grown on n-type GaAs. Room-temperature photocurrent spectra measured with the applied voltage U in the range 0 to −15 V showed a red shift characteristic of QCSE. As expected the shift increased with the applied voltage. Because of the broadening of the photocurrent spectra at room temperature the shift was not so prominent. To investigate the shift further, the substrate was etched away and transverse transmission measurements were made. These measurements showed more clearly the red shift.

PL measurements of $Zn_{1-x}Cd_xSe/ZnSe$ quantum wells by Babucke et al. [379] were discussed in section 5.4.3. Babucke et al. [379] also studied the effect of the electric field on the reflection and transmission spectra of these quantum wells. The measured and calculated reflection spectra of 10 quantum well structure under different applied fields are shown in Fig. 6.14. The red

6.6. EFFECT OF APPLIED ELECTRIC FIELD

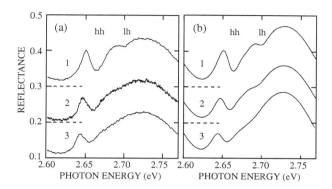

Figure 6.14: Reflection coefficient spectra of ZnCdSe/ZnSe 10 quantum well structure at 5 K at (1) 0 V, (2) –3 V and (3) –5 V. (a) shows experimental results and (b) shows calculated values [379].

Table 6.7: Excitonic parameters of the ZnCdSe/ZnSe 10 quantum well structure derived from the reflection spectra of Fig. 6.14 at 5 K. The symbol OS indicates oscillator strength.

U (V)	E_{hh} (meV)	OS_{hh} (a.u.)	$FWHM_{hh}$ (meV)	E_{lh} (meV)	OS_{lh} (a.u.)	$FWHM_{lh}$ (meV)
+12	2657	1.6	14	2699	0.9	20
0	2657	1.6	14	2698	0.9	22
–3	2652	1.3	16	2691	1	32
–5	2648	1.2	18	2681	1	40

shift of the hh feature caused by the quantum confined Stark effect is clearly seen. The Stark shift at $U = -5$ V is 9 meV. The shift measured by photocurrent spectroscopy is larger. The shift obtained in similar GaAs p-i-n multiple quantum wells is also larger [379]. Babucke et al. [379] suggested that the smaller shift in their experiments is probably due to the fact that the actual field in the quantum wells near the Schottky contact is smaller. At the applied voltage $U = -5$ V the change in the reflection is 25%. This value is more than twice the change due to field ionization of non-confined excitons. The reflection spectra calculated by the method of transmission matrices [379] for the multilayered structure is shown in Fig. 6.14(b). The known dielectric function for each layer was used. Field-induced corrections to the values of the dielectric function were taken into account. The transition energies E, the oscillator strength OS and the half-width FWHM were used as adjustable parameters. The best calculated values shown in Fig. 6.14(b) yielded values of the parameters shown in Table 6.7. The hh oscillator strength decreases with applied field because, as discussed earlier, electrons and holes are pulled in opposite directions and overlap of their

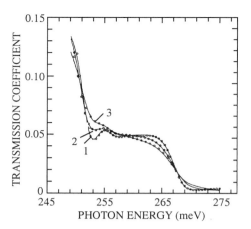

Figure 6.15: Symbols show the transmission coefficient spectra of ZnCdSe/ZnSe 30 quantum well structure at 300 K at different bias voltage U, (1) 0 V, (2) –15 V and (3) –30 V. The calculated values are shown by the solid line [379].

wave-functions decreases. The lh features are weak and parameters obtained from these features are less reliable.

Transmission spectra of the 30 quantum well structure measured at 300 K is shown by the symbols in Fig. 6.15. At zero field the hh exciton absorption peak is clearly visible. As the applied voltage U increases from 0 to –30 V, the transmission coefficient increases from 4.6 to 6.2% i.e. by more than 30%. The spectra calculated by the same method (as used for the reflection spectra shown in Fig. 6.14) are shown by the solid curves. The resonance line are considerably broadened. At room temperature practically there is no red shift or loss of oscillator strength. In the 30 quantum well structure the wells are on the back-side of the reverse biased Schottky contact. The field near the contact is partly compensated by the built-in field of the forward-biased contact (see Fig. 1(b) of Ref. [379]). Therefore the quantum wells in the low-field region do not contribute significantly to the total shift. The agreement between the observed and the calculated spectra is very good. The values of the parameters derived from these measurements for the 30 quantum well structure are shown in Table 6.8. The room temperature data are significantly different from those given in Table 6.7 at 5 K.

6.6.2.3 CdZnSSe multiple quantum wells

The visible spectral range is important for optical information processing systems. These systems employ two-dimensional arrays of surface-normal optical modulators [459]. The information is transmitted from one circuit to another by parallel optical beams. The information capacity depends on the area density of the modulators. The density depends on the physical size of the modulator. The minimum size is diffraction limited and increases approximately on the square

6.6. EFFECT OF APPLIED ELECTRIC FIELD

Table 6.8: Excitonic parameters of the ZnCdSe/ZnSe 30 quantum well structure derived from the transmission spectra of Fig. 6.15 at 300 K.

U (V)	E_{hh} (meV)	OS_{hh} (a.u.)	$FWHM_{hh}$ (meV)	E_{lh} (meV)	OS_{lh} (a.u.)	$FWHM_{lh}$ (meV)
0	2529	1.1	40	2570	0.32	40
−10	2527	1.1	44	2570	0.28	42
−15	2527	1.1	53	2570	0.16	46
−30	2527	0.9	60	2570	0.14	50

of the wavelength of the light used. The maximum pixel size can be reduced by a factor 3 by reducing the wavelength of the light from 850 nm to 500 nm. It is therefore interesting to investigate the electroabsorption of quantum wells which are used as active layers in Laser diodes. If structures similar to those in laser diodes can be used as modulators, the modulator can be integrated with a laser diode monolithically. Law et al. [459] investigated the effect of electric field on the absorption of the CdZnSSe/ZnSSe quantum wells used in laser structures. The $Cd_{0.3}Zn_{0.7}S_{0.06}Se_{0.94}$ multiple quantum wells consisted of 10.5 periods with 32.5 nm $ZnS_{0.06}Se_{0.94}$ barriers. Two thicknesses of the wells were used, 10 nm (sample 1) and 5 nm (sample 2). The n- and p-doped 100 nm $ZnS_{0.06}Se_{0.94}$ layers were used as cladding layers. The samples were pseudomorphic. They were free from dislocations. Room temperature photocurrent spectra were measured using light from a tungsten lamp dispersed through a monochromator. The spectra were normalized by the shortest wavelength response of the input optical system to account for bias dependent quantum efficiency. The normalized photocurrent spectra mimic closely the absorption spectra. Strong $n = 1$ hh exciton peak were observed at 512 nm in sample 1 and 507 nm in sample 2. The $n = 1$ lh feature was observed as a shoulder at 491 nm in sample 2. The values of FWHM of the hh feature were 42 meV for sample 1 and 76 meV for sample 2. Both widths are considerably larger than those observed in GaAs based quantum wells. The broadening occurs due to alloy concentration fluctuations, interface roughness and scattering of the excitons by the LO-phonons. A weak red shift and strong broadening of the hh absorption line occurred in sample 1 with increasing applied fields. This broadening is caused by field induced ionization of the excitons and reduction of their lifetime. Probably the red shift was partially masked because of the very large width of the spectrum at room temperature. The broadening of the spectra occurred in sample 2 also. However instead of the red shift, a blue shift was observed. Law et al. [459] mentioned that the reasons of the blue shift were not clear. The exciton Bohr diameter d_{Bohr} in bulk ZeSe is 6 nm. The thickness of sample 1 is bigger and that of sample 2 is smaller than the Bohr diameter. We have already discussed the work of Hartwit et al. [456] who also observed the blue shift in quantum wells which had thin periods so that the different wells interacted giving rise to the superlattice effect and mini-band formation. The effect of the electric field is to break the coupling between different wells and localize the excitons in the wells which causes the blue shift [456].

6.6.3 Effect of electric field on GaN

Effect of external electric field has not been studied as extensively in III-Nitrides as in II-VI semiconductors. Binet *et al.* [449] and Duboz *et al.* [450] have measured photo-current, optical absorption and transmission in MOVPE grown 1 μm WZ GaN layers at 330 K and at low temperatures. The normalized field $f = e \times E \times a/R$ was used in discussing the results. Here a is the exciton radius and R is the exciton Rydberg. The absorption edge energy decreased as expected. For normalized field $f = 0.36$ the shift was 20 meV. Exciton absorption and photo-current quenching with increasing field was also observed. Experimental results agreed with the values predicted by the theory of Dow and Redfield [448] only qualitatively. It was difficult to take into account the broadening of the edge due to thermal effects. The photo-current showed a different behavior at 80 K. An increase in the excitonic peak with applied field (compared to band to band transitions) was observed at low fields. As the field increased the photo-current saturated as predicted by the theory. However quenching of the excitons was not observed as sufficiently high fields could not be applied. A shift of 18 meV in the photocurrent was observed at the highest field used.

Gupta *et al.* [451] made the first measurement of the electro-optic effect in GaN at visible wavelengths in 1995. The results of the measurements gave a moderate value of the transverse electro-optic coefficient in GaN. This preliminary work showed that GaN and its alloys are promising materials for developing integrated-optic modulators for visible and near-UV wavelengths. Pearton *et al.* [452] fabricated quantum well modulator structures in GaN/InGaN and conventional III-V systems using a combination of ECR dry etching ($Cl_2CH_4H_2Ar$, BCl_3Ar or CH_4H_2Ar plasma chemistries respectively) and subsequent wet chemical etching of the buffer layer underlying the quantum wells. Wet etchants were employed for AlGaAs and InGaP, respectively. A new KOH-based solution was developed for AlN which is completely selective over both GaN and InGaN. An UV modulator based on the excitonic Franz-Keldysh effect was demonstrated by Duboz *et al.* [450].

6.7 Piezoelectric effect

6.7.1 Piezoelectric effect in zinc blende and wurtzite semiconductors

Zinc-blende semiconductors have two different atoms occupying the unit cell. They do not have inversion symmetry. In this respect they are different from the diamond-structure Si and Ge. Applying a strain to the zinc-blende crys-

6.7. PIEZOELECTRIC EFFECT

tals induces electric fields, known as the piezoelectric fields. III-V (including III-Nitrides) and II-VI semiconductors are piezoelectric [462, 463]. Strain in certain directions in these crystals produces a distortion of core valence charge which in turn produces an internal electric field known as the piezoelectric field. The direction and magnitude of the field are determined by the strain, crystal structure, symmetry and the piezoelectric constants [463]. Early studies of piezoelectric fields were made in InGaAs/AlGaAs superlattice. If ZB semiconductors are grown in the [111] direction, piezoelectric fields along the growth direction are generated. Smith and Mailhiot [462] considered the piezoelectric fields in type I $Ga_{0.47}In_{0.53}As$-$Al_{0.70}In_{0.30}As$ superlattice in which Ga alloy has a smaller bandgap, is under compression and forms the quantum well. The lattice mismatch between the quantum well and the barrier layers was 1.5%. Smith and Mailhiot [462] found many interesting results in the superlattice they studied. The strain induced electric field was 1.4×10^5 V/cm. Strain induced fields reduce the bandgap. Electronic wave functions and energy levels are changed by this field. The piezoelectric field can reduce the overlap between the electron and hole wave functions which reduces the oscillator strengths. The electric polarization is related to strain by the piezoelectric tensor. The piezoelectric tensor has only one independent coefficient, e_{14} [464]. In bulk semiconductors the strain induced piezoelectric field is small. It is screened by the free carriers. The piezoelectric field can be enhanced by growing strained layers of the zinc-blende semiconductors along a polar axis. If the layers are grown in (100) direction the strain has isotropic and tetragonal components. In such cases no piezoelectric field is induced. In the epilayers grown in (110) and (111) directions trigonal strain is built in and large piezoelectric field are induced. Experiments confirm that the optical properties of pseudomorphic InGaAs epilayers grown on (111) GaAs are considerably modified by the internally generated piezoelectric fields [465]. No modification of the properties occurs when the epilayers are grown on (100) oriented substrates. The piezoelectric coefficient is not constant, it depends on the isotropic strain [464]. The piezoelectric coefficients in II-VI semiconductors are much larger than in III-V semiconductors. The effect of piezoelectric field on the electronic properties is similar to the effect of externally applied field discussed in section 6.6. In quantum wells the piezoelectric field gives rise to built-in quantum confined Stark effect, and alters their electronic and optical properties.

The piezoelectric field in a strained quantum well grown in the [111] direction is given by [464],

$$F = \frac{2\sqrt{3}}{\epsilon\epsilon_0} \frac{c_{11} + c_{12}}{c_{11} + c_{12+4c_{44}}} e_{14} f_m. \tag{6.1}$$

Here f_m is the lattice mismatch as defined earlier. It is assumed that the quantum well is pseudomorphic and f_m is equal to the strain. The calculation of F for [110] direction is more tedious. In this case the polarization is within the plane and is easily screened as in the bulk material. If the barriers are unstrained, the potential profile is determined by the band offsets and the linear piezoelectric field. The transition energy between the first confined electron and hole levels is given by,

$$E(e_1 h_1) = E_g + E_{e1} + E_{h1} - eFL_w - E_b, \qquad (6.2)$$

where E_g is the bandgap of the strained quantum well material, L_w is the width of the well, E_b is the exciton binding energy, and E_{e1} and E_{h1} are the confinement energies of electrons and holes.

The potential profile of a quantum well and electron and hole wavefunctions were shown in Fig. 6.12. For wide enough quantum wells the electrons and the holes do not see the opposite walls of the well and their energies become independent of the thickness L_w. The transition energy depends linearly on well thickness and its slope gives the value of F. Cibert et al. [464] determined the value $F = 0.94 \times 10^5$ V/cm for CdTe/Cd$_{0.82}$Mn$_{0.18}$Te strained quantum well.

The piezoelectric coefficient e_{14} is not independent of the isotropic strain. Experiments performed on heterostructures subjected to hydrostatic pressure show that in the II-VI semiconductors, in which ionic contribution to polarization dominates, the absolute value of the piezoelectric effect increases with isotropic strain. On the other hand the coefficient decreases in semiconductors such as InGaAs where ionic contribution does not dominate [464].

6.7.2 Experiments on piezoelectric effect in II-VI quantum wells and superlattices

6.7.2.1 CdS/CdSe superlattices

Wurtzite CdS/CdSe superlattices grown on (111) oriented substrates show interesting piezoelectric properties. The relevant piezoelectric coefficients in these materials are larger than in the III-V semiconductors by a factor 3. Since the lattice-mismatch between CdS and CdSe is 5%, giant piezoelectric fields are generated in these superlattices [465]. In the presence of photo-excited carriers the piezoelectric fields are screened. Optical properties are therefore affected by the incident light. The fundamental absorption peak occurs below the band-edge of the wurtzite CdSe. Its actual position depends on the widths of the quantum well (L_w) and barrier (L_b) layers. If $L_w = L_b = 1.3$ nm, the absorption peak is observed at 740 nm. For thicker periods with $L_w = L_b = 5.4$ nm, the peak is observed at 950 nm. To understand these properties and the effect of photo-generated carriers, we show the band-diagram of the superlattice in Fig. 6.16. Under the influence of the piezoelectric fields the carriers generated by the incident light in the superlattice drift to the interfaces. The electrons are accumulated in the CdS barrier because the conduction band-edge is lower. The valence band-edge has lower energy in the CdSe well layer and holes are trapped here. The carriers screen the internal fields and modify the band-bending. Chen et al. [465] have studied two CdS/CdSe superlattice samples, designated as MH114 and MH95, grown on (111) GaAs. To reduce the mismatch strain, a 600 nm CdS buffer layer was grown on the substrate before depositing the superlattice structure. Each sample consisted of 20 periods of the superlattice. $L_w = L_b = 1.3$ nm for sample MH114 and $L_w = 4$ nm and $L_b = 2.5$ nm for sample MH95

6.7. PIEZOELECTRIC EFFECT

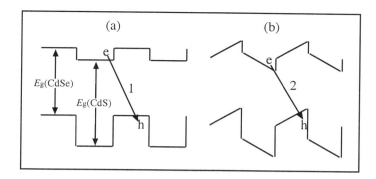

Figure 6.16: Band-structure (schematic) of the type II CdSe/CdS superlattice. (a) Flat band condition without the piezoelectric field and (b) Effect of the piezoelectric field. In the presence of large carrier densities the piezoelectric fields are screened and band structure becomes close to the flat band condition shown in (a). As the carrier density decreases, the band banding shown in (b) takes place. Recombination process in the superlattice are shown by arrows 1 and 2. Note the larger electron-hole energy separation in (a) than in (b).

were used. Time resolved PL spectra of both samples were studied using several values of delay time after the exciting pulse. Results of sample MH95 were reported in detail. The main peak due to the superlattice exciton was observed near 740 nm. This peak broadens and shifts to longer wavelengths as the initial delay increases from zero to 2 μs. In sample MH95 the total shift was about 5% of the bandgap. The shift was smaller in sample MH114. At short delay times the carrier density is large and they screen most of the piezoelectric field. The bandstructure is driven to the near flat-band condition. As can be seen in Fig. 6.16 the energy separation between the electron and hole levels is now large. At large time delays when many carriers have recombined and the remaining carrier density is small, the bands bend. The energy separation between the electron-hole levels is now reduced which gives rise to the observed red shift to the PL. This result is similar to that observed in Type I structures.

6.7.2.2 CdS/ZnSe superlattices

Bradley et al. [461] studied the PL of Type-II CdS/ZnSe strained layer superlattice grown on CdS buffers deposited on (111)A GaAs substrates. The samples were grown at two temperatures, 400°C and 300°C. TEM studies showed that the samples grown at the higher temperature were hexagonal and those grown at the lower temperature were cubic. Measurements were made in the range of power densities from 50 mW cm^{-2} to 20 kW cm^{-2}. Both samples showed blue shifts of the PL peaks with increasing excitation. The shift was 70 meV per decade of excitation intensity in both samples. Shifts in photoluminescence (PL) peak energies arise because photo-excited carriers screen the piezoelectric fields. Though no external strain was applied, piezoelectric fields were generated due to the strain in the layers. The existing red shift due to the piezoelectric field is reduced. The reduction of red shift appears as a blue shift. These results therefore provide evidence for the intrinsic Stark effect in these superlattices at visible wavelengths [461].

The piezoelectric coefficients of bulk cubic ZnSe are known to be much smaller than those for the wurtzite phase II-VI semiconductors. Therefore one would expect the shift in the cubic CdS-ZnSe superlattices to be much smaller. The shift of 70 meV per decade of excitation intensity is only slightly smaller than that observed in CdS-CdSe hexagonal superlattices of similar period. The piezoelectric coefficients change non-linearly with strain [461]. The effect of strain may change the relative values of piezoelectric coefficients for the two phases. In another publication Bradley *et al.* [460] examined the effect of excitation intensity on the PL of both the cubic and the hexagonal Type II CdS/CdSe and CdS/ZnSe superlattices. The excitation intensity was changed in the range 100 mW cm^{-2} to 1 MW cm^{-2}. The cubic samples were grown on (100) GaAs and the hexagonal samples on (111) GaAs. The blue shift in the PL of the hexagonal samples was very large, 200 meV for a 10 nm period of CdS/CdSe sample. The shift increased with the period of the superlattice linearly. The shift in the cubic samples was somewhat smaller than in the hexagonal samples but it was quite large in the cubic samples also. Bradley *et al.* [460] cited papers published by other groups who also observed such shifts in type II cubic superlattices. Since the cubic superlattices are non-piezoelectric the effect can not be attributed to the screening of intrinsic piezoelectric fields. The shift was attributed to the band-bending induced by the space charge. In type II superlattices strong space charge fields are created by the separation of the photo-generated carriers.

6.7.3 Piezoelectric effects in III-Nitrides

Recently many groups have studied the effect of piezoelectric fields on the band structure of WZ GaN and InGaN based devices. Field Emitter Arrays (FEA), Heterostructure Field Effect Transistors (HFETs) and quantum wells have been investigated [466]–[473]. Bernardini *et al.* [466, 467] have made *ab initio* investigations of the piezoelectric constants of the III-Nitrides using the Berry phase approach. They have also compiled the values of the piezoelectric constants for several II-VI and III-V semiconductors for both the ZB and the WZ structures. These values are given in Tables 6.9 6.10. We have used the same notation in these tables as in Ref. [466].

Bernardini *et al.* [466, 467] have surveyed the literature and found that the experimental values of the piezoelectric constants have not been measured in most cases. The only values available are for AlN, $e_{31} = 1.55$ C/m^2 and $e_{33} = -0.58$ C/m^2. which are in reasonable agreement with the values given in Table 6.9. There are some interesting features in the values given in the two tables. The piezoelectric constants of the III-Nitrides have the same signs as those of the II-VI semiconductors and different from those for III-V semiconductors (except InP). The magnitude of the piezoelectric constants are much larger than those

6.7. PIEZOELECTRIC EFFECT

Table 6.9: Piezoelectric constants e_{33} and e_{31} for wurtzite semiconductors (in units of C/m^2) from Ref. [466].

	e_{33}	e_{31}		e_{33}	e_{31}
AlN	1.46	−0.60	ZnO	0.89	−0.51
GaN	0.73	−0.49	BeO	−0.02	−0.02
InN	0.97	−0.57	–	–	–

Table 6.10: Piezoelectric constants e_{33} and e_{31} for zincblende semiconductors (in units of C/m^2) from Ref. [466].

	e_{33}	e_{31}		e_{33}	e_{31}
CdTe	0.03	−0.01	AlAs	−0.01	0.01
ZnS	0.10	−0.05	GaAs	−0.12	0.06
ZnSe	0.04	−0.02	InAs	−0.03	0.01
AlP	0.04	−0.02	AlSb	−0.04	0.02
GaP	−0.07	0.03	GaSb	−0.12	0.06
InP	0.04	−0.02	InSb	−0.06	0.03

for the III-V or the II-VI semiconductors (except for ZnO). AlN has the largest value among all the binary compounds [466].

Underwood et al. [463] have investigated theoretically and experimentally the effect of InGaN surface layer on the turn-on voltage of GaN FEA. In pseudomorphic strained InGaN layers the field F_z in the growth direction z is given by,

$$F_z = \frac{2d_{31}}{\epsilon_s \epsilon_0}(c_{11} + 2c_{13}^2/c_{33})\epsilon_{xx}, \qquad (6.3)$$

where d_{31} is the piezoelectric constant connecting the strain in the growth plane with the field in the z direction, C_{ij} are the elastic stiffness constants of InGaN, ϵ_{xx} is the strain in the x and the y directions, ϵ_s is the relative dielectric constant of InGaN and ϵ_0 is the vacuum permittivity. Strain is determined from the lattice mismatch between GaN and InGaN and is negative (compressive) for the InGaN films grown on GaN. The piezoelectric constants are 1.1×10^{-10} cm/V for GaN and 1.7×10^{-10} cm/V for InN [285]. The values of the piezoelectric constant for the InGaN alloys were determined by interpolation between the values of GaN and InN. Using a computer code, band structure of the InGaN/GaN was calculated. The piezoelectric effect was simulated by introducing an appropriate charge density (consistent with Eq. (6.3)) in the InGaN/GaN interface and the InGaN/vacuum interface. Different thicknesses and In molar fractions were considered. In the simulations GaN was assumed to be doped to an electron concentration of 2×10^{18} cm^{-3} and InGaN was undoped. A value 4.5 eV was used for the electron affinity of GaN. The electron affinity of the heterostructure was determined by subtracting the conduction band energy at the InGaN/GaN interface from the vacuum energy level at the InGaN/vacuum interface. The calculations showed that the piezoelectric field reduces the electron affinity considerably. The reduction of the electron affinity increases as the thickness of the InGaN layer increases. It also increases with the increase in the In molar fraction in the InGaN layer because larger In molar fraction produces larger strain. For large thickness and/or large In molar fraction, the electron affinity saturates at a low value. The minimum calculated value of the electron affinity was 0.9 eV.

Underwood et al. [463] also determined the electron affinity of the heterostructure experimentally. They fabricated an FEA by MOCVD. The thickness of the InGaN layer was 100 nm and In molar fraction was 0.2. The current–voltage and Fowler–Nordheim (F–N) characteristics were measured and plotted. The turn on voltage was found to be 70 V as compared to the turn-on voltage of 450 V GaN FEA without an InGaN layer. The slope of the F–N plot for the InGaN FEA was 0.24 times the slope for the GaN FEA. Because of processing differences, field enhancement effects are expected to be different in GaN and InGaN FEAs. It was difficult to separate out quantitatively the effect of the field enhancement and barrier lowering due to piezoelectric field. Comparing these results with the calculations, it appears that at least a significant part of the reduction in the turn-on voltage and change in the slope of the F–N plot must be due to the barrier lowering by the piezoelectric field.

Asbeck et al. [468] have estimated the piezoelectric charge density at (0001) AlGaN/GaN interfaces in GaN based HFET structures. The structures were grown on both the sapphire and the SiC substrates. The charge density 5×10^{13} cm^{-2} x_{Al} where x_{Al} is the Al mole fraction. The authors concluded that these charges will have a large effect on the performance of the devices.

Recently several groups have investigated piezoelectric fields and their effects on the optical properties of GaN based quantum wells. Wang et al. [469] have studied theoretically the influence of piezoelectric fields and many-body effects on the band structure, and the spontaneous- and stimulated-emission from the strained GaN/Al$_{0.2}$Ga$_{0.8}$N quantum wells. The thickness of the Al$_{0.2}$Ga$_{0.8}$N quantum well was 5 nm. The calculated threshold current density was 1 kA cm^2 and an intrinsic 3-dB modulation bandwidth was 11.7 GHz at 40 mW output power. This value of the threshold current is the theoretical limit. Deguchi et al. [470] have studied the emission spectra of heavily Si-doped InGaN multiquantum wells. They found that the spontaneous emission spectra showed potential fluctuations due to inhomogeneous distribution of In. Quantum-confined stark effect did not play a significant role in determining the optical characteristics because high carrier concentration screen the electric field. Muensit et al. [474] have measured the d_{33} piezoelectric coefficient of WZ GaN using a laser interferometer. The 1 μm thick GaN layer was grown by MOCVD. The spatial resolution of the techniques was 100 μm. The value of the coefficient is 2.0 ± 0.1 pm V^{-1}. Park and Chuang [471] have made self consistent calculations of the influence of piezoelectric effect and many body effects on the optical gain of GaN/AlGaN QW laser. Because of the change in the wave functions and energy levels by the internal piezoelectric field, the peak gain was red-shifted and was reduced. If the carrier density is high, $> 10^{19}$ cm^{-3}, the carriers screen the field and the effect of the internal Piezoelectric field becomes negligible.

Chichibu et al. [472] have made a thorough investigation of various factors which affect the optical emission from the InGaN quantum wells. They have taken into account the effects of quantum confined Stark effect (QCSE), quantum confined Franz-Keldysh effect (QCFK) and piezoelectric fields on the electronic transitions. If the piezoelectric field is strong, electrons-hole pairs are separated in the triangular potential wells formed on opposite sides of the well. This produces a Stokes-like shift in the emission. The oscillator strength becomes small in this case. This effect dominates if the quantum well thickness L is large. than the Bohr radius a_B. If the thickness of the quantum well (or radius of a quantum dot) is small, the carriers confined in the quantum well or the quantum dot do not suffer this reduction in the oscillator strength. In this case the carriers do not "see" the non-radiative recombination centers such as the threading dislocations and emission efficiency is high. If L is greater than the three-dimensional Bohr radius a_B, the effect of QCFK also becomes important. Chichibu et al. [472] suggested that research on cubic GaN quantum wells must be done to clarify the role of various effects in determining the optical properties of the quantum well lasers.

6.8 Effect of magnetic field on semiconductors

6.8.1 Magnetic polarons in diluted magnetic II-VI semiconductors

Borghs et al. [475] have described recent work on the growth and characterization of GaMnAs diluted magnetic semiconductor. The GaMnAs alloys are grown by MBE at low temperatures as Mn segregates if the temperature is more than 400°C. These alloys show a strong negative magnetoresistance effects up to 150 K. At temperatures higher than the growth temperature, Mn-As complexes are formed. As the temperature increases further, ferromagnetic MnAs precipitates out. Borghs et al. citeBorghs99 have given references to earlier work on III-V alloys containing large concentration of Mn. Considerable work on $Cd_{1-x}Mn_xTe$ and other diluted magnetic alloys was done in 1980s. Good reviews on this work have been published [476, and references given therein]. Several useful papers appeared in mid-1990s citeScholl93–citeFiederling98. Studies have been made in bulk alloys as well as in thin epilayers. The $Cd_{1-x}Mn_xTe$ alloy exhibits a variety of magnetic states as a function of Mn concentration and temperature. Its bandgap can be tuned by magnetic field [456]. Fiederling et al. [491] have given a good account of the properties of magnetic polarons in diluted magnetic or semimagnetic alloys containing Mn. Interesting properties of the alloys follow from a strong sp-d exchange interaction between spins of charge-carriers or excitons and magnetic moment localized on the transition metal ion. A spontaneous ferromagnetic alignment of spins of the magnetic transition metal Mn ion

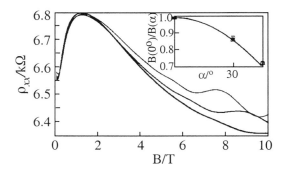

Figure 6.17: Magnetoresistance with Shubnikov de Haas oscillations for different values of angle α between the growth direction and magnetic field. The positions of maxima and minima divided by the positions at $\alpha = 0$ are shown in the inset. The solid line in the inset shows the $\cos\alpha$ dependence [477].

occurs as a result of the exchange interaction. The resulting center consisting of the localized carrier and the aligned spins is known as the magnetic polarons. Dietl [492] and Oka et al. [483] have also discussed effects of the exchange coupling between a charge carrier or an exciton and magnetic moment of the Mn ions in II-VI semiconductors containing Mn ions leading to the formation of the magnetic polarons (MP) or exciton magnetic polarons (EMP). The optical and magneto-optical properties of diluted magnetic semiconductors are modified by the formation of the MPs and EMPs. Usually carrier or exciton is trapped by an impurity or a defect, and the MPs or EMPs are localized. A free magnetic polaron (a delocalized carrier accompanied by a travelling cloud of polarized spins) is not expected to exist for the actual values of the coupling constants [492]. Carrier scattering by thermodynamic and static fluctuations affects significantly the energy of the magnetic polaron. If the carrier concentration is large (as obtained by optical pumping or heavy doping), the influence of the carrier liquid upon the Mn spins may lead to a ferromagnetic order, both in bulk and layered structures. Carrier-spin interactions are long range, and ferromagnetic ordering is not destroyed by the fluctuations [492].

6.8.2 Transport properties

Scholl et al. [477] studied two-dimensional Shubnikov de Haas (SdH) oscillation in modulation-doped 12 nm thick CdTe quantum wells with $Cd_{0.95}Mn_{0.05}Te$ (50 nm thick) barrier layers. The samples were grown on CdTe substrates with a 50 nm $Cd_{0.9}Mn_{0.1}Te$ buffer. The central part of the barrier layers was doped n-type with bromine. Magnetoresistance and Hall effect measurements were made in the temperature range 1.5–20 K. The SdH oscillations observed are shown in Fig. 6.17. The dependence on the angle α shown in Fig. 6.17 demonstrates that the electron gas is 2D. Assuming a circular Fermi level, the oscillation period yielded a value $(9 \pm 2) \times 10^{11}$ cm^{-2} of the 2D carrier density. The value of the 2D carrier density determined from the Hall effect measurements was 1.2×10^{11} cm^{-2} in each well. The value of the carrier concentration was also determined using the PL and PL excitation measurements. The difference between the PL peak and the PL excitation peak was taken to be the value of the Fermi energy. This value of Fermi energy gave a value of 8×10^{11} cm^{-2} for the carrier density. The agreement between the values determined by the three different methods is reasonable. The Hall mobility was 2×10^2 cm^2/Vs. The effective mass of the 2D carriers derived from the temperature dependence of the oscillation amplitude is $0.1 m_0$. The temperature dependence of the back-ground of magnetoresistance was investigated. Above 10 K, the behavior of the samples was CdTe-like and below 10 K it was CdMnTe-like.

Oscillations of PL intensity in the applied magnetic field have also been observed [480]. The oscillations were attributed to transitions between the Landau levels. There was evidence that the Landau level energy also oscillates. The energy oscillates because the dielectric constant is a function of the density of the electrons near the Fermi level. The dielectric constant determines the self-energy of the electron. Magnetoresistance and PL intensity were measured simultaneously in one sample. The minima in the magnetoresistance and the PL intensity coincided perfectly.

6.8.3 Magnetic and Optical properties
6.8.3.1 Bulk crystals

Stankiewicz et al. [487] have measured light induced changes in magnetization of Ga doped n-type $Cd_{1-x}Mn_xTe$ (x=0.01, 0.05) single crystals at T=2 K and 5 K and in magnetic fields of up to 0.5 T. Hall effect measurements were made in a temperature range 1.6 to 300 K. Initially magnetization increases with applied magnetic field up to H = $\sim$ 2000 Oe and then it saturates. The incident light increases the concentration of shallow donors. The light was below the bandgap energy. The changes in magnetization correlated with the increase in the concentration of light induced shallow donors. The authors used bound magnetic polaron theory to explain the temperature and magnetic field changes in the magnetization on illumination. The change in magnetization is a measure of the concentration of the bound magnetic polarons.

Photoluminescence and reflectivity of $Cd_{0.9}Mn_{0.1}Se_{0.3}Te_{0.7}$ crystals in a magnetic field of up to 5 T at T=4.2 K have been studied by Bryja et al. [489]. Two lines were observed in the luminescence spectra. Comparing these results with the previous studies in $Cd_{1-x}Mn_xTe$ and $Cd_{1-x}Mn_xSe$ the lines were identified as being due to free exciton X and exciton bound to neutral acceptor A^0X. The effect of magnetic field on the PL spectra was small. The authors mentioned that influence of high alloy disorder on the bound magnetic polaron was probably responsible for this small effect. In a later publication, Bryja et al. [490] studied bound magnetic polaron effect in cubic $Zn_{0.1}Cd_{0.89}Mn_{0.01}Te$ and wurtzite $Cd_{0.99}Mn_{0.01}Se_{0.3}Te_{0.7}$. The PL of both samples was investigated. Measurements were made at T=2 K and with magnetic field in the range $B = 0 - 5$ T. The cubic samples were studied in Faraday configuration and the wurtzite samples in Voigt configuration. In both samples optical transitions of free excitons X and of exciton bound to neutral acceptor A^0X were observed. In cubic sample the A^0X complex is destabilized in magnetic field as in ternary compounds containing Mn but at higher fields. In the wurtzite sample the complex still existed at 5 T. A third line with energy above the free exciton was observed in the wurtzite sample. The line was attributed to the transitions of the carriers trapped at the composition and potential fluctuations in the alloy.

6.8.3.2 ZnTe and ZnSe epilayers

Diamagnetic shift and Zeeman splitting of the exciton bound states caused by an external magnetic field in ultra-thin strained MBE-grown epilayers of ZnTe and ZnSe have been studied by Lee et al. [482]. Magnetic fields up to 6 T were used. The structure in the continuum due to formation of Landau levels was also investigated. The transmission spectra at 1.5 K was measured of both samples. Strong $1s$ peaks were observed at 2.3789 eV and 2.3804 eV in ZnTe and 2.8028 in ZnSe. The peaks due to $2s$ states were observed at higher energies. The $3s$ states were resolved only in ZnSe. In the presence of external magnetic field the transmission spectra became very complex due to Zeeman splitting, the diamagnetic shift of the excitons and formation of Landau levels. The experimental results were analyzed taking into account these factors and also the splitting of the exciton state by the biaxial strain. Landau levels were calculated using $8 \times 8 k \cdot p$ Hamiltonian. Using the eigenstates of the 8×8 Hamiltonian, matrix elements of the transitions between the hole and electron Landau levels were determined. Thus oscillator strengths of the transitions in the continuum above the fundamental gap were calculated. In the calculation of magnetic field dependence of the Landau level transition energies, the heavy-hole valence band edge was taken as an adjustable parameter. The best fit was obtained for heavy hole bandgap = 2.398 eV for the strained ZnTe sample. Taking into account the effect of strain, the bandgap of the unstrained ZnTe came out to be 2.3934 eV. The corresponding values for ZnSe are 2.8250 eV for the strained and 2.8246 eV for the unstrained ZnSe. The exciton binding energy was estimated to be 12.4 meV on ZnTe and 22.2 meV in ZnSe. There were small but systematic deviations of the calculated values from the observed values due to electron-hole correlation energies which were not included in the calculations.

6.8.3.3 Quantum wells and superlattices of diluted magnetic alloys

$CdTe/Cd_{1-x}Mn_xTe$ quantum wells

In narrow non-magnetic quantum wells with semi-magnetic barriers low dimensional EMPs are formed because the exciton wavefunctions penetrate the barrier layers [479]. Magnetic properties of the interface can be studied by investigating EMP energies and Zeeman splittings. CdTe/CdMnTe quantum wells are inter-

6.8. EFFECT OF MAGNETIC FIELD ON SEMICONDUCTORS

esting because if the Mn concentration is high, the band offsets are large. At low temperatures a spin-glass phase of the Mn containing alloy is formed. However large Mn concentration causes broadening of the PL line and determination of the Zeeman splitting becomes difficult. Mackh et al. [479] have studied PL and PL excitation spectra of cubic CdTe/Cd$_{1-x}$Mn$_x$Te quantum wells with high Mn concentration ($0.4 \leq x \leq 0.8$). For $x = 0.78$ a Stokes shift 35 meV is observed. The shift is large because the excitons are localized due to fluctuations in the well width and alloy composition. The EMP formation also contributes, to a minor extent, to the shift. The excitation spectrum showed clearly the hh and lh peaks with a separation of 125 meV due to strong quantum confinement. The EMP peak was also detected in the PL excitation spectrum. The localization potential of the excitons was large because of the large composition and potential fluctuations. This made it possible to detect the EMP peak in the excitation spectrum. The excitation spectra were measured under right- and left-circularly polarized light. As the magnetic field increased the energy difference between the EMP line and the right-circularly polarized (σ^+) excitation light decreased while the difference between the EMP energy and left-circularly polarized (σ^-) light increased. σ^--excitation produces higher energy excitons antiparallel to the magnetic field direction. Most of the excitons in the antiparallel direction relax to the parallel direction and form EMPs parallel to the field direction. The difference between the excitation energy and the EMP energy is due to both the EMP formation and Zeeman splitting. Zeeman splitting can be determined by the energy difference between the EMP energies taken under σ^+- and σ^--polarized lights. The Zeeman splitting can also be determined from the observed hh peaks in the excitation spectra. The Zeeman splitting determined by the two methods agreed well. At 1.6 K and for a 6ML (2 nm) quantum well with ($x = 0.78$) the Zeeman splitting was ~ 8 meV at $B = 0.5$ T. The splitting increased sublinearly with B to ~ 40 meV at $B = 4$ T. EMP energy was determined for a 4 ML quantum well as a function of Mn concentration. The energy was about 5 meV for $x = 0.2$ (this value was taken from earlier work) and 20 meV for $x = 0.8$. The magnetic susceptibility of bulk Cd$_{1-x}$Mn$_x$Te decreases as x increases because of the antiferromagnetic pairing of the neighboring Mn ions. The penetration of the exciton wavefunction in the barrier also decreases with x. Therefore one would expect the EMP energy to decrease with increase of Mn concentration. Mackh et al. [479] attributed the observed behavior to the interface structure. The nearest neighboring Mn ions are partially missing in the interface. Intermixing at the interface also occurs. However it was difficult to model the effect of the interface structure on the EMP energy.

Cd$_{1-x}$Zn$_x$Te/Cd$_{1-x'-y}$Zn$_{x'}$Mn$_y$Te quantum wells

Magneto-optical properties of Cd$_{1-x}$Zn$_x$Te/Cd$_{1-x'-y}$Zn$_{x'}$Mn$_y$Te single quantum wells have been investigated by Takeyama et al. [478]. The composition was $x = 0.07$, $x' = 0.04$, and $y = 0.48$. Four Cd$_{1-x}$Zn$_x$Te quantum wells with widths = 13, 19, 40, and 90 Å were fabricated. The width of the diluted magnetic semiconductor Cd$_{1-x'-y}$Zn$_{x'}$Mn$_y$Te barrier layers was 450 Å. PL measurements were made in the temperature range ~ 4 to ~ 115 K. Each quantum well exhibited characteristic PL lines. The 40 and 90 Å wells exhibited two closely spaced PL lines. Takeyama et al. [478] suggested that the low energy line from each of these two wells is due to an exciton bound to a donor. Magneto-photoluminescence was measured with applied magnetic fields up to 3.5 T. The σ^+-component of the PL line shifted to low energy with the applied magnetic field. The shift to low energy reflected a large Zeeman splitting of the PL. A larger shift was observed in wells with smaller widths. As the thickness of the well decreased, the wave-function of the exciton penetrated deeper into the magnetic barriers and gave rise to a larger splitting. Analysis of the experimental results yielded a binding energy of ~ 27 meV for the excitons.

$Cd_{0.9}Mn_{0.1}Te/Cd_{0.6}Mg_{0.4}Te$ quantum wells

Merkulov et al. [488] have studied extensively the magnetic polaron states of two-dimensional excitons in $Cd_{0.9}Mn_{0.1}Te/Cd_{0.6}Mg_{0.4}Te$ quantum wells. The thickness of the quantum well was 1.8 nm. The 1.6 K PL of the structure was dominated by a line due to recombination of excitons localized at monolayer fluctuations of the thickness of the quantum well. The energy of the magnetic polaron measured by the selective excitation of the localized excitons was 10 meV. Measurements were made in both Faraday and Voigt geometries. At moderate in-plane magnetic fields, magnetic-polaron formation leads to a lowering of the system symmetry. The magnetic moment of the polaron thus formed contains a component normal to the quantum well plane, the moment is not parallel to the external magnetic field. The polarization characteristics of the luminescence from magnetic polaron states is affected by the spontaneous lowering of the symmetry.

$Zn_{1-x}Mn_xSe/ZnSe$ multiple-quantum-wells

Klar et al. [357] studied extensively the excitonic and electronic transitions in $Zn_{1-x}Mn_xSe/ZnSe$ multiple-quantum-well structures. The width of the ZnSe quantum wells was in the range 29 to 91 Å. The thickness of the $Zn_{1-x}Mn_xSe$ barrier layers was considerably larger. The value of x in the barrier layers was in the range 0.18 to 0.26. Reflectivity, photoluminescence excitation, and spin-flip Raman spectroscopy measurements were made at liquid helium temperatures. The applied magnetic fields were up to 7.5 T. The three techniques give complementary information about the excitonic and electronic transitions in the quantum wells and in the barrier layers. Values of the three quantities, the energy splitting between the first light and heavy-hole quantum-well transitions in zero-magnetic field, the energy splitting between the two polarization components of the first heavy-hole quantum-well transition (at 3 T) and the saturation Raman shift for spin-flip scattering in the conduction band of the quantum well were calculated for different thicknesses of the wells. These quantities are sensitive to valence band offset, strain and to the interface quality. To take into account the interface mixing, Mn concentration of $0.9x$ and $0.1x$ were used in the cation layers immediately adjacent to the interface. The strain was calculated assuming a common in-plane lattice constant $a_L = [(a_{ZnSe} + 2a_{ZnMnSe})]/3$. The values of the valence band offset for the $Zn_{0.75}Mn_{0.25}Se/ZnSe$ were determined

6.8. EFFECT OF MAGNETIC FIELD ON SEMICONDUCTORS

for the best fit of the experimental data with the theory. The effects of interface roughness and strain were taken into account. The best value of valence band offset was $(20 \pm 10)\%$ of the bandgap difference between $Zn_{0.75}Mn_{0.25}Se$ and ZnSe.

6.8.3.4 Digital alloys

Very short-period superlattices are known as the artificial crystals of digital alloys. Kutrowski et al. [486] fabricated digital alloys using non-magnetic CdTe and magnetic $Cd_{1-x}Mn_xTe$ epilayers. The authors studied the magneto-optical properties of the samples and of the random alloys of the same average composition. The magnetic behavior of the samples grown in the $\langle 100 \rangle$ was different from that of the random alloys of the same average composition. The spin-splitting was different in the two cases. The behavior of the digital alloys grown in the $\langle 120 \rangle$ direction was also different. The binding energy of the exciton magnetic polaron in the alloys grown in the $\langle 120 \rangle$ direction was greater by 10-20% due to anisotropy of the heavy hole mass. In these digital alloys (grown in the $\langle 120 \rangle$ direction), no modification of the spin-splitting was observed. Fiederling et al. [491] also studied optical properties of the exciton magnetic polarons in digital alloys of diluted magnetic semiconductors using magneto-optical methods. The samples consisted of $Cd_{0.2}Mn_{0.8}Te$ random alloy barriers and $Cd_{1-x}Mn_xTe$ digital alloy as quantum wells. Magnetic field of up to 7 T was applied parallel to the growth axis (Faraday geometry). Some experiments were performed in the Voigt geometry i.e. the magnetic field applied perpendicularly to the growth axis. As in the work of Kutrowski et al. [486], samples grown on both (120)- and (100)-oriented substrates were studied. Photoluminescence excitation by and photoluminescence measurements selective excitation were made. No difference in the magnetic properties, as measured by the giant Zeeman splitting of excitonic states, was observed between (120)- and (100)-oriented structures. However the exciton magnetic polaron energy was found to be larger by 10-40

6.8.3.5 Effect of magnetic field on the properties of Quantum dots

Oka et al. [483] studied the dynamics of excitons in the quantum-dots and quantum-wells of diluted magnetic II-VI semiconductors using time-resolved luminescence and transient non-linear optical spectroscopy. The evidence of formation of EMP in CdMnSe quantum dots was obtained. In asymmetric double quantum wells of $Cd_{1-x}Mn_xTe/ZnTe$ the carriers and excitons tunnel in the picosecond time range. The observed ultrafast dynamics and the enhanced magnetic-polaron effect are characteristic excitonic properties of the diluted magnetic II-VI low dimensional structures. Magnetic polarons have been studied theoretically by the well known method of Baldereschi and Lipari by Bhattacharjee et al. [484]. Properties of the magnetic polarons were studied in two magnetization limits, saturation and linear. Unlike the previous theoretical work, the valence-band degeneracy was taken into account explicitly. It was shown that the theories which neglect the valence band degeneracy underestimate the binding energy of the MPs. The available experimental results on EMPs containing quantum dots in CdMnSe were explained satisfactorily by the theory.

6.8.3.6 EMP bifurcation and asymmetric quantum wells

Several theories predict that symmetry of the exciton magnetic polaron (EMP) formed in quantum structures with non-magnetic quantum well and diluted magnetic barriers will be spontaneously broken [485]. This phenomenon is also called the EMP bifurcation. The bifurcation shifts the wave-function of the EMP to one side of the interface. Whether or not EMP bifurcation exists is still being debated [485]. So far no clear observation of the EMP bifurcation exists. EMP wave-function can move to the interface for other reasons e.g. the EMP may be bound to an impurity which is located near the interface. Takeyama et al. [485] have studied the effect of magnetic field (up to 15 T) on the observed PL of in $Cd_{1-x}Mn_xTe$-$CdTe$-$Cd_{1-y}Mg_yTe$ single quantum wells at different temperatures. The quantum well is asymmetric because one barrier is semi-magnetic and the other is non-magnetic. Two PL lines were observed and attributed to symmetric and asymmetric exciton states. The co-existence of both symmetric and asymmetric states suggest that there is a finite barrier between the two states. The observation is similar to exciton EMP bifurcation formed on one of the magnetic barriers in a symmetric quantum well. Takeyama et al. [485] have suggested that co-existence of both symmetric and asymmetric states should be taken into account in the theory of the bifurcation problem.

6.8.3.7 Cyclotron resonance measurements

Since the mobility in II-VI semiconductors is low, it is difficult to perform cyclotron resonance measurements in these semiconductors [493]. The recent advances in producing very high magnetic fields has made it possible to make the cyclotron measurements in low mobility materials. Imanaka and Miura [493] have determined the effective mass of electrons in $Zn_{0.75}Cd_{0.25}Se$/ZnSe quantum wells using cyclotron resonance measurements at magnetic fields up to 150 T. The quantum wells were modulation doped with Cl. The measurements were made in the temperature range 20 K to 300 K. The normalized effective mass for a 9 nm thick well was 0.126 at 20 K. The effective mass increased linearly with temperature to ~ 0.131 at 100 K. It became practically independent of temperature beyond 100 K. In n-type CdS the effective mass decreases gradually with increasing temperatures. The origin of the temperature dependence of the effective mass is not clear. The effective mass also depended on the width of the well. At 100 K and for 3 nm well, the effective mass increased to 0.14.

Chapter 7

Strained layer optoelectronic devices

7.1 Conventional-III-V semiconductor lasers

7.1.1 Suppression of Auger recombination by strain in semiconductor lasers

Auger recombination affect adversely the performance of InGaAs based visible lasers and other long wavelength lasers. Auger processes are intrinsic and can not be suppressed by improving the quality and purity of the sample. Early predictions that strain can suppress the Auger recombination and reduce the threshold current were made by Yablonovitch and Kane [51]. In the following years strain played an important role in improving the performance of InGaAs lasers grown on GaAs substrates. Excellent reviews of the subject have been written by Morkoç et al. [12] and by O'Reilly et al. [42]. At the present time there is a great interest in the study of mid-IR (2 to 5 μm) InGaSb-based lasers. There is a transparency window in the 2-5 μm range in the atmosphere. Therefore mid-IR devices are useful for several applications discussed in section 7.5.6. Detrimental effect of Auger recombination increases as the temperature increases and bandgap decreases. Therefore the harmful effect of Auger recombination is more serious in mid-IR lasers. The mid-IR lasers do not yet operate at room temperature because of the limitation due to Auger recombination. Mid-IR lasers using Sb based III-V semiconductor strained layers have been fabricated and studied recently. In this section we review the theoretical and experimental work that has been done to suppress the Auger recombination in InGAs- and GaSb-based lasers.

The difference of electron and hole quasi-Fermi levels $F_c - F_v$ must be larger than the bandgap of the semiconductor for the laser action to take place [51]. The threshold carrier density and the threshold current have minimum values if the holes have the same density of state effective mass as the electrons, i.e. they

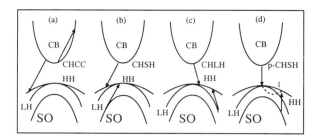

Figure 7.1: Electronic transitions leading to Auger recombination are shown. (a) electron-electron-Auger recombination, (b) and (c) hole-hole-Auger recombination and (d) energy and/or momentum are conserved with the involvement of phonons [51].

are not heavier. Strain is also used to reduce the effective mass and density of sates in the valence band.

The four processes leading to Auger recombination are shown in Fig. 7.1. Auger recombination increases rapidly as the density of injected carrier increases. Strain reduces the density of state effective mass of the holes and also the threshold carrier density. If lasing takes place at a lower carrier density, Auger recombination is reduced decreasing further the level of injection at which the lasing occurs. In the modified band structure conservation of momentum and energy (a necessary condition for Auger recombination to occur) can not take place and Auger recombination is reduced.

Calculated band structure of a 3.0 nm InAs quantum well with InP barriers is shown in Fig. 7.2 [51]. These calculations are important because they predicted in 1988 that strain could be used to suppress the Auger recombination. The dashed curve in Fig. 7.2 is for the valence band of the unstrained InAs. At the zone center the bandgap of the strained quantum well is the same as that of the unstrained InAs. Strain decreases the bandgap but the decrease is compensated by its strain induced increase. The most important result of the calculations is that the hole mass at the zone center is reduced in the strained layer. The hole mass increases in the off-center positions but here the density of the holes in this position is small. In the modified band structure it is difficult for the electron-hole Auger recombination to take place. The modified band-structure also suppresses the intervalence band absorption (recombination of an electron from the spin-split-off band with a free heavy hole) which competes with the laser gain. Both of these advantages are due to strain, the shape of the hole bands is not affected so much by the confinement [51]. These improvements predicted by the calculations [51] have been amply confirmed by the subsequent experimental work on the strained layer InGaAs based devices [12, 42, 495].

Grein et al. [496] made similar calculations to determine the reduction of Auger recombination in mid-IR laser structures. Calculated band structure of a 1.7 nm InAs/3.5 nm $In_{0.25}Ga_{0.75}Sb$ strained superlattice is shown in Fig. 7.3. The bands are highly non-parabolic. In the strained superlattice an electron

7.1. CONVENTIONAL-III-V SEMICONDUCTOR LASERS

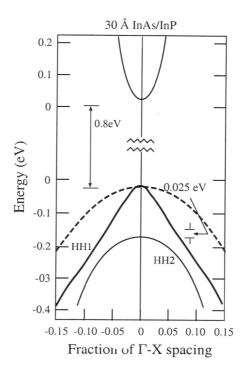

Figure 7.2: The (calculated) effect of strain and confinement on the band structure of a 3.0 nm InAs quantum well with InP barriers [51].

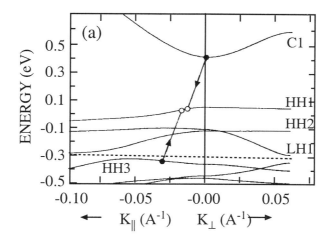

Figure 7.3: Calculated band-structure of a strained layer 1.7 nm InAs/3.5 nm $In_{0.25}Ga_{0.75}Sb$ superlattice is shown. Bands are highly non-parabolic [496].

near the bottom of the conduction band can not Auger recombine with a hole with zero $K_\parallel$ because of the absence of states an energy gap below the top of the HH band shown by the dotted horizontal line [496]. Therefore Auger recombination can take place only as shown by the arrows. The density of holes shown by open circles is not very large and therefore the Auger recombination is suppressed. The curvature in the conduction band for $K_\perp$ suppresses the electron-electron Auger processes.

7.1.2 Pump-probe and other methods for measurement of Auger lifetimes

In the pump-probe method, an unfocused free electron laser beam is used to inject (or pump) a large density of carriers in to a semiconductor. Band filling (Dynamic Moss-Burstein shift) causes bleaching and high transmission of above bandgap radiation. A part of the laser beam passed via a delay line is used as a probe. The transmitted probe signal, measured as a function of probe delay by Ciesla *et al.* [497], is shown in Fig. 7.4(a) for two values of photon

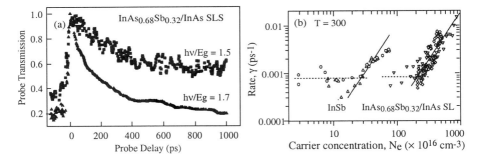

Figure 7.4: (a) Transmission of an InAsSb/InAs superlattice after high injection of carriers as a function of probe-delay time is shown. (b) Recombination rates in a strained InAsSb/InAs superlattice are shown as a function of carrier concentration. Recombination rates in an unstrained InSb (which has a similar bandgap) sample are also shown. Symbols show the measured values, dotted lines show the calculated Shockley–Read type rates and solid lines show the calculated Auger rates [497].

energies. The effective pump energy of 350 mJ/cm^2 per micro-pulse and probe energy of 10 mJ/cm^2 per micro-pulse were used in the experiments [497]. In the initial stages the transmission of light increases due to filling of the bands by the carriers. After the injected carrier density becomes sufficiently high, the injection is stopped and the transmission is measured after different delay times. As the probe-delay increases, the transmission of the probe decreases because carrier densities decrease due to Auger recombination. Values of carrier density at different times are determined using the observed transmitted signal intensities. Values of the Auger recombination rates and Auger coefficients are calculated using the carrier densities as a function of time. Measured and calculated values of the Auger recombination rate as a function of the carrier density are shown in Fig. 7.4(b). The recombination rate in the unstrained InSb (which has a comparable bandgap) is also shown for comparison. The Auger recombination rate in the strained superlattice is about two orders of magnitude smaller. Dotted lines show that Shockley type of recombination are negligible except at the lowest carrier densities.

Youngdale et al. [498] have measured the Auger life-time in (3.9 nm)/(2.5 nm) InAs/Ga$_{0.75}$In$_{0.25}$Sb, superlattice. The bandgap of the superlattice is 140 meV. Steady state photo-generated carrier density Δn versus laser light intensity was determined at 77 K by photoconductivity measurements. A frequency doubled CO_2 laser was used for injecting the carriers. For large carrier densities ($\Delta n > 2 \times 10^{17}$ cm^{-3}) measured lifetime values agreed with the calculated Auger values. The lifetime values were up to two orders of magnitude larger than in unstrained Hg$_{0.77}$Cd$_{0.23}$Te with a similar bandgap (140 meV) [498].

The Auger recombination in the lattice matched InGaAs/InP laser diodes and also in the strained InGaAsP quantum wells by the differential carrier lifetime technique [499] have been measured. The transparency carrier densities were 1×10^{18} cm^{-3} for the strained device and 1.5×10^{18} cm^{-3} for the unstrained device. Auger recombination rate was reduced from 1×10^{27} cm^{-3}s^{-1} in the unstrained structure to 0.2×10^{27} cm^{-3}s^{-1} in the 1.8% strained structure.

7.1.3 PL of superlattices containing Sb in the active layers

Characteristics of the InAsSb/InAlAs superlattice active region emitting at 2.65 μm have been studied by Grietens et al. [22]. The characteristic temperature T_0 of these laser diodes was rather small, 28 K. Recent efforts have been to optimize structures to obtain higher values of T_0 [21]. Van Hoof et al. [21] fabricated symmetrically strained InAs$_{0.9}$Sb$_{0.1}$/In$_{0.85}$Al$_{0.15}$As superlattices on InAs substrates. The superlattices consisted of 100 periods of 2.5 nm InAs$_{0.9}$Sb$_{0.1}$ and 2.5 nm In$_{0.85}$Al$_{0.15}$As layers. The structures were characterized by RHEED, PL and x-ray techniques. The quality of the layers was very good. Integrated intensity of the PL as a function of temperature is shown in Fig. 7.5. It is seen from this figure that T_0 for the IMEC 177 InAs$_{0.9}$Sb$_{0.1}$/In$_{0.85}$Al$_{0.15}$As sample is 300 K, somewhat smaller than the value 360 K for the Hughes sample. The integrated intensity of the PL as a function of temperature for the superlattices with improved structure is shown in Fig. 7.6. The structure included InAs spacers of different thicknesses between the two superlattice periods. Values of several parameters derived from the data of Fig. 7.6 are shown in Table 7.1. Both the emission wavelength and T_0 change with the thickness of the spacer. Best values of T_0 are obtained with the thickest spacer i.e. the spacer with 12 nm thickness. The value $T_0 = 446$ for this sample is the highest. The value of T_0 for bulk InAs is 120 K. Hoof et al. [21] implemented an 8-band model to calculate accurately the band structure of the superlattices. Structures with a larger InAs spacer showed the highest laser gain. The wavefunctions for the electrons also show increased confinement with increasing InAs spacer layer thickness.

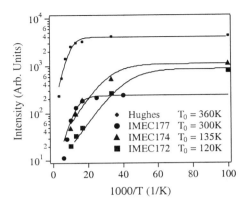

Figure 7.5: Integrated PL intensities versus 1000/T for an $InAs_{0.9}Sb_{0.1}/In_{0.85}Al_{0.15}As$- and two $In_{0.9}Ga_{0.1}As/GaSb$-superlattices, and an InAs/InAsSb SL laser grown at Hughes Research Laboratories (see text) [21].

PL of $InAs_{0.91}Sb_{0.09}/In_{0.87}Ga_{0.13}As$ superlattice matched to InAs substrate have been measured at 4 K by Kurtz and Biefeld [500]. The measurements were made with several values of externally applied magnetic fields. Similar studies were also made for strained pseudomorphic $InAs_{0.91}Sb_{0.09}$ quantum wells separated by 50 nm InAs barriers. From the slope of the PL versus magnetic field plots, the m_h/m_0 values of 0.015–0.024 were derived for the strained sample and 0.031–0.037 for the unstrained structures. As mentioned earlier the reduction in hole mass reduces the threshold current as well as the Auger recombination.

7.1.4 $InAs_{0.9}Sb_{0.1}/In_{0.85}Al_{0.15}As$ strained layer superlattice mid-IR lasers

Grietens et al. [22] and Hoof et al. [21] have fabricated and studied mid-IR strained layer superlattice lasers emitting at 2.65 μm. The lasers were grown on an InAs substrate. The active region consisted of 200 periods of 2.5 nm $InAs_{0.9}Sb_{0.1}$ and 2.5 nm $In_{0.85}Al_{0.15}As$ superlattice. The active region was surrounded by superlattice cladding layers. The claddings consisted of 400 periods of 2.5 nm $AlAs_{0.16}Sb_{0.84}$ and 2.5 nm InAs layers. The active region was left undoped. The p-type cladding was formed by doping both AlAsSb and InAs with Be. The n-type cladding was formed by doping the InAs layers with Si and leaving the AlAsSb layers undoped. Using standard lithographic techniques the grown material was processed into 100 μm wide broad-area lasers. Mesas were etched and covered with an isolating Aluminum-oxide. Ohmic contacts were made by the deposition of TiW/Au. Individual devices 230 μm long were mounted on Cu heat sinks and installed in a liquid-nitrogen dewar for low-temperature measurements. The characteristics of the laser diode are shown in Fig. 7.7. Pulsed laser emission was observed up to 120 K using 2 μsec pulses at 10 kHz repetition rate. Intensity of the emitted light versus input current at

7.1. CONVENTIONAL-III-V SEMICONDUCTOR LASERS

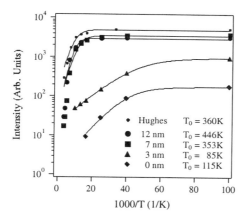

Figure 7.6: Integrated PL intensities versus 1000/Temperature for various $InAs_{0.9}Sb_{0.1}/InAs/In_{0.85}Al_{0.15}As/InAs$ superlattices with varying thickness of the InAs layers and an InAs/InAsSb SL laser grown at the Hughes Research Laboratories.

80 K is plotted in Fig. 7.7(a). The laser action starts at about 200 mA at this temperature. At 80 K the laser shows light emission at 2.65 μm (Fig. 7.7(b)). The threshold current density versus temperature is plotted in Fig. 7.7(c). At 80 K the threshold current is 80 mA/cm^2. Light output versus current at 80 K is plotted in Fig. 7.7(d). From threshold measurements between 80 K and 120 K the characteristic temperature was calculated as $T_0 = 28$ K. The highest measured power output was 110 μWatt/facet using a 1.1 A injection-current. This corresponds to a peak efficiency of $\sim 4\%$. Somewhat similar structures were also by Zhang et al. [501]. The stimulated emission by optical pumping was at 3.4 μm. Their experiments show that lasers based on InAs/InAsSb type II superlattices can cover a wavelength range from 3 μm to 10 μm.

7.1.5 Summary of this section

Strain modifies the awkward bandgaps of *natural* semiconductors and changes heavy holes into light holes at the zone center. It suppresses Auger recombination. In the layers under compression, the hole mass at the zone center is reduced by a factor up to 5. The holes are still heavy away from the zone center but here their density is reduced. Reduction in the hole mass makes it possible to satisfy the laser condition and the transparency condition at a carrier density smaller by a factor up to 10. This reduces the threshold carrier and current densities by about the same factor. Lower carrier density results in reduced Auger recombination. Strain also suppresses the inter-valence band absorption. The shape of the conduction band can be changed such that electron-electron Auger recombination is suppressed. Laser diodes have been designed and fabricated based on the above advantages of strain. Expected high performance has been

Table 7.1: Characteristic temperature T_0, activation energy ΔE and the adjustable parameter c_r for four samples of Fig. 7.6.

	Hughes 514	IMEC 233	IMEC 234	IMEC 235
SLS	InAs InAsSb InAsSb	$InAs_{0.9}Sb_{0.1}$ InAs (12 nm) $In_{0.85}Al_{0.15}As$	$InAs_{0.9}Sb_{0.1}$ InAs (7 nm) $In_{0.85}Al_{0.15}As$	$InAs_{0.9}Sb_{0.1}$ InAs (3 nm) $In_{0.85}Al_{0.15}As$
λ μm	3.27	3.84	3.50	3.30
I_0 (A.U.)	4784	2929	3390	929
c_r	27.7	71.9	64.0	44.9
T_0 (K)	360	446	354	84.8
ΔE (meV)	31	37	29.5	7

achieved.

7.2 ZnSe-based light emitters and other devices

7.2.1 Light Emitting Diodes

Early work on the ZnSe-based optoelectronic devices has been reviewed by several authors [53, 189, 411]. ZnSe and ZnS have bandgaps of 2.7 eV and 3.7 eV respectively. Bandgaps of zinc telluride and CdSe are smaller than the bandgap of ZnSe. Therefore, for fabricating light emitting devices, ZnCdSe (10 to 20% Cd) or ZnTeSe ($\sim$ 10% Te) are used as active layer materials and wider bandgap ZnSe or ZnSSe as the cladding layers [181]. By alloying ZnSe with Cd or Te in the active layer, the laser wavelength can be tuned in the blue to green region of the spectrum. Refractive index of the ZnCdSe layers is larger than that of the ZnSSe barrier layers. This produces the desired wave-guiding action. The photons emitted by the radiative recombination are confined in the active layer region. Recently quaternary alloys containing Mg or Be have been used for cladding layers. The quaternary alloys are discussed later.

The first blue p-n LED was a ZnSe homojunction diode, reported in 1988 by Yasuda et al. [494]. The p-doping was obtained by co-doping with Li and N. Several authors reported double heterojunction LEDs in 1990s [181, 502, 503]. The double heterostructure LEDs fabricated by Eason et al. [181] showed very good characteristics. The LEDs were fabricated by MBE on (100) ZnSe substrates. A 2 to 3 μm thick ZnSe:Cl layer was first grown on the substrates. The active layer was deposited on the ZnSe:Cl layer. The active layer for the blue diode consisted of 5- to 10-nm thick ZnSe/$Zn_{0.9}Cd_{0.1}$Se quantum wells. For the green diode the active layer was a single 100 nm thick $ZnTe_{0.1}Se_{0.9}$ layer. One micron thick p-type ZnSe:N layers were grown on the active layers in each case. The contact to the p-ZnSe was made using HgSe/ZnTeSe layers.

7.2. ZNSE-BASED LIGHT EMITTERS AND OTHER DEVICES 215

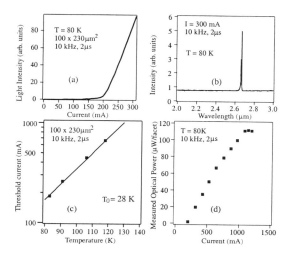

Figure 7.7: Characteristics of a mid-IR laser fabricated in our laboratory. (a) Output power versus injected current for a 100×230 μm^2 device. (b) Emission spectrum demonstrating stimulated emission at 2.65 μm. (c) Threshold current versus temperature and (d) Optical power versus input current.

Current-voltage characteristics showed that the contact to the p-ZnSe layer was ohmic and its resistance was low. The LEDs were packaged in a standard T-1 3/4 clear-epoxy lamp configuration for testing. Electroluminescence intensity versus wavelength of the blue diode is plotted in Fig. 7.8(a). The spectrum is sharply peaked at 489 nm and the FWHM is 72 meV. The external quantum efficiency is 1.3%, 30 times larger than that of the commercially available SiC LEDs. At the operating current of 10 mA and voltage of 3.2 V, the output power was 327 μW. The brightness (luminous efficiency) of the blue diode was 1.6 lm/W. At the time of publication it was the brightest blue LED fabricated using II-VI semiconductors. For comparison light output of a blue GaN/InGaN

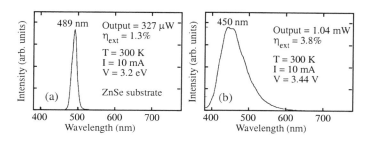

Figure 7.8: Light output characteristic (a) of the blue ZnSe/ZnCdSe LED fabricated by Eason et al. [181] and (b) of the blue GaN/InGaN LED fabricated by Nakamura et al. [504].

diode fabricated by Nakamura et al. [504] is shown in Fig. 7.8(b). The spectrum of the GaN/InGaN diode is peaked at 450 nm. The diode produces 1.04 mW at 10 mA corresponding to an external quantum efficiency of 3.8%. These are very impressive values, ~ 3 times larger than the corresponding values for the ZnSe/ZnCdSe diode. The luminous efficiency of the GaN/InGaN diode is also much larger, 3.6 lm/W as compared to 1.8 lm/W of the ZnSe/ZnCdSe diode. However the spectrum of the GaN/InGaN LED is very broad with an FWHM of 510 meV, because the active region was doped with Zn. Zn produces a deep level emission at ~ 450 nm. To the naked eye the output appears whitish-blue because of the large FWHM.

At the same operating current and voltage (as the blue diode) the green LED produced 1.3 mW at a peak wavelength of 512 nm (not shown in the figure). The FWHM of the green LED was 245 meV a relatively large value due to the presence of Te in the active region. This corresponds to an external quantum efficiency of 5.3%. The brightness (luminous efficiency) of the green diode was 1.7 lm/W. This diode is 50 times brighter than the commercial GaP LED which produces not green but yellow-green light at 555 nm. In fact in 1995 this diode was the brightest diode made from any semiconductor emitting at 512 nm.

The degradation of output power Φ was studied as a function of time at two current densities, $J = 15$ A/cm^2 and $J = 50$ A/cm^2. In both cases the decay of output power Φ could be described by the equation:

$$\Phi = \Phi_0 \exp(-t/\tau), \tag{7.1}$$

where Φ_0 is the output power before degradation begins and τ is the time at which output power decays to Φ_0/e. Values of τ were 675 h for $J = 15$ A/cm^2 and 350 h for $J = 50$ A/cm^2. If the lifetime can be further increased, 100 μW at 1 mA devices can be commercialized. There is a substantial market for low power green LEDs [181].

7.2.2 Photo-pumped lasers

We have seen in section 5.4.3 that probability of radiative recombination of electrons and holes confined in quantum wells increases and efficient emission of light occurs. In the lasers, active layers are generally quantum wells placed between p- and n-doped large bandgap layers. Electrons and holes are injected into the quantum wells where they recombine and give rise to light emission. Typical widths of the quantum wells are 5 to 10 nm. As discussed earlier, at these widths the transition energy (which is equal to the energy of the emitted photons) is determined not only by the bandgap of the well material but also by the quantum confinement effects. If the quantum well is strained, the transition energy also depends on the magnitude and kind of strain (i.e. whether the strain is compressive or tensile). Lasing has been observed by electron-beam pumping in bulk crystals and by current injection in homojunction. However threshold power in the first case and threshold current in the second case are high and efficiencies are low. Most work on lasers diodes has been on double heterostructures, involving quantum wells as active layers.

7.2. ZNSE-BASED LIGHT EMITTERS AND OTHER DEVICES 217

Considerable work has been done on photo-pumped lasers [505]-[507]. Nakanishi et al. [505] fabricated a photo-pumped ZnSe/ZnSSe blue semiconductor laser in 1991. The threshold at room temperature was very low, only 10.5 kW/cm^2. This is one of the lowest threshold observed in the optically pumped lasers and is close to the theoretically calculated threshold. The threshold was low because the active region was directly excited by the 450 nm light from a dye laser. This excitation is more effective than the excitation by the 337 nm or 335 nm light from N$_2$ or Nd:YAG lasers. If excited by the short wavelength 337 nm or 335 nm light, a part of the energy is wasted in exciting the barrier layers. The lasing action was observed up to 400 K with a characteristic temperature $T_0 = 124$ K. At 400 K the differential quantum efficiency decreased to 77% of its value at room temperature. Okuyama et al. [506] reported probably the first ZnSe-ZnMgSSe double-heterostructure photopumped blue lasers in 1992. The threshold excitation intensity was 150 kW/cm^2 at room temperature. Polarization measurements confirmed the lasing action. The ZnMgSSe layer was lattice matched to GaAs (100) substrate. The bandgap difference between the cladding layer and the active layer was more than 0.3 eV. In a later paper [507] Okuyama et al. reported a photo-pumped laser that showed improved characteristics. The threshold intensity now was 105 kW/cm^2. The laser emitted at 464.5 nm. The laser worked up to 500 K with a characteristic temperature $T_0 = 170$ K. The high temperature performance was a result of large bandgap difference between the cladding and the active layer. These results showed that ZnMgSSe is a useful material for the cladding layer of blue (and green) light emitting devices.

Toda et al. [508, 509] fabricated photo-pumped lasers on GaAs substrates by MOCVD. The laser structure consisted of (1.5 μm ZnMgSSe buffer)/(60 nm ZnSe)/(150 nm ZnMgSSe) layers. The quality of ZnMgSSe layers was very good. Mirror-like surface and near band-edge PL was observed from the ZnMgSSe layers containing up to 13% Mg. The cladding layers of the laser structures contained 10% Mg and 15% Se. Characteristics of the laser were measured using a N$_2$ laser. The pulse width and emission wavelength of the exciting light were 5 ns and 337 nm. Room temperature threshold was 70 kW/cm^2. Below threshold, the PL was broad (FWHM = 150 nm) and peaked at 450 nm. Above threshold the emission was at 465 nm with FWHM = 15 meV. The emission was strongly TE polarized, confirming that the emission was stimulated. The threshold value was comparable to that obtained with the MBE grown lasers.

Recently (1998/1999) several groups have fabricated and studied the characteristics of photo-pumped lasers. Ivanov et al. [510] deposited dot-like CdSe islands on ZnSe by MBE. The values of the thickness of the CdSe layers (which produced the islands) were between 0.75 and 3 mono-layers. The corresponding peak positions of the PL varied in the range 2.8 eV to 2.3 eV. The best performance was obtained when the CdSe thickness was 2.8 monolayers. The PL peak position at this thickness was at 2.49 eV. The actual structure of the photo-pumped lasers was 0.5-μm Zn$_{0.92}$Mg$_{0.08}$S$_{0.14}$Se$_{0.86}$ layer followed by a 0.1-μm ZnS$_{0.12}$Se$_{0.88}$/ZnS superlattice, 10 nm ZnSe quantum well, 0.1-μm ZnS$_{0.12}$Se$_{0.88}$/ZnS superlattice, and top ZnMgSSe layer. A 2.8 monolayer thick CdSe layer was inserted at the center of the 10 nm ZnSe quantum well. TEM pictures showed clearly self organized CdSe islands in the quantum well layer. Lasing action was studied using 8 ns pulses from a nitrogen laser. At room temperature the threshold intensity was less than 4 kW/cm^2 (a very low value) as compared to the value 19 kW/cm^2 for a laser without the CdSe layer insertion. The emission showed a small blue shift at the threshold. The threshold intensity was also measured at different temperatures in the $\sim$ 4 K to 300 K range. The increase of the threshold intensity with temperature was not monotonous, it showed a shallow minimum at $\sim$ 100 K. The laser with CdSe layers showed no degradation when operated at 25 times the threshold for 24 hours. Gurskii et al. [511] studied optical properties and lasing of ZnMgSSe/ZnSe-, ZnMgSSe/ZnSSe/ZnSe-, and ZnMgSSe/ZnMgSSe/ZnSe-based single- and multiple-quantum-well heterostructures grown by metalorganic vapor phase epitaxy. The effect of well width on the optical properties was studied in detail. Laser action under transverse optical pumping was obtained only if well widths were $L_w \geq$ 4 nm and optical confinement factor $\Gamma >$ 0.04. Best results were obtained with ZnMgSSe/ZnMgSSe/ZnSe SCH with an active layer thickness of 40 nm. The lasing emission was TE polarized and peaked at $\sim$ 2.67 eV. The lowest threshold intensity $I = 10 - 30$ kW/cm^2 at T = 78 K was achieved. The lasing was observed up to 577 K, a significant improvement over the previously observed maximum value of 500 K [511].

Optically pumped ZnSe based lasers have also been fabricated on InP substrates [513, 512]. The cladding and waveguiding layers lattice matched to GaAs substrates can be grown using ZnMgSSe and ZnSeS alloys. However there is a significant lattice mismatch between ZnCdSe quantum wells and GaAs substrate. The quantum wells are therefore strained. The thickness of the wells must be kept small in order to keep the layers pseudomorphic. If the thickness is not sufficiently small, the strain can relax by the creation of misfit dislocations during the operation of the laser. This affects adversely the long-term reliability of the lasers. In order to solve this problem, Guo et al. [513] and Zeng et al. [512] fabricated optically pumped lasers on InP substrates by MBE. Room-temperature optically pumped emission in the red, green, and blue was obtained from ZnCdMgSe/ZnCdSe quantum well laser structures. A 4 nm thick ZnCdSe quantum well (E_g = 2.2 eV) was embedded in 0.2 μm thick graded composition Zn$_x$Cd$_y$Mg$_{1-x-y}$Se layers [512]. The grading of the composition was such that the bandgap of the layer varied continuously from 2.7 to 3.0 eV. Zn$_x$Cd$_y$Mg$_{1-x-y}$Se alloy with E_g = 3.0 eV was used for the cladding layers. To avoid oxidation of Mg in the cladding layer, a 10 nm ZnSe cap layer was used. The refractive index versus bandgap relation –0.7/eV, valid for ZnMgSSe alloy, was used for ZnCdMgSe alloy also [513] to calculate the optical confinement factor. The calculated confinement factor was 1.39%. A dye laser operating at 460 nm was used for excitation. The laser emission was green at 512 nm. The width of the emission line was 5 nm. The threshold power density was 160 kW/cm^2. The change in the threshold power density with temperature was measured and a value of 150 K was determined for the characteristic temperature T_0. Since the effective masses, band offsets and refractive indices for the ZnCdMgSe alloy were not known, it was not possible to optimize the design.

7.2. ZNSE-BASED LIGHT EMITTERS AND OTHER DEVICES

These studies [513, 512] show that the ZnCdMgSe alloys and InP substrates are potentially useful as alternatives to ZnMgSSe alloys and GaAs substrates. This work demonstrates the potential of these materials for use in integrated full color display devices. This is the only material system which allows the whole laser structure to be pseudomorphic on one single substrate (i.e. without a relaxed buffer layer) and which emits in whole of the visible range [512].

Jeon et al. [514] fabricated a photo-pumped vertical-cavity surface-emitting laser (VCSEL) operating in the 480–490 nm range. The separate-confinement heterostructure (SCH) consisted of three ZnCdSe QWs cladded by ZnSSe and ZnMgSSe inner and outer confinement layers. Jeon et al. [514] removed selectively III-V semiconductor and deposited low-loss multilayer stacks of SiO_2/HfO_2 as reflectors. The cavity Q factor was about 400. The electron-hole pairs were injected in the ZnCdSe active layer by a dye laser. Lasing was observed up to 200 K. Far-field measurements showed that spatial coherence was good; typical beam divergence was $\sim 4°$.

7.2.3 ZnSe-based electron-beam pumped lasers

Several papers have appeared on the electron-beam-pumped lasers [515]-[519]. Electron-beam-pumped lasers are important for high brightness, high resolution Red-Blue-Green (RGB) laser projection application [101, 516]. The military applications of interest are flight simulator visuals, command center large screen projection and possibly cockpit display. Laser radar and communication are also important applications. Some of the commercial applications include electronic cinema, teleconference display, university auditorium display, laser microscopy, and optical computing [516].

At liquid nitrogen temperature emission of coherent and bright blue light at 450 nm [101] is obtained from ZnSe by electron beam pumping. However sensitivity of eye to this wavelength is low and therefore very high power outputs are required. This problem can be solved by shifting the wavelength to 470–480 nm range where the eye sensitivity is higher. ZnSe does emit at 473 nm at room temperature. However an increase of temperature of the crystal reduces the output power and working life of the system. Akhekyan et al. [101] have grown bulk single crystals of $ZnSe_{1-x}Te_x$ ($x \leq 0.03$), $Zn_{1-x}Cd_xSe$ ($x \leq 0.15$), and $Zn_xCd_{1-x}S$ ($x \leq 0.35$) solid solutions and investigated their optical properties to determine whether these crystals can be used as electron-beam-pumped lasers in the 470–480 nm range. The crystals were grown from the vapor phase by the closed-tube sublimation method. The diameter of the crystals was 50 mm and thickness, 20 nm. Cathodoluminescence measurements showed deep levels in the $ZnSe_{1-x}Te_x$. Lasing action in the other crystals was observed at higher pumping rates. The emission was shifted to longer wavelengths by the required amount.

Kozlovsky et al. [518] fabricated room temperature (RT) VCSEL based on ZnCdSe/ZnSe superlattice structure. Several samples containing different number of quantum wells were fabricated. The heterostructures were characterized by cathodoluminescence (CL), photoreflection (PR), and phototransmission

(PT) techniques. Stokes shift increased and new lines in CL and PT spectra appeared as the number of the wells increased. Output power up to 2.2 W in single longitudinal mode with emission at $\lambda = 493$ nm was obtained by electron beam pumping.

Full color TV projectors based on II-VI electron-beam pumped lasers have been developed [515, 517]. Cathode ray tubes with phosphor screens are generally used in projection televisions. The phosphor screens are not ideally suited for this purpose. Since the emission spectrum is wide, the chromaticity is poor. Phosphor emission follows approximately cosine law and divergence is high. The emission saturates at large electron-beam intensities. Laser emission with electron beam pumping has high spectral purity and high brightness. Nasibov et al. [515] demonstrated a projector system in which the phosphor screen was replaced by electron beam-pumped screen. The screen consists of three units, one for the blue, one for the green and one for the red color. At $T = 80-100$ K, the three units consisted of $Zn_{0.15}Cd_{0.85}S$, $CdS_{0.8}Se_{0.2}$ and $CdS_{0.6}Se_{0.4}$. At room temperature, alternative materials ZnSe. CdS, and $Zn_{0.4}Cd_{0.6}Se$ were used. The colors obtained at room temperature are 475 nm (blue), 530 nm (green) and 630 nm (red). Because of the alloy used for the red region, the efficiency of red emission was lower. For room temperature operation, the laser screen was water cooled. The efficiency was about 10 lm/W. The quality of the projected pictures was very good. In order to reduce the threshold and improve room temperature performance Kozlovsky et al. [517] used quantum well structures. A significant improvement in the threshold was obtained.

Bonard et al. [519] fabricated electron-beam-pumped ZnCdSe/ZnSe (Cd concentration was between 15 and 25%) graded index SCH (GRINSCH) lasers and studied their degradation using TEM and cathodoluminescence (CL) methods. The degradation was monitored in real time with a submicron resolution of the CL detection mode. TEM studies were made before as well as after the degradation. The authors found that the degradation occurs due to the formation of dark spot defects. The dark spot defects are related to electron-bombardment-induced dislocation loops in the quantum well. It was suggested that a climb process of dislocations induced by non-radiative recombination is involved in the degradation process.

7.2.4 ZnSe-based laser diodes

7.2.4.1 Structure and performance

Extensive work on the ZnSe-based laser diodes has been done by 3M company [133, 520, 521], Purdue-Brown group [134, 502, 522], Philips Research Laboratory [523], Sony Corporation Research Center [409, 509, 524] and Photonics Laboratory of Samsung [411]. As in the case of III-V semiconductors, ZnSe-based heterojunction lasers greatly outperform the homojunction devices. The first electrically pumped heterostructure laser diode was reported in 1991 [133] by the 3M company. The active layer of the laser was a ZnCdSe quantum well. ZnSe was used for barrier layers and $ZnS_{0.03}Se_{0.97}$ for outer cladding layers.

7.2. ZNSE-BASED LIGHT EMITTERS AND OTHER DEVICES

$ZnS_{0.03}Se_{0.97}$ layers are lattice matched to GaAs substrate. Concentration of Cd in the active layer was 20% and thickness of the quantum well was 10 nm. The thickness of the ZnSe barrier layer was 500 nm. The p-doping was achieved by using plasma activated nitrogen. The doping level was low, 2×10^{17} cm^{-3}. A film of polymide was used to suppress current spreading. The contact to the p-ZnSe layer was made using Au metallization. The laser was gain guided. The design of this first device was not optimized. The top p-contact resistance was high. The thickness of the ZnSe layers was more than the critical thickness. Concentration of dislocations was 10^7/cm^2. Confinement of the carriers was also weak. In spite of these deficiencies, 1 mm cavity mounted p-side up emitted 100 mW under pulsed operation at 77 K. Threshold current density was 320 A/cm^2 and differential quantum efficiency was > 20%. The refractive index of the active layer was reduced by the influx of electrons and holes and therefore laser experienced anti-guiding effect [53, 133]. The difficulty of confining the optical field in the active layer can be solved by reducing the thickness of the light-guiding layer and using alloys with lower refractive index. Subsequently 3M company reported the fabrication of improved laser diodes [520, 521]. The buried ridge laser design reported in 1993 lased at 9 V [520]. This threshold voltage is still high, it is about 3 times larger than the bandgap of the active layer. In 1994 the buried DH laser was reported [521] which had a low pulsed threshold current of 2.5 mA. This threshold current is comparable to that obtained with GaAs based lasers. However the lifetime of the laser was only a few seconds.

Pseudomorphic ZnCdSe/ZnSSe structure can be fabricated with small Cd and S concentrations. However optical confinement in these structures is poor. In the laser fabricated with these layers, pulsed operation at room temperature was possible only with reflective end coatings. The lasers emit light in the range 500 to 530 nm depending on the concentration of Cd in the ZnCdSe quantum well. To improve the carrier confinement and to design lasers for emission in the shorter wavelengths, it is necessary to find a semiconductor for cladding layers with wider bandgap and lattice matched to GaAs. ZnMgSSe with appropriate concentrations of Mg and S is a suitable quaternary for this purpose [409]. The use of the quaternary allows independent control of bandgap and lattice constant of the epilayer. Epitaxial layers of ZnMgSSe lattice matched to GaAs can be grown over a large range of bandgaps (2.78 to 4.5 eV) [409]. The introduction of the ZnMgSSe quaternary made it possible to grow pseudomorphic quantum wells and cladding layers lattice matched to GaAs substrate. ZnSe/ZnMgSSe is a type I heterostructure with approximately equal band-offsets for the conduction and the valence band. The refractive index of ZnMgSSe is smaller than that of ZnSe. Therefore both carriers and photons can be confined in ZnSe quantum wells with ZnMgSSe barrier layers. Operation of photo-pumped ZnSSe/ZnMgSSe DH lasers up to 500 K has been demonstrated by several authors (see the previous subsection) confirming the type I band alignment and refractive index differences mentioned above.

Another quaternary being used in ZnSe-based lasers is BeMgZnSe (see references in section 7.2.4.2). The quaternary BeMgZnSe can be grown lattice matched to GaAs as well as to silicon. The bandgap can be tuned from 2.8 eV to more than 4 eV. The ternary BeMgSe is lattice matched to GaAs if Be concentration is 2.8%. The sticking coefficients of the three metals Be, Mg, and Zn are high and high quality epilayers of the quaternary and ternary alloys can be grown over a wide range of temperatures. The width of x-ray diffraction curves are small confirming the high quality of the epilayers [406]. In contrast it is more difficult to grow ZnMgSSe epilayers because sticking coefficient of S is very sensitive to the surface temperature. The surface temperature keeps changing during the growth because the surface emissivity changes as the growth proceeds. The incorporation of Be does not introduce deep levels which show that during the growth the high temperature of the Be effusion cell does not cause impurity problems. However heating BeTe above 100°C under normal atmosphere deteriorates the surface. Laser structures containing Be are discussed later in section 7.2.4.2.

For the fabrication of a laser diode high p- and n-doping are necessary. Under the limited p-doping of ZnMgSSe (see chapter 6), 77-K CW operation of blue LD has been demonstrated [506, 526]. The threshold current density was 225 A/cm^2 and the emission wavelength was 447 nm. The separate confinement heterostructure (SCH) laser diode with $Zn_{1-x}Mg_xS_ySe_{1-y}$ cladding layers were fabricated in several laboratories in 1993 (see the review by Nurmikko and Ishibashi [367]). A schematic structure of the edge emitting SCH laser diode and band gap of the constituent layers are shown in Fig. 7.9. 77 K pulsed laser operation of the

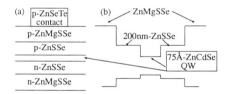

Figure 7.9: Schematic structure of (a) the ZnCdSe QW laser diode with ZnSSe wave-guiding layer and $Zn_{1-x}Mg_xS_ySe_{1-y}$ cladding layers and (b) bandgaps of the constituent layers (see Fig. 4 of Ref. [409]).

ZnCdSe/ZnSe/ZnMgSSe SCH diodes was reported by Sony group [509, 409] in 1995. The heterostructure consisted of 6 nm $Zn_{0.8}Cd_{0.2}$ quantum well, 120 nm ZnSe optical guiding layer, 60 nm $Zn_{0.93}Mg_{0.7}S_{0.1}Se_{0.9}$ inner cladding layer 500 nm $ZnS_{0.06}Se_{0.94}$ outer cladding layer and 20 nm p-ZnSe contact layer. The nitrogen doping in ZnSe was 1×10^{18} cm^{-3}. The TE-polarized laser line at 473.3 nm with a threshold of 0.9 kA/cm^2 was observed. With uncoated facets, pulsed blue laser-emission at 462.7 nm was observed from the SCH diodes mounted p-side down on a heat sink [409]. By reflection coating at the facets (70% at the front facet and 95% at the rear facet), CW operation at room temperature could be obtained. For green (523.5 nm) lasers the composition of the alloy layers was adjusted so that the bandgap of the ZnCdSe active layer was 2.54 eV and that of ZnMgSSe was 2.94 eV. For fabricating the blue lasers, the bandgaps of the active and cladding layers were set at 2.65 eV and 3.03 eV respectively. The stimulated CW emission was observed at 489.9 nm at room temperature. The threshold current was 1.4 kA/cm^2 at 6.3 V. For other similar lasers, the power output was 834 mW for pulsed operation and 30 mW for CW operation. The characteristic temperature T_0 was as high as 126 K.

Grillo et al [522] also used ZnMgSSe quaternary alloy for cladding layers in the lasers. By using ZnMgSSe alloy a confinement factor of 0.017 was obtained. Graded ZnSeTe layer was used for p contact which reduced the threshold voltage to 7 V. The lifetime of the laser under pulsed operation at room temperature was 1 h. Philips Research Laboratories reported pulsed operation of the ZnSe-based laser up to 394 K. Threshold current was measured as a function of temperature. The characteristic temperature near room temperature was $T_0 = 150$ K. At RT the pulsed peak power was 500 mW. The external quantum efficiency per facet was 0.56 W/A at 85 K and 0.446 W/A at 394 K. The combined dislocation and stacking fault density was 10^6 cm^{-2} and the lifetime at room temperature was 10 min.

Toda et al. [509] fabricated the first MOCVD grown laser. The laser consisted of a single ZnCdSe QW, ZnSe optical guiding layers, and ZnMgSSe/ZnSSe double stacked cladding layers. The structure was grown on n-type GaAs substrate. The laser operated under pulsed conditions at 77 K. The emission wavelength was 473 nm. The threshold current density varied in the range 0.9 to 1.8 kA/cm^2.

Itoh et al. [528] have reported considerably improved lifetime of the laser diode. The laser structure was grown on Si-doped (100) GaAs substrate by MBE. A Si-doped GaAs buffer layer was grown under As rich conditions. The structure of the gain-guided SCH laser diode was similar to that shown in Fig. 7.9. The epitaxial layer sequence was: 0.25 μm GaAs buffer, 30 nm n-ZnSe, 150 nm n-ZnSSe, 1 μm n-ZnMgSSe, 10 nm n-ZnSSe, ZnCdSe (35% Cd) QW, 10 nm p-ZnSSe, 1.5 μm p-ZnMgSSe, 1.6 μm p-ZnSSe, and 0.1 μm n-ZnSe [528]. For making p-contact, p-ZnSe/ZnTe MQW and 50 nm p-ZnTe layers were grown in the central area at the top. These layers were surrounded by an insulator layer to avoid current spreading. Finally Pd/Pt/Au metallization was used as the p-electrode. The light output versus current and voltage characteristics of the diode are shown in Fig. 7.10. The laser emitted at 514.7 nm. The threshold current under room temperature CW operation is 32 mA and the threshold current density is 533 A/cm^2 at 11 V. The threshold voltage is higher than the previously reported values. The main feature of this laser is the improved lifetime as discussed in section 7.2.4.2. Recently further improvements in the process technology have been made and lifetime of laser diode has increased to 389 h. The aging characteristics of this diode are also discussed in section 7.2.4.2.

7.2.4.2 Degradation and reliability

Short working lifetime of the ZnSe-based LDs prevents the commercialization of the blue-green ZnSe-based laser diodes [529]. Extensive work has been done

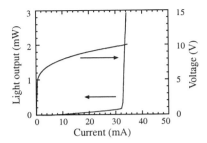

Figure 7.10: Light output power versus current and voltage of the laser diode in CW operation at room temperature [528].

by many groups world wide to understand the causes of the degradation and to improve the lifetime. Reviews of this work have been published by Ishibashi [409], by Nurmikko and Ishibashi [367] and by Gunshor et al. [529]. Two kinds of defects are observed in the degraded LDs. Arrow-head-like triangular dark regions and dark line defects. TEM studies show that the triangular dark regions are clusters of dislocations. The dislocations nucleate in the active region at the threading dislocations. The threading dislocations are produced at the stacking faults which exist at the interface in the as grown structures [409]. The main cause of the defects at the interface was identified to be the poor quality of the II-VI/GaAs substrate interface. The pre-existing defects propagate and multiply. They accelerate the degradation and reduce the lifetime of the laser diodes. Experiments show that the lifetime is inversely proportional to the pre-existing defect density. It was found that the interface quality can be improved substantially by introducing a GaAs epitaxial buffer layer between the substrate and the II-VI epitaxial layer [367, 409, 528]. The buffer layer reduces the concentration of pre-existing defects and improves the lifetime. The one hour lifetime mentioned earlier was obtained by improving the interface and reducing the defect density to $10^5/\text{cm}^2$.

The degradation rate is roughly proportional to I^{-n} where I is the operating current and n has values between 1 and 2. Since the operating current I is smaller in the LEDs, their lifetimes are larger than those of the LDs. Ishibashi [409] showed that the defect density $N(t)$ varies with time t (for which the devices have worked) as follows.

$$N(t) = \text{const.} t^\gamma. \tag{7.2}$$

If the defects in the active layer serve as the defect generation sites for additional defects (due to weak bond-length), value of γ is more than 1 and the increase of $N(t)$ is superlinear in time. In this case the degradation is catastrophic. On the other hand, if the degradation is by a diffusion like process, the value of γ is less than one and increase of $N(t)$ is sub-linear. Experiments showed that value of γ is less than 1. Ishibashi [409] concluded that the degradation is not due to weak bond-strength but due to pre-existing point defects. Both the pre-existing stacking faults and threading dislocations associated with them, and the point defects are responsible for the short lifetimes [529, 530, 367]. During laser operation the threading dislocations propagate and multiply in the active quantum well layer. Point defects act as non-radiative recombination centers. The heat generated by the recombination process helps in generation of additional defects [367, 530]. Kim et al. [411] have investigated etch-pit configuration and their relation to the defects in ZnCdSe QW laser structures using a new etchant and TEM. They found that each defect, stacking faults, threading dislocations or point defects, have specific etch-pit configuration.

The identification that pre-existing macroscopic defects, particularly that the stacking faults are responsible for the short lifetime of the laser diodes has motivated research leading to the reduction of pre-existing defects from 10^5-10^6 to 10^3-10^4 [367, 411, 528, 531]. The reduction in the macroscopic defect density is obtained by growing high quality GaAs buffer layer between the II-VI heterostructure, and by irradiating the GaAs surface with Zn before growing the ZnSe epilayer [528]. This led to the increase of lifetime of the laser diode to 101 hours under CW operation at room temperature in 1996 [528, 531, 532]. This is a significant improvement in the reliability. Further improvements in the lifetime have been made recently [137, 533]. Kato et al. [137] fabricated ZnCdSe/ZnSSe/ZnMgSSe single quantum well SCH laser diodes. The authors studied the effect of different values of VI/II ratio used during the growth of the ZnCdSe active layer. The active layers were grown with several values of the VI/II ratio in the range 0.8 to 3.5. The threshold current density was 431 A/cm^2, the operating voltage was 5.3 V and the emission peak was at 514 nm. These characteristics were independent of the VI/II ratio used during the growth of the active layer. However the lifetime depended on the ratio used. It increased as the ratio increased from 0.8 to 2.2, was maximum at 2.2 and then started decreasing. From these results it can be concluded that the point defects responsible for the degradation of the device are related to Zn rich conditions, i.e. Se vacancies or Zn interstitials. The aging characteristic of the diode (active layer grown with VI/II ratio = 2.2) is shown in Fig. 7.11. The figure shows that the lifetime under CW operation increased to an all time high value of about 400 hours at room temperature. The same group studied [533] the effect of the stripe-width on the operating voltage of a similar diode. The active layer was again grown in VI-rich conditions. There was a substantial decrease in both the threshold current and the operating voltage when the stripe-width was increased from 3 μm to 20 μm. The minimum operating voltage for 20 μm stripe width was 3.5 V and threshold current density was 310 A/cm^2. The lifetime of the laser diode was 300 h at 4 V.

It has been suggested that the laser performance can be improved significantly if the laser diodes are fabricated on ZnSe substrates [534, 535]. Recently Nakanishi et al. [534] have fabricated SCH ZnCdSe/ZnSe/ZnMgSSe laser structures on conductive ZnSe substrates by molecular beam epitaxy. Continuous-wave laser operation at room temperature was achieved at a wavelength of 527.9 nm (2.349 eV). The laser operated at a very low threshold current density of 222 A/cm^2 and threshold voltage of 5.4 V. However the lifetime was only 74 s

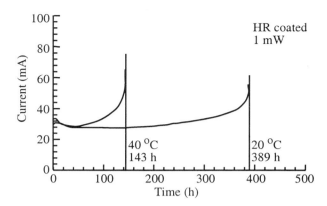

Figure 7.11: Aging characteristics of the laser diode under CW operation at 20°C and 40°C respectively and under constant 1 mW output power [137].

at a constant light output power of 2 mW. Wenisch et al. [535] have also fabricated SCH ZnCdSe/ZnSe/MgZnSSe laser diodes on Al doped 700 μm thick ZnSe substrates by MBE. Stable room temperature operation of planar green laser diodes at 512 nm was achieved on ZnSe substrates, which are conductive in the first 200 μm from the top due to post-growth aluminum doping. The threshold current density was 900-1000 A/cm^2 and the operating voltage was 14 V. The lifetime in pulsed operation exceeded 1 h. In an effort to reduce the stacking faults that originate at the II-VI/III-V interface, Albert et al. [536] fabricated MgZnSSe/ZnSSe/CdZnSe-quantum well laser diodes on Te terminated GaAs buffer layers (grown on GaAs) as substrates. They also fabricated the diodes using the conventional Zn terminated substrate surfaces. They obtained a low threshold current density of 350 A/cm^2 for a gain guided laser structure. The degradation studies show that during the first 4 hours the degradation occurs by an increase in the width of the dark spots. Later the degradation occurs by darkening of the areas between the dark spots. The output power was significantly larger in these diodes than in the diodes grown on Zn terminated substrates. However the lifetime of the lasers did not increase beyond several hours. Faschinger et al. [537] have also shown that quality of ZnSe epilayers improves substantially if the layers are grown on Te terminated GaAs.

Another route to avoid degradation and increase lifetime is to use alloys containing Be. Both Cd and Be harden the lattice and should make the layer resistant to defect creation. The bond in Be-chalcogenides is not as ionic as in other II-VI semiconductors. Landwehr et al. [] have plotted the bond energy (cohesive energy per bond) of II-VI and III-Nitride semiconductors (see Fig. 1.4 in chapter 1). The bond energies of Be-chalcogenides are in general much larger than those of other II-VI semiconductors. In some cases they approach the values as high as those for III-Nitrides. Be-chalcogenides should be stable against the creation of defects because of larger covalency in the bond and larger bond energy.

7.2. ZNSE-BASED LIGHT EMITTERS AND OTHER DEVICES

The first laser diode using Be containing ZnSe based semiconductors was reported by Waag et al. [138]. The commonly used structure with a top p contact and n-doped (100) GaAs substrate was used. Normally used ZnTe-ZnSe contact is not lattice matched to ZnSe and strain can create misfit defects. The contact to the top p-cladding layer was made using lattice-matched graded BeTe-ZnSe layer. The active region consisted of 7 nm BeZnCdSe layer. For the composition of the active layer used the emission wavelength was at 530 nm at room temperature and 508 nm at 77 K. For cladding layers the quaternary BeMgZnSe and for waveguide layers the ternary BeZnSe were used. The room temperature bandgap of both layers was 2.9 eV. Be instead of S (normally ZnMgSSe alloys are used for cladding layers) provides better control of composition and lattice constant of the epilayers. For 77 K pulsed operation, the threshold current was 80 mA (240 A/cm^2). At higher temperatures the threshold current increased rapidly, at 150 K it was 500 mA (1.5 k/cm^2). The reason for this rapid increase of the threshold current is not clear. The operating voltage was high, 20 V. The series resistance was high because the p-contact was not optimized, it was only lightly doped p-type. Output power of more than 35 mW per facet was obtained. Lasers containing Be compounds have been fabricated by other workers [406, 510, and references given therein]. More recently a lifetime of 57 hours of CW operation at room temperature of Be containing laser diode has been demonstrated (unpublished work of Miller, Baude, Hauger, Haase and Grillo, quoted in Ref. [406]). Legge et al. [538] have fabricated a ridge waveguide SCH laser using Be containing II-VI alloys and grown on n-type (100) GaAs substrate. The active layer was a 4 nm thick $Zn_{0.65}Cd_{0.35}Se$ single layer. 10/15 $ZnSe/Be_{0.04}Zn_{0.0.96}Se$ superlattices embedded in $Be_{0.06}Mg_{0.06}Zn_{0.88}Se$ layers were used as waveguides. Fabry-Perot laser structures with cleaved facets were realized. The smallest stripe width was 1 μm and the lengths were between 400 and 800 μm. The output characteristics of a laser diode with 1.5 μm wide stripe were discussed in detail. The laser was operated with 2 μs pulsed current injection and with 0.05% duty cycle. The emission spectrum peaked at 521 nm. The threshold current was 15 mA, differential quantum efficiency was 0.021 and the power output was up to 10 mW per uncoated facet. The laser operated up to 371 K. For a wider stripe-width, the lasing was observed up to 413 K. Both the near field and far field patterns were recorded. For stripe widths < 3 μm, lateral single mode emission was obtained. These initial studies show that Be-chalcogenides can be used in laser structures with reasonable performance. The motivation for using Be-chalcogenides was to improve long term reliability and working lifetime of the lasers. No significant increase in lifetime has yet been demonstrated. However the high temperature performance of the lasers [538] discussed above is impressive.

The lifetime of about 400 hours of the ZnSe-baser laser diodes achieved by the Sony group [137] is the highest at the time of this writing (December 1998). However the lifetime of a few hundred hours is not sufficiently large, a value of some 10,000 hours is required for commercialization of the devices. Considerable work will have to be done before the II-VI laser diodes can reach the market place.

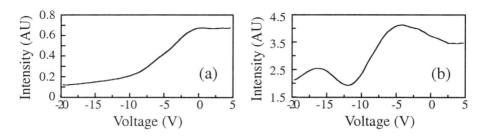

Figure 7.12: Transmission versus voltage for the planar waveguide at (a) 490 nm and (b) at 514 nm ([457]).

7.3 Other II-VI semiconductor applications

7.3.1 Modulators and switches

Thompson et al. [457] investigated properties of ZnSe/ZnCdSe multiple quantum well p-i-n structures which are useful for fabricating modulators. The i region of the p-i-n consisted of the ZnSe/ZnCdSe multiple quantum wells surrounded on the two sides by n- and p-doped ZnSe. The extinction ratio is defined as the ratio of high transmission (at low voltage) and low transmission (at high reverse voltage). The on/off or extinction ratio for the lateral transmission was measured with argon ion laser light at 496, 501 and 514 nm [457]. The observed transmission at 496 nm with applied voltage U is shown in Fig. 7.12(a). The extinction ratio is 6. Similar result was obtained with 501 nm light but the extinction ratio was 4. The result with 514 nm light (below the bandgap) shown in Fig. 7.12(b) is different. In addition to a small change in electro-absorption, Fabry-Perot-like peaks are seen in the transmission versus voltage curve. This observation shows that in addition to the intensity modulation, a phase modulation also occurs. The net result is a combined effect of the electro-absorption change and a transverse electro-optic effect in which the refractive index changes with the applied voltage. Intensity modulation was also observed in a room temperature laser diode structure with ZnMgSSe cladding layers. The active layer was a single ZnCdSe quantum well. The extinction ratio at 501 nm was 9. This large ratio is the result of stronger confinement in this structure because of the quaternary cladding layers. Electro-optic effect was not observed in this structure. This result shows that monolithic integration of modulators and laser diodes is possible.

More recently Babucke et al. [458] reported fabrication and characteristics of ZnSe-based waveguide modulators and switches for the blue-green region of the spectrum using MBE. These devices are useful for the integrated optics in the short wavelength region of the spectrum. The devices were planar waveguide ZnSe/ZnMnSe heterostructures grown on n-doped GaAs substrates with a 0.5 μm ZnSe buffer layer. A 4 μm thick ZeSe waveguiding layer was sandwiched between two ZnMnSe cladding layers, each layer being 0.6 μm thick. Mn con-

7.3. OTHER II-VI SEMICONDUCTOR APPLICATIONS

centration was 23% in the top layer and 30% in the bottom layer. The cleaved structures were 1 to 5 mm long. AuTi Schottky contact at the top and AuGe n-ohmic contact at the bottom were used. The Schottky contact was used for applying an electric field parallel to the growth direction. Two oppositely directed fields were produced in the depletion layer, one by the Schottky contact and the other at the ZnSe/GaAs heterojunction. Since the ZnMnSe cladding layers were quite thick, the fields did not penetrate the waveguiding layer in the absence of an external applied field. External field was applied parallel to the growth direction. Optical measurements were made by focusing light from a dye laser or Xe lamp with a microscope perpendicularly on a (110) cleaved surface of the waveguiding layer. A second microscope objective was used to collect the transmitted light from the opposite facet. Because of the lattice mismatch between ZnSe and ZnMnSe layers the device had a built-in strain and optical anisotropy even in the absence of an electric field. The built-in strain makes the active layer birefringent and causes modulation of the intensity of transmitted light at $U = 0$, where U is the applied voltage. The birefringence spectra of a 1.2 mm device was plotted for values of applied voltage U in the range –50 V to 50 V. As expected the spectra (plotted as a function of photon energy in the range 2.4 to 2.6 eV) showed maxima and minima. The spectra shifted to higher photon energies for negative voltages and lower energies for positive voltage. The contrast ratio (ratio of transmissions with and without eternal field) was determined for different photon energies. The ratio was strongly dependent on the photon energy. Switching occurred in both directions, from dark to bright or from bright to dark. Maximum contrast ratio was 80:1 at a photon energy of 2.485 eV.

To use the structures described above as switches and modulators, the top electrode layer was replaced by a stripe electrodes. The AuTi stripes were 200 μm wide and the spaces between the tripe were 7 μm wide. A revere bias is applied between the AuGe bottom contact and the AuTi stripe electrodes at the top. The refractive index below the stripes is reduced and this causes lateral optical confinement of the TE-polarized waves in regions under the spaces. The effect is absent for the TM-polarized light. The authors demonstrated that this structure was capable of various modes of operation. One example is a spatial switch between adjacent channels. An electric field is applied between two top electrodes. Switching into another spatial channel was accomplished by reversing the polarity. The switching occurs as a result of asymmetry of the field under the two electrodes. The device exhibited large switching contrasts and short switching times. Schottky contact was found to be convenient and effective for applying the control voltage.

7.3.2 Optical detectors

Optical detectors with high efficiency in the visible range of spectrum are still not available. Bandgap of Si is in the infrared and is indirect. The quantum efficiency of the Si detectors in the blue and green is very low. To improve the efficiency of the Si detectors, the visible radiation is down-converted to infrared by using coatings on the back-side of the detector [537]. However the quantum efficiency is still low. Since the wide bandgap II-VI semiconductors have direct bandgap and the gap energy can be tuned over a wide range, they are ideally suited as detector materials in this spectral range. Faschinger et al, [537] fabricated and characterized PIN ZnSe photo-diodes using MBE. The n-doping was 1×10^{18} cm^{-3} and the p-doping was 5×10^{17} cm^{-3}. Contact to the p-layer consisted of a highly p-doped ZnTe layer and a semitransparent thin gold layer. The contact was not ohmic, there was a Schottky barrier. However since the photodiode works under reverse bias, the Schottky contact is forward biased and is not so much of a problem. The efficiency of the ZnSe detector is compared with that of the Si detector in Fig. 7.13. The efficiency of the Si detector is

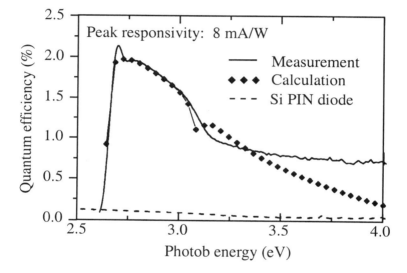

Figure 7.13: Measurement and calculation of the quantum efficiency of a ZnSe photodetector as a function of the photon energy. For comparison the efficiency of a commercially available Si detector is also plotted [537].

very low. At its energy gap the efficiency of ZnSe detector is 2%. The peak responsivity is 8 mA/W at 2.7 eV. The performance of the diode is not as good as expected. Model calculations of the spectral response were made and results were compared with the observed response as shown in Fig. 7.13. The calculated results agree well with the observed values at lower energies. To bring about this agreement, the diffusion current had to be neglected in the calculations. This result suggests that the intrinsic part of the diode is absent. Chlorine was diffusing into the intrinsic region making it n-type. $C - V$ measurements confirmed that this was indeed the case. The lifetime and the diffusion lengths of the minority charge carriers was also low. The performance can be improved by using better design and better quality materials. The performance of the device can be extended to longer wavelengths by using ZnCdSe alloys.

7.4. III-NITRIDE LIGHT EMITTING DIODES

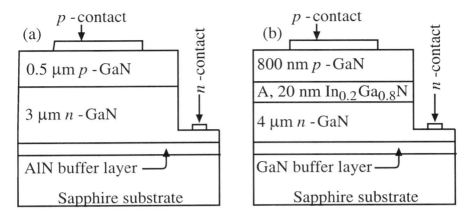

Figure 7.14: Structure of the first blue GaN p-n -based LEDs. (a) Structure of the first blue GaN p-n junction diode fabricated in 1989 [72]. (b) Structure of the DH GaN/InGaN blue diode. A is the active layer [77].

7.4 III-Nitride Light Emitting Diodes

7.4.1 Fabrication and performance

The first GaN p-n junction diode was fabricated by Amano et al. [72] after they discovered that Mg dopants can be activated as acceptors by LEEBI. The structure of the diode is shown in Fig. 7.14(a). The n-type layer was undoped 3 μm GaN film grown on AlN buffer/sapphire substrate. It had 2×10^{17} cm^{-3} electrons. 500 nm p-type layer was grown on the n-type layer with a doping of 2×10^{20} cm^{-3} Mg atoms. After the LEEBI treatment the hole concentration was 2×10^{16} cm^{-3}. The resistance of the p-layer changed from $> 10^8$ Ωcm to 35 Ωcm. The hole mobility was 8 cm^2/Vs. The EL of the LED showed two peaks, one in the UV region at 370 nm and the other in the blue region at 430 nm. Amano et al. [72] showed that the electrons injected into the p-region recombine with holes to produce the blue emission. The blue light originates in the Mg doped region. The near band-edge emission at 370 nm is emitted by the injection of holes into the n-region. The EL peak wavelengths and other performance characteristics of the LEDs are given in Table 7.2. Nakamura et al. reported similar structure LED with improved performance in 1991 [539]. LEDs with different hole and electron concentrations were fabricated. In a typical case, a Si-doped n-type GaN layer ($n = 5 \times 10^{18}$ cm^{-3}) was grown on GaN buffer deposited on a sapphire substrate. A Mg-doped p-type layer ($p = 5 \times 10^{18}$ cm^{-3}) was grown on the n-type layer. Its EL was at 430 nm. At higher current densities a weak UV EL at 390 nm was observed. A weak deep level emission at 550 nm was also observed. The UV and deep level emissions occur due to recombination in the n-type GaN layer [539]. At 20 mA the output power of GaN LED was 42 μW (see Table 7.2). The performance of the LEDs with lower

Table 7.2: Performance of GaN based LEDs in the visible range of the spectrum. The abbreviations used are: Cl for color, GN for GaN, AG for AlGaN, IG for InGaN, V for violet, B for blue, G for green and Y for yellow. η_e indicates external quantum efficiency and P is the output power at 20 mA.

Cl	λ (nm)	Semiconductor	P	η_e (%)	Ref.
V&B	370, 430	p-GN:n-GN	–	–	[72]
V&B	390, 430	p-GN:n-GN	42^a μW	0.18	[539]
B	440	GN/IG/GN-DH	125 μW	0.22	[77]
B	450	AG/IG/AG-DH^b	3 mW	5.4	[83]
B-G	500	AG/IG/AG-DH^c	1.2 mW	2.4	[540]
G	520	IG SQW/AG	3 mW	6.3	[541]
B,G,Y	d	AG/IG SQW/AG	d	d	[542]
V(B)^e	e	AG/IG SQW/IG/AG	5.6(4.8) mW	9.2(8.7)	[543]

^{a}For the blue color.
^{b}Co-doped with Zn and Si, $x = 0.06$.
^{c}Co-doped with Zn and Si, $x = 0.23$.
^{d}The color, P and η_e depend on the In content (see text).
^{e}In content in the QW layer was 0.09 for the V and 0.2 for the B diode. λ was 405 nm and 450 nm for the V and the B diodes. The numbers in the brackets are for the blue diode.

hole concentrations was inferior.

In 1993 Nakamura et al. [77] fabricated the first p-GaN/n-InGaN/n-GaN *double heterostructure* (DH) LED. The DH structure consisted of a 4 μm Si doped GaN layer, a 20 nm Si doped In$_{0.2}$Ga$_{0.8}$N layer and a 4.8 μm Mg-doped GaN layer (see Fig. 7.14(b)). Au and Al were used for contacts to p- and n-layers. The peak of the EL was in the blue region at 440 nm with FWHM = 20 nm. The EL was due to recombination of electrons and holes in the In$_{0.2}$Ga$_{0.8}$N layer. In 1994 Nakamura et al. [504, 540, 544] reported (DH) LEDs with In$_x$Ga$_{1-x}$N active layers and AlGaN barrier layers. The color of the EL of these diodes could be changed from blue to blue-green by changing the In content in the active layer. The active layer was co-doped with Zn and Si. Structure of a typical DH diode with $x = 0.06$ is shown in Fig. 7.15(a) [83]. Ni/Au contact to p-layer and Ti/Al contact to the n-layer were used. The power output and external quantum efficiency η_e of this DH LED with 20 mA forward current are 3 mW 5.4% respectively. The characteristics of this DH diode are given in Table 7.3. The EL spectra of the diode as a function of forward current are shown in Fig. 7.15(b). For application to traffic lights, a bright blue-green color at longer wavelengths is required. By using In content $x = 0.23$, EL at 500 nm at 20 mA forward current was obtained [540]. The power output of this diode was 1.2 mW and $\eta_e = 2.4\%$ (see Table 7.3). This was the first blue-green GaN based LED [540].

Further improvements in the performance of the LEDs were made by using

7.4. III-NITRIDE LIGHT EMITTING DIODES

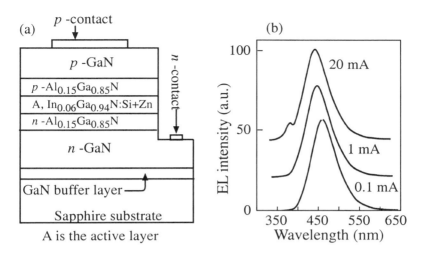

A is the active layer

Figure 7.15: (a) Structure of $Al_{0.15}Ga_{0.85}N/In_{0.06}Ga_{0.94}N/Al_{0.15}Ga_{0.85}N$ DH LED. The active InGaN layer is co-doped with Zn and Si. (b) EL spectra of the LED under different forward currents [83].

an InGaN *quantum well* as an active layer [542, 543]. The characteristics of a QW diode with one p-AlGaN barrier between the QW and the p-GaN layer [541] is shown in Table 7.2. The In content in the active layer was 45%. The color of the EL was green at 520 nm, the power output was 3 mW and η_e was 6.3% [541].

The structure of a QW diode with AlGaN barriers on both sides is shown in Fig. 7.16(a) [542]. The peak wavelength of the EL was changed by changing

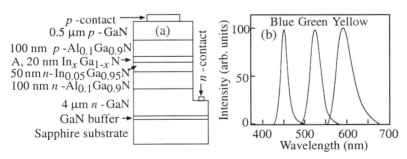

Figure 7.16: (a) Single $In_xGa_{1-x}N$ QW LED. A is the active layer. Light of blue, green or yellow color can be obtained by changing the value of indium in the range $0.2 < x < 0.7$. For green color $x = 0.43$. (b) Electroluminescence of 3 GaN based single quantum well LEDs at 20 mA forward current [542].

the In concentration x in the $In_xGa_{1-x}N$ QW layer. Typical EL spectra of the diodes with three values of x are shown in Fig. 7.16(b). The values of x for the three diodes were 0.2 for the blue, 0.43 for the green and 0.7 for the yellow diode respectively. The FWHM increases as the value of x increase presumably because the crystal becomes inferior with larger values of In concentration. The output power of the three LEDs is shown in Fig. 7.17(a) as a function of forward current. The power output increases linearly with the lower values of the forward

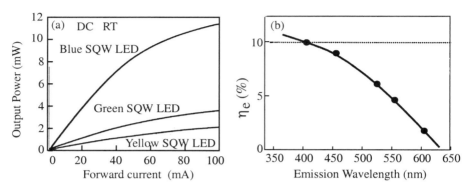

Figure 7.17: (a) The output power of blue, green and yellow SQW LEDs as a function of forward current [83]. (b) The external quantum efficiency as a function of the emission wavelength of InGaN SQW [83].

current. It tends to saturate at higher values of the current. The power output of the longer wavelength diodes with larger values of x is smaller. This is demonstrated clearly in Fig. 7.17(b) where η_e is plotted as a function of emission wavelength. Nakamura et al. [542, 83] suggested that this is due to the thermal and lattice mismatch between the active and barrier layers which gives rise to a large tensile stress in the active layer. The mismatch and the stress increase with the In concentration. Nakamura et al. [541] also noted that the peak wavelength of the $In_xGa_{1-x}N$ QW active layer is longer as compared to its value for the bulk $In_xGa_{1-x}N$ with the same value of x. For $x = 0.45$, the peak wavelength of the QW LED was at 520 nm whereas it was at 495 nm for the bulk alloy. It was felt that the tensile stress in the active layers is responsible for this shift [541] also.

In 1995 Nakamura et al. [543] also fabricated $In_xGa_{1-x}N$ SQW violet and blue LEDs with much better performance. The top layer was p-GaN followed by p-$Al_{0.3}Ga_{0.7}N$ and $In_xGa_{1-x}N$ QW active layer. On the other side of the QW active layer were the n-$In_{0.02}Ga_{0.98}N$, n-$Al_{0.3}Ga_{0.7}N$ and n-GaN layers. The value of In concentration x was 0.09 for the violet and 0.2 for the blue LED. The power output and efficiency values of the violet and blue diodes are given in the last row of Table 7.2.

Several other groups have also fabricated LEDs using III-Nitrides. Sun et al. [217] used low pressure (LP) MOVPE to grow Mg doped $In_{13}Ga_{87}N$ LED structure on (111) $MgAl_2O_4$ spinel substrates [217]. The active layer consisted of Mg-doped $In_{0.13}Ga_{0.87}N$ layer. The LEDs showed emission at 510 nm with a FWHM of 60 nm. At forward current of 20 mA, the output power was 200 μW

7.4. III-NITRIDE LIGHT EMITTING DIODES

Table 7.3: Performance of II-VI, III-V and SiC LEDs. Y-G is used for the yellowish-green color and other abbreviations used are the same as in Table 7.2.

Color	λ (nm)	Semiconductor	P	η_e (%)	Ref.
Blue	489	ZnSe-based	0.3 mW	–	[83]
B	–	SiC	7 μW	0.02	[545, 83]
Y-G	555	GaP	–	0.1	[546]
Y-G	570	AlInGaP	–	1	[546, 547]
G	512	ZnSe-based	1.3 mW	–	[181]

and the external quantum efficiency was 0.3%. This work demonstrated that high quality epilayers doped with Mg can be grown with LPMOVPE on (111) MgAl$_2$O$_4$ spinel substrates.

Recently LEDs have been fabricated using MBE. An LED consisting of a 3 μm GaN:Si ($n \sim 5 \times 10^{18}$ cm^{-3}) epilayer and a 0.7 μm GaN:Mg ($p \sim 3 \times 10^{17}$ cm^{-3}) epilayer was fabricated by Grandjean and Massies [225] using MBE. Circular mesas of 320 μm diameter were formed by dry etching. Ni/Au (300 Å/ 1000 Å) ohmic contacts were deposited by photolithographic technique on the n- and p-type layers. The RT electroluminescence peak was at 390 nm due to DAP recombination. The turn on voltage was 3 V. The reverse leakage current was very low.

Sun et al. [473] have studied the band edge emission from InGaN/GaN MQW light emitting diodes. Both spontaneous and stimulated emission were observed. The peak-wavelength of the stimulated emission shifts to higher energies as the well thickness decreases. The observed results were compared with the quantum-mechanical calculations of the sub-band energies and band offsets were calculated. In the case of GaN/In$_{0.13}$Ga$_{0.87}$N interface, the band discontinuities are 130–150 eV for the conduction band and 245–220 eV for the valence band. The ratio of the conduction to valence band discontinuity is 0.5–0.7 in agreement with the results obtained by photoemission studies [285].

7.4.2 Comparison with other LEDs

The values of η_e and other characteristics of LEDs based on other semiconductors are given in Table 7.3. Table 7.3 shows that LEDs with good power output in the wavelength region < 570 nm have not been fabricated using conventional III-V semiconductors. The power output of AlInGaP LEDs is good in the yellowish-green region at 570 nm. If the alloy composition is changed to obtain emission in the green region, 510 to 530 nm, the efficiency of the LEDs decreases to low values. Before the GaN devices were developed, blue LEDs could be fabricated using only SiC as no other stable semiconductor with sufficiently large bandgap was available. The power output of the SiC diodes is low as shown in Table 7.3.

A comparison of external quantum efficiencies η_e of commercially available

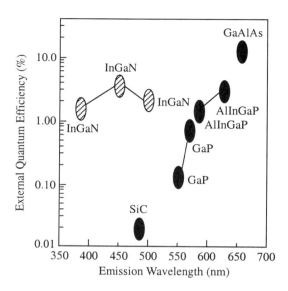

Figure 7.18: External quantum efficiencies as a function of the peak wavelength of various commercially available LEDs [83].

LEDs is shown in Fig. 7.18. This figure shows that there are no LEDs which emit with a peak wavelength below 550 nm with $\eta_e > 0.01$. This is the reason why the III-V Nitrides have become indispensable for making LEDs in the short wavelength region of the visible spectrum. II-VI semiconductor based LEDs have shown high efficiencies but their lifetimes are rather short and they are not available commercially. They have not reached the market place. The relatively short life is caused by the multiplication and propagation of defects in these materials [83].

7.5 GaN based Lasers

7.5.1 General remarks

The threshold values depend strongly on the quality of the sample and varies from sample to ample. In addition to the purity and structural quality, cavity length and edge facets play an important role in determining the threshold. It is difficult to form high quality facets in GaN grown on sapphire. Efforts have been made to improve the facets by mechanical polishing [548]. A reduction in the threshold due to polishing was observed. However mechanical polishing is a time consuming process. The difficulty of cleaving can be avoided by growing the nitride layers on spinel $MgAl_2O_4$ substrates [330].

Observations of Fabry-Perot cavity modes in stimulated emission have been reported [330]. It is difficult to calculate the mode spacing and interpret the observed spectra because the refractive indices of the nitrides are not known

7.5. GAN BASED LASERS

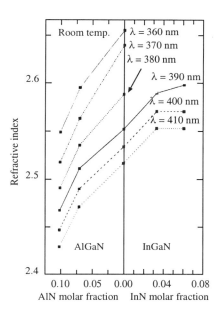

Figure 7.19: Values of refractive index of GaN and its alloys at different wavelengths [81].

accurately. Recently Akasaki and Amano [81] have reported the calculated values of the refractive indices. The values for wave vector of light perpendicular to the c axis are shown in Fig. 7.19. This figure shows that the refractive index decreases as the bandgap increases. This shows that both carrier and optical confinement can be obtained in the InGaN QW with wider-gap GaN or AlGaN barrier layers.

The density of states for both the conduction band and the valence band are large in GaN. As discussed earlier, the effective mass of electrons perpendicular to the c-axis is $\sim 0.22 m_0$. The hole effective mass in the top three Γ valence bands (shown in Fig. 5.3) has values 1.76, 1.76 and $0.16 m_0$ parallel to the c axis and 0.16, 1.61 and $0.14 m_0$ perpendicular to the c axis. The carrier density at which GaN becomes transparent at room temperature is 1.0×10^{19} cm^{-3}. It is 4 times and 2 times larger than the corresponding values for GaAs and ZnSe [81].

7.5.2 Optically pumped III-Nitride lasers

As mentioned earlier, numerous groups have reported fabrication and performance of optically pumped lasers [548]-[551]. Typical emission spectra of a GaN/AlGaN SCH laser is shown in Fig. 7.20. Individual spectra are vertically displaced for clarity. Observed spectral narrowing, superlinear increase of the emission intensity with exciting power, and suppression of broad emission are the signatures of the stimulated emission. The measured threshold for stimu-

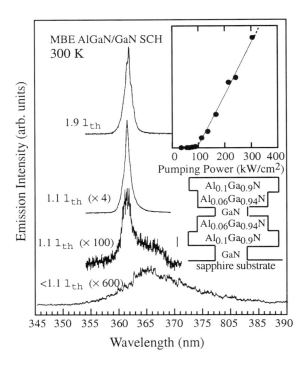

Figure 7.20: Emission spectra of optically pumped GaN/AlGaN SCH. Individual spectra are displaced vertically for clarity. Pumping power is denoted by l_{th} in kW/cm^2. The threshold pumping power is 90 kW/cm^2. Layer structure is also shown. Inset shows the emission intensity as a function of pumping power [548].

lated emission is $\sim$ 90 kW/cm^2. This laser did not have a cavity and interference modes were not observed. By attaching high reflectivity dielectric mirrors to the laser high contrast Fabry-Perot fringes were observed. In some cases very narrow single mode-like peaks of laser emission with a value $\sim$ 0.3 nm of FWHM were observed. The lowest measured threshold power remains at 27 kW/cm^2 achieved by Akasaki et al. [551, 81] in 1995.

7.5.3 III-Nitride laser diodes

The first short wavelength (490 nm) laser diode was made using ZnSe based II-VI semiconductors in 1991 [552]. However as discussed earlier the degradation problem of II-VI devices has not yet been solved. The maximum lifetime of the II-VI laser is still about 100 hours. The first III-Nitride injection laser was reported by Nakamura et al. [553] in 1996 and by Akasaki et al. [86] also in 1996. The laser diode fabricated by Akasaki et al. [86] consisted of a InGaN SQW as an active layer and emitted at 376.0 nm. This is the shortest wavelength

7.5. GAN BASED LASERS

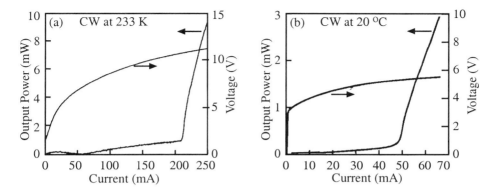

Figure 7.21: (a)The $I - V$ curve and output power per coated facet of a ridge geometry InGaN MQW violet laser with CW operation at 233 K [556, 82] (b) Same for an improved laser diode with CW RT operation lifetime of 100 hours [92].

laser diode fabricated so far. The laser diode fabricated by Nakamura et al. [553] consisted of multiple quantum wells of InGaN in the i-region capped on both sides by the GaAlN layers. The capping layers confined both the injected carriers and the emitted radiation in the active i-region. The i-region formed a wave guide in the edge-emitting Fabry-Perot laser structure. The mirror facets for the laser cavity were formed by etching the III-V nitride layers without cleaving. The pulsed threshold current density was 4 kA/cm^2. Resistance of the i-layer was high which required large heat dissipation. It was necessary to drive the laser with low duty cycle microsecond pulses. The wavelength of the emitted light was 417 nm. The laser produced a power output of 215 mW at 2.3 A at RT. Soon afterwards the lasers grown on (111) MgAl$_2$O$_4$ were reported [554, 555]. The emission wavelengths of these lasers were 410 and 417 nm. The laser action was confirmed from the power output versus current curves and $I - V$ curves and also from the TE-polarization. The performance of the laser diodes improved rapidly after these first reports. CW operation at 233 K was reported a few months later [82, 556]. High reflection facet coatings consisting of quarter-wave TiO$_2$/SiO$_2$ multilayers were used which reduced the threshold current. The power output and $I - V$ curves for this laser are shown in Fig. 7.21(a). The lasing started at a threshold current of 210 mA which corresponded to 8.7 kA/cm^2. The lifetime of this laser diode was 30 min [82]. The lifetime for RT CW operation of the device was $\sim$ 1 s. At the forward current equal to threshold current, a sharp peak at 413.1 nm with small side structure was observed. When the forward current was increased by 5%, the laser emission peak at 413.3 nm became dominant. The spectral width of the EL was 0.1 nm. With further increase of the forward current the side-structure disappeared and only a single laser emission peak at 413.8 nm remained.

The rapid pace of development of the laser diodes continued throughout 1997

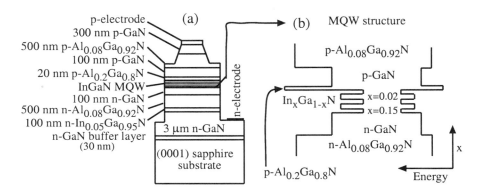

Figure 7.22: (a) The structure of the InGaN MQW laser diode. (b) Structure of the MQW and layers on both sides of the MQW is shown in details. The thickness of each of the three well layers was 3.5 nm and that of the barrier layers, 7nm. The figure is constructed based on Fig. 1 of Ref. [92].

and 1998. Room temperature CW operation of InGaN MQW laser diode was demonstrated in February 1997 [557]. The laser emission was at 400.23 nm. The lifetime of the laser was 24 to 40 min. One month later the lifetime for RT CW operation increased to 27 hours [558]. The threshold current density was 3.6 kA/cm^2. In June a laser diode with CW operation lifetime of 100 hours at RT was achieved [92]. The output power and $I-V$ characteristics of this diode are shown in Fig. 7.21(b) and the structure of this diode is shown in Fig. 7.22. Mg and Si were used for p- and n-type doping. The n- and p-type GaN layers were used for light-guiding layers and n- and p-type AlGaN cladding layers were used for confining the carriers. Ridge geometry was formed by etching as shown in the figure. The cavity length of this laser was 150 μm. The threshold current was 50 mA which corresponded to a threshold current density of 8.8 kA/cm^2. The operating voltage at the threshold current was 5.5 V, significantly lower than the values needed by earlier laser diodes [84]. The reduction of the operating voltage was obtained by optimizing the doping profile and improving the contacts. Also unlike the previous lasers, the InGaN layers in the MQW were doped with Si. The exact mechanism by which Si doping of the MQW layers improves the performance of the diodes is not known. The temperature of activation of the Mg acceptors was changed from 700 to 600°C. The annealing was done after evaporation of Ni-Au contacts. The low temperature annealing gave lower contact resistance. The lifetime of these diodes varied between 40 and 100 hundred hours.

Two important changes in the structure of the laser diode were made in the following months. (1) To avoid the problem of cracking of thick AlN layers, $Al_xGa_{1-x}N$/GaN modulation doped strained layer superlattices were used as cladding layers; and (2) The structure was grown on ELOG substrates (see section 2.5.2) [85, 214]. Except for these changes, the design of the diode was similar to that shown in Fig. 7.22. The lifetime test of the diode with these

7.5. GAN BASED LASERS

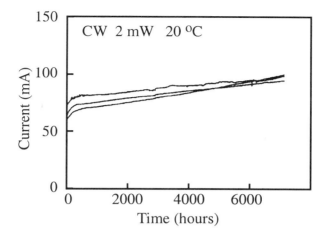

Figure 7.23: Operating current as a function of time under a constant output power of 2 mW per facet. The laser diodes were grown on the ELOG substrate and had modulation-doped AlGaN/GaN superlattices as the cladding layers [85].

modifications is shown in Fig. 7.23 where current for constant power output is plotted as a function of time. The laser diodes with this improved structure operated without significant degradation for 6000 hours. The extrapolated lifetime of the diode was more than 10,000 hours [85, 214]. Recently the life time has increased to $\sim 10,000$ hours and Nichia has started shipping engineering samples to customers (see the web site http://www1a.mesh.ne.jp.nichia/index-e.htm).

7.5.4 Mechanism of laser emission

The emission and PC spectrum of the laser diode (shown in Fig. 7.22) are shown in Fig. 7.24. There are three features in the PC spectrum, at 345 nm (3.59 eV), 360 nm (3.44 eV) and 381 nm (3.25 eV). These features are due to absorption in the AlGaN cladding layers, in the GaN light guiding and InGaN barrier layers and in the InGaN well layers respectively [92]. The laser emission from the well layers is at 399 nm (3.11 eV). Thus emission line is situated at the low energy tail of the exciting spectrum. The Stokes shift is quite large, 140 meV. This suggests that laser emission occurs at regions of local high In concentration. These regions act as quantum dots.

The delay time of the laser emission was measured as a function of the operating current [92]. The lifetime of the minority carriers was found to be 3.5 ns from these measurements. The threshold carrier density was $\sim 1.8 \times 10^{20}/cm^3$, a rather large value [92]. As mentioned earlier, the reason for this large value is probably the large density of states because the effective masses are high. However the density of states in a quantum dot should be small and lasing should occur at lower values of carrier concentration. The differential

242   CHAPTER 7. STRAINED LAYER OPTOELECTRONIC DEVICES

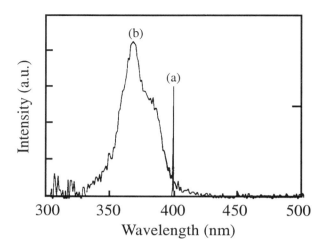

Figure 7.24: (a) Emission and (b) PC spectra of the InGaN MQW laser diodes [92].

gain coefficient was 5.8×10^{-17} cm^2 [92]. One would normally expect a higher differential gain from quantum dot lasers.

Fig. 7.25 shows the emission spectra of the laser diodes at two different operating currents at RT. The cavity length of this laser diode was 150 μm. It was assumed that the different closely spaced peaks at lower operating currents arise due to different modes of the diode. At 53 mA operating current a narrow lasing emission at 400.2 nm (3.098 eV) is obtained. The mode separation of the spectrum in Fig. 7.25(b) is 0.125 nm which gives a value of the effective refractive index as 4.3 [92]. The values of refractive index given in Fig. 7.19 are smaller.

7.5.5   Comparison of ZnSe- and GaN-based lasers

Both II-VI and III-Nitrides semiconductors support the heterostructure technology. ZnSe is nearly lattice matched to GaAs and therefore high quality epilayers of ZnSe and ZnSe-based alloys can be grown on GaAs substrates. Because of large lattice mismatch between GaN and substrates the quality of the layers is not as good. Pseudomorphic or lattice-matched ternary and quaternary alloys make it possible to fabricate quantum wells with sufficient optical and carrier confinement in both cases. It is not difficult to dope ZnSe and related alloys n-type. Cl forms shallow donors in these materials. Similarly GaN can be easily doped n-type with Si. It is more difficult to dope both ZnSe and GaN p-type. The p-type doping of ZnSe is achieved using plasma assisted nitrogen radicals. For the p-type doping of GaN Mg is used. Mg is activated either by low energy electron beam irradiation or by high temperature annealing. Making ohmic contacts is difficult for both p-type ZnSe and GaN. Good p-contacts are obtained

7.5. GAN BASED LASERS

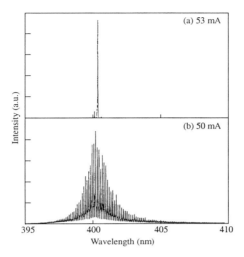

Figure 7.25: Emission spectra of CW InGaN MQW laser diode with two different operating currents at RT [92].

by using a heavily nitrogen doped ZnTe layer. For best contacts a compositionally graded layer from p-type ZnSe to heavily doped p-type ZnTe layer is used. Contacts to p-GaN continue to be leaky high resistance Schottky contacts. The first ZnSe based laser was reported by 1991 [133]. The firsts GaN laser diode was reported in 1996 [86, 553]. The degradation problem in ZnSe devices could not be solved and maximum life time achieved is only 400 h. The GaN devices have shown 10,000 h working life time and have been commercialized. Efforts are being made to increase the lifetime of ZnSe devices by using Be-chalcogenide based semiconductors. Be-chalcogenides have large bond-strengths. Extensive work is being done on these chalcogenides with the hope that long lifetime of the lasers will be achieved [138, 408]. High initial performance of the II-VI lasers is not difficult to obtain. Typically, the threshold current is 6.9 mA, the threshold voltage is as low as 3.3 V and the output power is 87 mW in CW room temperature operation. Lasing emission at wavelength of 485 nm has been reported [560]. However high performance mentioned above is obtained only at wavelengths in the range 500-540 nm [560]. The shortest emission wavelength of GaN based laser diode is 376 nm [81]. The emission wavelength of the GaN laser is close to 400 nm [85].

7.5.6 Applications of LEDs and LDs

Important applications of mid-IR lasers are remote sensing and pollution monitoring, short-link high bandwidth optical communication, sensing trace gases, night vision devices, research and medical diagnostics, and IR counter measures for defense applications. There is a large demand of LEDs and LDs emitting in the short wavelength region (i.e. in the green to ultraviolet region) of the spectrum. Violet, blue and green light emitting diodes (LEDs) are required for full color display, laser printers, high-density information storage and for under-water optical communication. Aerial packing density of information in a compact disc is inversely proportional to the square of the wavelength of the laser light [62]. By reducing the laser wavelength to 360 nm the information that can be stored in a compact disc can be increased by a factor ~ 4 [62] as compared to the present technology. Short-wavelength sources will enable us to produce compact and high density optical storage system and a high speed laser printing system. They will be very useful in designing and developing instruments for medical diagnostic purposes [80]. All solid state full color flat-panel display system can be produced using these sources. Full color display using LEDs was demonstrated by the Akami Electric Company Ltd. in Japan in 1996 [82]. For red color, GaAlAs LEDs were used. Successful use of LEDs for traffic lights has also been demonstrated [83]. The saving of electric power by replacing incandescent lamps with the LEDs will be enormous.

Chapter 8

Transistors

8.1 InGaAs transistors

8.1.1 Early work

The demand for high frequency wireless data transfer is increasing very rapidly. Some of the present and projected requirements are [24]: Mobile telephones (900 MHz and 1.8 GHz), Communication through low orbit satellite systems (20, 30 and 60 GHz), Automotive industry (collision avoidance radar work at ~ 77 GHz), High resolution radar for airport surveillance (planned to work at 94 GHz) and space observation systems (140 GHz). GaAs based MESFETs (Metal-Semiconductor FETs) and HEMTs or MODFETs (High Electron Mobility Transistors or Modulation Doped Field Effect Transistors) are the key components for high frequency data transfer systems. High frequency, low power consumption and excellent noise performance are important features of GaAs based FETs. With the advent of strained layer epitaxy, HEMTs containing strained pseudomorphic InGaAs active layers were developed. These transistors, designated as pseudomorphic or PM-HEMTs, are fabricated on GaAs substrates. InGaAs alloys with high In concentration have considerably higher mobility and larger intervalley separation in the conduction band than GaAs. For several years InGaAs strained layer devices dominated the HEMT technology [12]. The main limitation of this technology is that a large fraction of indium can not be built in the layers. If indium concentration is large critical thickness becomes very small. For best performance 30 to 40% mole fraction of InAs is required in InGaAs layers. An alternative technology based on InP substrates uses lattice matched $In_{0.53}Ga_{0.47}As$ layers. An InGaAs layer with 53% indium is lattice matched to InP. The HEMTs based on InP substrates show higher operating frequencies because of the increased mobility and conduction band discontinuity. Fig. 8.1 shows the evolution of the values of cut-off frequency f_t of both GaAs-substrate and InP-substrate devices. The lowest frequency data point in Fig. 8.1 is for an AlGaAs/GaAs HEMT. The top two points are for the transistors with more than 53% In in the active layers. These

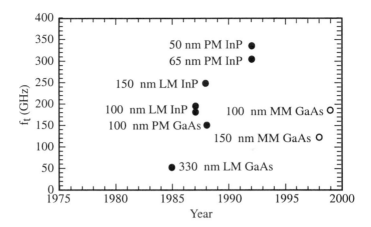

Figure 8.1: Evolution of reported values of cut-off frequencies f_t. The data shown by filled symbols is taken from Ref. [24]. 100 nm MM GaAs data point is from Ref. [561] and 150 nm MM GaAs point is from Ref. [562].

transistors are pseudomorphic grown on InP substrates. Fig. 8.1 shows how the frequency has continuously increased starting from the GaAs active layer and going to pseudomorphic (PM) InGaAs active layer on GaAs substrate, InGaAs (53% In) active layer lattice matched to InP substrate and finally pseudomorphic InGaAs (In content > 0.53) active layer with InP substrates. The highest frequency has been obtained with transistors grown on InP substrates. However GaAs substrate has several advantages over InP substrate. GaAs is less fragile. Large wafers of GaAs (up to six inch diameter) are commercially available at a reasonable price. If InP substrate is used, mole fraction of In in the active layer can not be changed if lattice matching is to be preserved. For best performance 30 to 40% mole fraction of InAs in InGaAs is required. It is therefore important to develop a technology which can produce HEMTs on GaAs substrates and perform as well as those fabricated on the InP substrates. A new structure called the lattice-matched metamorphic HEMT or LM-MM-HEMT has been developed to meet these requirements. Schematic structure of a typical MM-HEMT is shown in Fig. 8.2. In the LM-MM-HEMTs the concentration of In can be varied and yet strain free lattice matched structures can be grown on GaAs substrates. MM-HEMTs with 30% to 50% In have been studied by several groups for low noise and power applications. Large values of conduction band offsets in this system improve the maximum current density and reduce the gate leakage current due to holes generated in the channel by impact ionization. The effects are small because of a relatively large bandgap of InGaAs (~ 1 eV). Similarly large bandgap (~ 2 eV) of the InAlAs layers improves the turn-on voltage [562]. Breakdown voltage generally limits the performance of the power devices. Large bandgaps of both the InGaAs and the InAlAs layers improve the breakdown voltage.

8.1. INGAAS TRANSISTORS

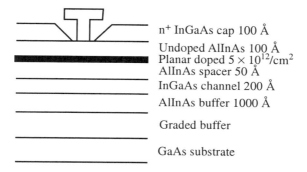

Figure 8.2: Schematic cross section of MM-HEMT investigated by Happy et al. [564].

8.1.2 Recent work

Extensive work is being done on the LM-MM-HEMTs at the present time [561]-[563] The heterostructure HEMT is shown in Fig. 8.2. The structure consists of $In_yAl_{1-y}As/In_xGa_{1-x}As$ layers. Values of x and y are so chosen that both layers are lattice matched to top layer of the graded relaxed buffer layer and are unstrained. The graded buffer consists of $In_zAl_{1-z}As$ or $In_zGa_{1-z}As$ layers where $0 \leq z \leq x$ [564]. The in-plane lattice constant of the top layer of the buffer is lattice matched to the epilayers constituting the transistor. The graded buffer is grown at low temperatures. Under optimized growth conditions concentration of the threading dislocations in the top part of the buffer is low.

Happy et al. [564] have analyzed by Monte Carlo method the effect of In concentration on the performance of the device. The active layers were 200 Å InGaAs layers. The gate length was $L_g = 0.25$ μm for all devices. Calculations were made for several values of Indium mole-fraction in the range 0.3 to 0.6. Transport properties were studied in detail. The calculations were made assuming a quasi-2D channel since at room temperature quantum-size effects in the 200 Å channel are not significant. Charge control law (relation between sheet carrier density in a layer and applied voltage) was calculated by self-consistent solutions of Schrödinger's and Poisson's equations. The modeling of the dc, ac, noise and high frequency performance of a device with 0.25 μm gate length was performed using these solutions. The calculated values of the low field 300 K mobility in the InGaAs layers are plotted in Fig. 8.3. Experimental values quoted by Happy et al. [564] and those recently measured by Cordier et al [561] are shown in the figure. 77 K data of Cordier et al. [561] are also included. Fig. 8.3 shows that mobility increases as InAs mole fraction increases. The increase in mobility is due to decrease in the effective electron mass and increase in the intervalley separation $\Delta_{\varepsilon\Gamma L}$. At a given gate voltage the maximum sheet carrier density in the $In_xGa_{1-x}As$ channel decreases slowly as x increases. This decrease is associated with the decrease in the conduction band offset. The values of conduction band offset used in the calculations were 0.78 eV, 0.66 eV,

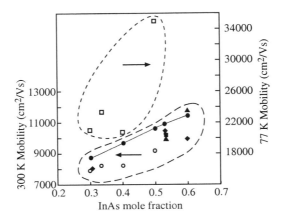

Figure 8.3: 77 K and 300 K low field mobility versus InAs mole fraction in $In_xGa_{1-x}As$. Solid curve with filled circles show the Monte Carlo values calculated by Happy [564]. Filled diamonds, filled squares and filled triangles are the literature values of experimental mobility quoted by Happy [564]. Open circles (300 K) and open squares (77 K) show the mobilities measured recently by Cordier [561].

0.51 eV and 0.44 eV for $x = 0.3$, 0.4, 0.53, and 0.6 respectively. In their experimental investigations Cordier et al. [561] also found that the sheet carrier density in the channel is not very sensitive to the In mole fraction. Calculated $C - V$ plots [564] show that at low x there is a large plateau in the values of C which improves the linearity of the device. These results show that a trade off between the transport properties and high carrier density is necessary for optimizing the device performance. Calculated dc drain current I_{dc} and extrinsic transconductance G_m are plotted in Fig. 8.4. As x decreases the maximum dc drain current increases. This increase correlates with the increase of the carrier density mentioned earlier. The transconductance G_m has a maximum value of about 1.3 S/mm. G_m is also not very sensitive to the indium concentration x. Calculated values of the extrinsic current gain cut off frequency f_t and maximum oscillation frequency f_{max} are given in Table 8.1. This Table shows that

Table 8.1: Cut-off and maximum oscillation frequencies of the 0.25-μm gate length MM-HEMT [564]. $I_{ds} = 50$ mA/mm and $V_{ds} = 1.5$ V.

In (%)	f_T GHz	f_{max} GHz
30	94	246
40	98	261
53	100	268
60	104	285

both f_t and f_{max} increase with In concentration. The ratio f_t/f_{max} remains

8.1. INGAAS TRANSISTORS

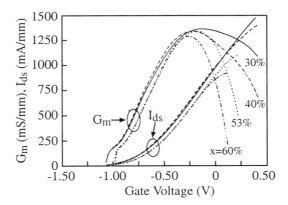

Figure 8.4: Calculated values of extrinsic transconductance G_M and source-drain current I_{ds} for $V_{ds} = 1.5$ V are shown as a function of gate voltage [564].

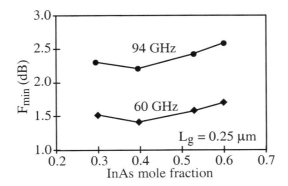

Figure 8.5: Minimum noise figure F_{min} versus InAs mole fraction for $I_{ds} = 50$ mA/mm and $V_{ds} = 1.5$ V [564].

close to 2.65. Since the gate length and extrinsic parasitics are the same for all devices, the improvement in the frequencies must be related to the increase in mobility shown in Fig. 8.3. Happy et al [564] calculated and compared the performance of 0.1-μm gate length MM-HEMT and LM-HEMT transistors. For $x = 0.60$ in the active layer of the MM-HEMT the values $f_{max} = 500$ GHz and $f_T = 180$ GHz were obtained. The frequency values for the LM-HEMT were similar.

Minimum noise figure F_{min} of the 0.25 μm MM-HEMTs were calculated for several different values of x (see Fig. 8.5). F_{min} was calculated at two frequencies, 60 GHz and 94 GHz. For $x = 0.3$ F_{min} was 1.5 dB at 60 GHz and 2.3 at 94 GHz. These values decreased to 1.4 dB and 2.2 dB respectively at $x = 0.4$. The values started increasing as x increased further.

Results of several experimental studies [561, 563, and references given there

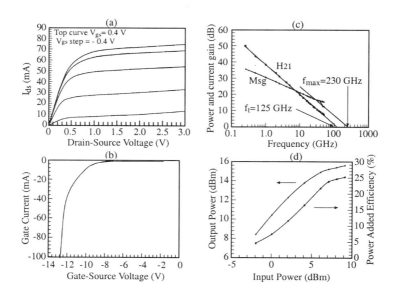

Figure 8.6: (a) $I-V$ characteristics of a 0.15×100 μm transistor. (b) Gate diode characteristic of the same device, (c) Power and current gain against frequency for the same transistor at $V_{ds} = 1.5$ V and $V_{gs} = -0.8$ V, and (d) Output power and Power Added Efficiency (PAE) versus input power at 60 GHz of a 0.2×150 μm transistor. The figures are taken from Ref. [562].

in] are consistent with the numerical calculations of Happy et al. [564]. Zaknoune et al. [562] fabricated and studied a $In_{0.3}Al_{0.7}As/In_{0.3}Ga_{0.7}As$ MM-HEMT grown on GaAs substrate. The gate length of the transistor was 0.15 μm. The characteristics of the transistor are shown in Fig. 8.6. The maximum drain current was 750 mA/mm at a gate-source voltage of +0.4 V. The extrinsic transconductance was 660 mS/mm and the output conductance had a low value of 30 mS/mm. The forward turn-on voltage was 0.9 V and gate breakdown voltage was a high -13 V (measured at 1 mA/mm). The values of cut-off frequency $f_t = 125$ GHz and maximum oscillation frequency $f_{max} = 230$ GHz were determined. Power measurements were made at 60 GHz. An output power of 240 mW/mm with 6.4 dB power gain and a power added efficiency of 25% were obtained [562]. A significant improvement in the breakdown voltage is obtained by reducing the In content from 50% to 30%.

Cordier et al. [561] made extensive experimental investigations of the In-GaAs/InAlAs MM-HEMTs. The Schottky barrier height of Ti/Pt/Au contact on InAlAs was 1.3 V. The effect of In content on the source-drain breakdown voltage is shown in Fig. 8.7. In a 0.1 μm MM-HEMT the values of f_t were 160, 195, and 187 GHz for $x = 0.33$, 0.40 and 0.50 respectively. The lower value of f_t in the $x = 0.5$ device was due to parasitic elements. The value of f_{max} was 400 GHz for $x = 0.33$ device.

The development of InAlAs/InGaAs high electron mobility transistors on

8.2. II-VI SEMICONDUCTOR TRANSISTORS

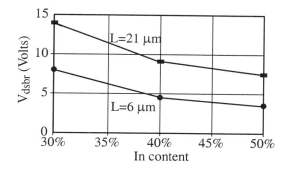

Figure 8.7: Dependence of source drain breakdown voltage on indium content in InAlAs/InGaAs HEMTs for contact distance $L = 6$ mum and 21 μm [561].

high-quality GaAs substrates (metamorphic HEMTs) is of primary interest for high frequency devices. The work described in this section shows that an InAs mole fraction of 0.30 to 0.40 is the optimum composition range for manufacturing high gain low noise MM-HEMTs on GaAs substrates. In this range of composition and for the dimensions investigated, the performance of MM-HEMT structures is similar to that obtained for lattice-matched HEMTs on InP substrates [563, 564]. Recently Van der Zanden et al. [563] have fabricated the first metamorphic microwave monolithic integrated circuit. The circuit is a coplanar Q-band high gain amplifier. The amplifier used a single-stage dual-gate MM-HEMT. The small signal gain at 40.5 GHz was 17.5 dB.

8.2 II-VI semiconductor transistors

The first Field Effect Transistor (FET) using ZnSe was fabricated by Studtmann et al. [565] in 1988. It was a ZnSe/n-GaAs heterostructure FET. An undoped 100 nm ZnSe layer was deposited on a 400 nm n-type GaAs layer. The n-type GaAs layer was deposited on a 1.5 μm GaAs buffer layer/(100) semi-insulating GaAs substrate. Two samples with different dimensions were fabricated. Gate width/length of the two samples were 45 μm/45 μm and 45 μm/2 μm. The transistors were isolated by wet chemical etching. Source and drain were thermally evaporated 88% Au-12% Ge alloys. The alloy layers were 300 nm thick and annealed at 450°C for 90 s. The gate consisted of 300 nm Al layer. The source, drain and gate were defined by lift off technology. Characteristics of both the long channel (45 μm) and short channel (2 μm) devices were measured. The long channel transistor showed depletion mode characteristics with complete pinch off and current saturation. The breakdown between the gate and the drain was at about 42 V. The transconductance at zero gate voltage was 3.5 mS/mm. After correction for series resistance this value became 5.1 mS/mm. The transconductance of the short channel device was 6.5 mS/mm without correction for series resistance and 13.5 mS/mm after the correction was applied. The transconductance was independent of the gate voltage. Lee et al. [566] fabricated a depletion mode Au/ZnSe/InP heterojunction gate field-effect transistor by MOCVD. The current-voltage and other characteristics of the transistor were measured. Gate leakage current was small and breakdown voltage was high. For a 2 μm gate-length a transconductance of 9.5 mS/mm was obtained.

Wang et al. [567] reported fabrication and characteristics of a depletion mode metal-semiconductor field effect transistor (MESFET) fabricated using Cl-doped $ZnS_{0.07}Se_{0.93}$ active layer. $ZnS_{0.07}Se_{0.93}$ alloy is lattice matched to GaAs. Therefore the transistor was grown directly on semi-insulating GaAs substrate; no buffer layer was necessary. The thickness of the $ZnS_{0.07}Se_{0.93}$ layer was 400 nm and the doping level was in the $10^{16} - 10^{17}$ cm^{-3} range. Mesa structure was used to isolate the devices. Cr/In/Cr was used to form the source and drain ohmic contacts. The source and drain contacts were annealed at 380°C in N_2 for 18 s. Au Schottky contact was used as gate. The gate was formed by recess etching and self-aligning techniques. Source, drain and gate were defined by lift-off technology. Values of gate width to length ratio were $W(\mu m)/L(\mu m) = 200/20, 200/4$ and $200/2$ respectively. For the 2 μm device, the turn-on voltage was 1.75 V, pinch-off voltage was 13 V and transconductance was 8.73 mS/mm. The breakdown voltage without a gate-drain bias was 28 V. Accurate modeling of the devices was difficult because reliable values the mobility were not available. The series resistance was probably high because the ohmic contacts were not of low resistance.

Before concluding the discussion of II-VI FETs, mention should be made of the use of II-VI layers to passivate III-V semiconductor layers which allows fabrication of improved III-V FETs. Dauplaise et al. [568] deposited thin CdS layers on InP before depositing an insulator to fabricate metal-insulator-semiconductor transistors. In the insulator/InP structures, the interface-state density was reduced from 5×10^{11} to 5×10^{10} eV^{-1}cm^{-2} by inserting the thin CdS layers at the interface. The InP did not become P deficient after the deposition of the insulator. Depletion mode MIS transistors made with a CdS treated interface showed superior characteristics as compared to the transistors fabricated without the CdS treatment.

Heterostructure Bipolar Transistors (HBTs) using ZnSe have also been fabricated. Tseng et al. [569] fabricated ZnSe/Ge/GaAs and ZnSe/InGaAs/GaAs HBTs. The lattice mismatch between different layers and the substrate was small. The structures were grown by MOCVD. The valence band offset in both ZnSe/Ge/GaAs and ZnSe/InGaAs/GaAs is large. As compared with the AlGaAs/GaAs HBT, the transistors showed lower base resistance and lower contact resistance. S-parameter measurements were made at 77 K. The values of unity-gain cut-off frequency for the ZnSe/Ge/GaAs HBT is $f_T = 6$ GHz and of the maximum oscillation frequency it is $f_{max} = 13$ GHz at a collector current density of 1.8 kA/cm^2. These transistors will be useful for application in high-speed digital circuits.

The ZnSe/Si blue light-sensitive bipolar transistor has a high-potential for application in photo-receivers that require short-wavelength detection. Recently

8.3. III-NITRIDE BASED TRANSISTORS

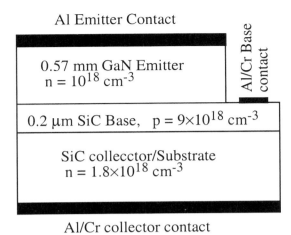

Figure 8.8: Schematic cross-section of the GaN/SiC heterojunction bipolar transistor [79].

Wen [570] has reported a ZnSe/Si HBT. An unintentionally n-doped ZnSe layer was used as an emitter layer. The layer was deposited by vapor-phase epitaxy. For a transistor with a base-width of 5 μm and a base-to-emitter doping ratio of 50, the common-emitter current gain was 10. The base could be doped heavily because of the large valence band offset at the interface. The device can be used as a phototransistor in the visible range of the spectrum. At $\lambda = 470$ nm (blue light), the responsively was 1.3 μA/lx. The dark current was very small, 1.8 pA/μm^2.

8.3 III-Nitride based transistors

8.3.1 GaN/SiC HBT

Pankove *et al.* [79] have fabricated a heterostructure bipolar transistor (HBT) using p-type 6H SiC as the base and n-type GaN as the emitter. The structure of the device is shown in Fig. 8.8. The (collector current)-(collector to base voltage) curves are shown in Fig. 8.9. The curves in Fig. 8.9 were measured at room temperature in common-base mode. For small voltages the current is practically independent of the collector to base voltage. The absence of the Early effect is due to very heavy doping in the base and the collector. The breakdown voltage was estimated to be about 10 V. High leakage and low breakdown voltage were attributed to heavy base and collector doping and large (0.25 cm^2) base-collector junction area. The collector current is very close to the emitter current. This shows that recombination in the base is small and current gain is high. The measured value of the current gain at room temperature was 10^5. The base transport factor was 0.999987. The emitter efficiency γ was 0.999999. The high

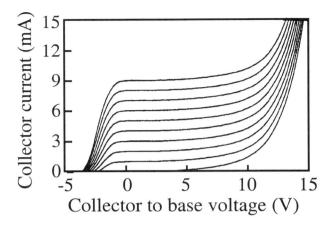

Figure 8.9: Common base characteristics of a GaN/SiC HBT. The emitter current varies from 0 for the bottom curve to 9 mA for the top curve in equal steps [79].

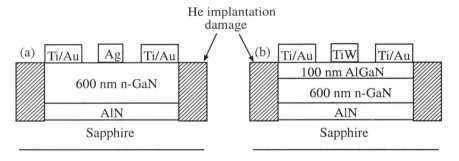

Figure 8.10: (a) Cross section of the first reported GaN MESFET [78]. (b) Cross section of the first reported AlGaN MODFET [571].

temperature and power handling capability of the transistor were good. It could be operated at 535°C with a current gain of 100. The transistor could sustain a power density of 30 kW/cm^2. The temperature dependence of the current gain showed that the valence band discontinuity between GaN and SiC is 0.43 eV.

8.3.2 Field Effect Transistors on sapphire substrate

Several reviews have appeared on the GaN based FETs [55, 61, 96]. The first MESFET was fabricated by Khan *et al.* [78] in 1993. The structure of the MESFET is shown in Fig. 8.10(a). The 300 K electron mobility in the n-type GaN layer was 350 cm^2/Vs. Transconductance of the device was 23 mS/mm. Soon afterwards the same group reported the first AlGaN/GaN Heterostructure FET (HFET) [571]. The structure of the HFET is shown in Fig. 8.10(b). The

8.3. III-NITRIDE BASED TRANSISTORS

Table 8.2: Experimental values of f_t (GHz), f_{max} (GHz), of AlGaN/GaN FETs grown on sapphire substrates. The abbreviations used are: PO for power output, H for HFETs and M for MODFETs.

Transistor L_g (μm)	f_t	f_{max}	PO (W/mm)	Ref.
H AlGaN/GaN,1	11	35	–	[571]
H AlGaN/GaN,1	18.3	–	–	[572]
H AlGaN/GaN,0.25	36	–	–	[573]
H AlGaN/GaN,0.12	46.9	103	–	[95]
M AlGaN/GaN,1	8	27.2	1.57	[574]
M AlGaN/GaN,0.2	50	92	1.7	[575]
M AlGaN/GaN,0.7	–	–	2.57	[576]

Table 8.3: Experimental values of transconductance g_m (mS/mm), product $Pr = f_t \times L_g$ (GHz μm), peak drain current I_{dsp} (mA/mm) and drain-gate breakdown voltage V_B (V) of GaN based FETs grown on sapphire substrates. The abbreviations used are: H for HFETs and M for MODFETs.

Transistor L_g (μm)	Pr	V_B	g_m	I_{dsp}	Ref.
H AlGaN/GaN,1	–	–	28	–	[571]
H AlGaN/GaN,1	18	–	–	–	[572]
H AlGaN/GaN,0.25	8	–	–	–	[573]
H AlGaN/GaN,0.12	5.6	–	–	–	[95]
M AlGaN/GaN,2	–	90	186	490	[577]
M AlGaN/GaN,1	–	340	140	400	[578]
M AlGaN/GaN,1	8	–	–	> 500	[574]
M AlGaN/GaN,0.2	10	–	240	800	[575]
M AlGaN/GaN,0.7	–	200	–	1000	[576]

AlGaN layer was 100 nm thick and contained 0.14 molar fraction of Al. The layer was grown on top of a 600 nm n-type GaN layer. None of the layers was intentionally doped. At 300 K the sheet charge density was 1.2×10^{13} cm^{-2} and the Hall mobility was 560 cm^2/Vs. The values of f_t and f_{max} were 11 and 35 GHz respectively. Transconductance of the device was 28 mS/mm. The experimental characteristics of the MODFETs are compiled in Tables 8.2 and 8.3.

Binari [579] fabricated a high temperature GaN MESFET with Pt/Au Schottky gate in 1995. The structure of this transistor is shown in Fig. 8.11. The forward characteristics of this transistor were exponential over many decades of current and the ideality factor was 1.2. The reverse leakage current was 1 mA/mm at a gate voltage of −90 V. A maximum transconductance of 23 mS/mm was measured. The pinch-off characteristics were very good. For $V_{DS} = 10$ V

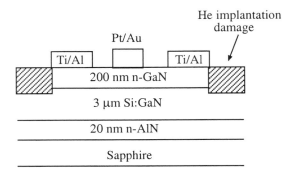

Figure 8.11: Cross section of a Si doped GaN MESFET. The doped channel layer is grown on a semi-insulating unintentionally-doped GaN buffer layer [579].

and $V_{GS} = -11$ V, I_{DS} was 10 μA. The estimated power output from this device was 1.2 W/mm. The device was tested at high temperatures. The transconductance was 10 mS/mm at 400°C and 8 mS/mm at 500°C. The pinch-off characteristics were good up to 400°C but degraded at higher temperatures because of the increased gate leakage.

GaAs MESFETs have played a dominant role in high speed microelectronics. Extensive efforts have been made to find a suitable gate dielectric so that Metal-Insulator-FET (MISFET) can be fabricated with GaAs [55]. GaAs MISFETs have been fabricated using deposited insulators. However these devices show large hysteresis loss [55]. In principle AlN can be used as an insulator and MIS technology with III-Nitrides is feasible. GaN MISFETs with deposited Si_3N_4 have been fabricated [579]. The maximum transconductance of the device was 16 mS/mm and the pinch-off voltage was −50 V. The cut-off and maximum oscillation frequencies, f_t and f_{max}, had values 5 GHz and 9 GHz respectively.

AlGaN/GaN modulation doped FETs (MODFETs) have been studied extensively in the last three years. Si doped GaN/AlGaN heterostructures were grown by LPMOVPE by Knap et al. [324] to determine whether or not the electrons at the interface form a 2D gas. Knap et al. [324] studied cyclotron resonance and quantum Hall effect in the electron gas confined at the interface. The experiments confirmed that the electron gas was 2D. The experiments also demonstrated that abrupt and good quality interfaces can be obtained in these heterostructures. In MODFETs, conduction band discontinuity should be large to confine the electrons in the channel effectively. In AlGaAs devices, this can be achieved by using a larger Al molar fraction in the barrier layer. However a large Al concentration causes problems due to the formation of donor-complex (DX) centers [55]. The work reviewed in this section shows that AlGaN/GaN MODFETs do not suffer with these difficulties. A 50% Al molar fraction has been used in the barrier layers with good performance of the transistors [576].

Khan et al. [95] fabricated AlGaN/GaN HFETs which showed $f_t = 46.9$ GHz and $f_{max} = 103$ GHz at room temperature. High performance MODFETs were fabricated by Fan et al. [577] (see Table 8.3 for comparison of different

8.3. III-NITRIDE BASED TRANSISTORS

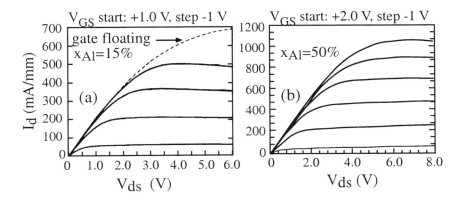

Figure 8.12: (a) $I-V$ characteristics of a $L_g = 1$ μm $Al_{0.15}Ga_{0.85}N/GaN$ MODFET fabricated on a 2 μm insulating GaN buffer [574]. (b) $I-V$ characteristics of a $L_g = 0.7$ μm MODFET with high (50%) Al molar fraction in the barrier layer [576]. This transistor was also grown on a 2 μm insulating GaN layer.

transistors). The MODFETs were fabricated by reactive MBE. In these MODFETs the gate length was 2 μm, sheet electron density was 3×10^{12} cm^{-2} and mobility was 700 cm^2/Vs. The threshold voltage was –1 V. The peak drain current was 490 mA/mm and the transconductance was 186 mS/mm. These values were measured at $V_{ds} = 4$ V. The breakdown voltage of 90 V given in the table is for the gate-drain spacing of 1 μm. The high performance was obtained by improved material quality and suppression of the leakage currents.

MODFETs with impressive performance have also been fabricated by Wu et al. [574]-[578]. In the early devices [578], the doping in the barrier layers was high so that the sheet charge density was as high as 8×10^{12} cm^{-2} with a high electron mobility, 1500 cm^2/Vs. The transconductance of a device with 1 μm channel length was 140 mS/mm. The highest reported breakdown voltage, 340 V for a 1 μm gate length MODFET with 3 μm gate-drain spacing was reported by Wu et al. [578] in 1996. In a subsequent publication the same group [404, quoted in [61]] was able to reduce the contact resistance from 3 Ω mm to 0.44 Ω mm. The transconductance of the device increased to a high value of 179 mS/mm. With increase in the drain voltage, the electron mobility degraded due to channel heating. The devices were grown on sapphire substrates and the thermal conductivity of sapphire is low.

The devices fabricated by Wu et al. [578] in 1996 had poor sub-threshold characteristics because the crystal quality of GaN was not very good and because as compared to gate-length the channel-thickness was large [574]. The RF output resistance was therefore large. In their later work, Wu et al. [574] have added an insulating GaN buffer layer between the channel and the sapphire substrate. This improved the output conductance and at the same time gave high current levels and high gate-drain breakdown voltage. The $I-V$ characteristics of a transistor grown on the insulating layer are shown in Fig. 8.12(a). The drain saturation current was higher than 500 mA/mm. The peak value of the transconductance is $g_m = 160$ mS/mm at $I_d = 200$ mA/mm. The calculated value of the intrinsic transconductance was 240 mS/mm. The combined source and the drain resistance was 4.6 Ω-mm. The reduction in the resistance gave the cut-off frequency $f_t = 8$ GHz. The other characteristics did not degrade e.g. the pinch off was hard and breakdown voltage was 100 V for the 1 μm gate device. The breakdown voltages were 150 and 220 V for 2 and 3 μm gate-lengths. The frequency f_t was measured at different drain voltages for a fixed value of the drain current. The frequency f_t was practically independent of the drain bias. This result is important and is strikingly different from the behavior of GaAs devices where the frequency degrades rather sharply with increase in the drain voltage [574]. Soon afterwards Wu et al. [575] made further improvements in the design of the MODFETs. Table 8.3 shows that as the gate-length L_g decreases, the product $Pr = f_t \times L_g$ also decreases. This is contrary to the results based on model calculations [580]. At $L_g < 0.3$ μm, the pad and gate line fringe capacitances do not remain negligible in comparison with the active input capacitance. This leads to a decrease in the RF input current to the intrinsic gate and gives a lower value of f_t. To remedy this problem, a thinner (20 nm) but more heavily doped AlGaN barrier layer was used [575]. This increased the device active input capacitance and reduced the source and drain contact resistances. The device showed considerably improved performance as shown in Tables 8.2 and 8.3. In particular the CW output power density of 1.7 W/mm at 10 GHz exceeded the previous record for any semiconductor MODFET. One year later Wu et al. [576] showed that Al molar fraction in the barrier layer can be increased from 15% to 50% with out degrading room temperature mobility. Higher Al molar fraction increases the breakdown field, the conduction band discontinuity and the piezoelectric effect. A high saturation velocity and a large carrier density co-exist in the channel. The initial measurements suggested that higher Al content makes the metal-GaAlN interface more stable. The $I - V$ characteristics of the high Al content transistor are shown in Fig. 8.12(b). The characteristics of the device are shown in Tables 8.2 and 8.3. The product $Pr = f_t \times L_g$ increased from 11 for 15% Al to 15.3 for 50% Al, a considerable improvement.

8.3.3 FETs on SiC substrate

Early work on GaN based MODFETs grown on SiC substrates has been discussed by Duboz and Khan [61]. For high power operation, thermal conductivity of the substrate becomes an important factor. The thermal conductivity of sapphire is low, 0.5 W/cm-K. SiC has excellent thermal properties, its thermal conductivity is 4.9 W/cm-K [581]. The lattice mismatch between SiC and Nitrides is smaller than that between sapphire and the nitrides. This should improve the material quality of the nitride layers grown on SiC substrates. Since SiC is conducting and can be doped, it can act as a second gate. Simulation of device RF power density performance including the effect of self heating have been made recently [582]. The simulations have been made for devices fabri-

8.3. III-NITRIDE BASED TRANSISTORS

Table 8.4: Experimental values of f_t (GHz), f_{max} (GHz), of GaN based FETs grown on SiC substrates. The abbreviation M is used for MODFET.

Transistor L_g (μm)	f_t	f_{max}	Ref.
M AlGaN/GaN,1	6	11	[201]
M AlGaN/GaN,0.25	53	58	[585]

Table 8.5: Experimental values of transconductance g_m (mS/mm), product $Pr = f_t \times L_g$ (GHz μm), peak drain current I_{dsp} (mA/mm) and Drain-gate breakdown voltage V_B (V) of GaN based FETs grown on SiC substrates. The abbreviation M is used for MODFET.

Transistor L_g (μm)	Pr	V_B	g_m	I_{dsp}	Ref.
M AlGaN/GaN,1	–	100	70	–	[201]
M AlGaN/GaN,0.25	–	100	229	1430	[585]

cated on both the sapphire and the SiC substrates. These simulations show that AlGaN HFETs fabricated on SiC substrate will achieve the highest RF power densities and lower degradation rates. As discussed below, AlGaN/GaN MODFETs grown on SiC substrate have already shown much better high power and high temperature performance.

AlGaN/GaN MODFETs on n-type 6H SiC substrate were fabricated by Binari et al. [201] in 1997. The layer sequence was 150 nm AlN, 3 μm semi-insulating GaN and 50 nm $Al_{0.15}Ga_{0.85}N$. The nominal gate-length was 1 μm (see Tables 8.4 and 8.5). The measured characteristics were: $g_m = 70$ mS/mm and $V_B = 100$ V. The device had a hard pinch-off. The value of f_t and f_{max} were 6 GHz and 11 GHz respectively. Sub-threshold characteristics were nearly ideal. Recently several authors have fabricated and studied performance of the MODFETs on the SiC substrates [583]-[586]. Ping et al. [585] have fabricated the MODFETs with 0.25, 0.5 and 1 μm gate-lengths. The observed characteristics of the 0.25 μm gate-length device are shown in Tables 8.4 and 8.5. The peak drain current of 1.43 A/mm is a considerable improvement over the values obtained with sapphire substrates. The $I - V$ characteristics showed a very small negative differential resistance. The small negative differential resistance is caused by self heating and is larger in devices grown on sapphire substrates because of the poor substrate thermal conductivity. The threshold voltage was -6.7 V for the 0.25 gate-length device. Gaska et al. [584] have compared the performance of the MODFETs grown on sapphire and on conducting 6H SiC substrates. The gate-lengths of the devices were in the range 1.7 to 2 μm. The devices grown on 6H SiC substrates operated with DC power up to 0.8 MW/cm^2. The measured thermal impedance was low, only 2°C mm/W, smaller by a factor $\sim$ 20 than for the GaAs power FETs. The drain currents were up to 1.35 A/mm and $V_B > 100$ V. The highest extrinsic gain was 185 mS/mm. When the substrate was used as a second gate, the threshold voltage was negative and high. The maximum transconductance was 25 to 45 mS/mm at voltages –10 to –15 V. The self-heating causes a faster decay of the source drain current after the voltage pulse is applied. The decay was studied with a time resolution of a few nano-seconds. The decay rate in the devices grown on the sapphire substrates was much faster. The high performance of the devices grown on SiC substrates shows that these devices have a great potential for applications as high power microwave, millimeter wave, and switching transistors [584].

In another publication, Gaska et al. [583] studied the 2D electron transport at the AlGaN/GaN interface. The mobilities are considerably enhanced in structures grown on SiC substrates. The reported values for highest 77 K mobility in the heterostructure grown on sapphire is 4000 cm^2/Vs [583, and references given therein]. Gaska et al. [583] obtained a value 8583 cm^2/Vs at 77 K in structures grown on SiC. At 4.2 K a mobility of $\sim 10,500$ cm^2/V s was observed. The sheet charge density was 10^{13} cm^{-2} at 77 K and 0.97×10^{13} cm^{-2} at 4.2 K. The scattering by defects becomes relatively more important at cryogenic temperatures. The fact that the improvement in the mobility is larger at low temperatures shows that the concentration of dislocations and other defects in the layers is smaller as compared to the concentration in the layers grown on sapphire substrates. Sullivan et al. [586] have demonstrated the high power capability of the AlGaN/GaN MODFETs grown by MOCVD on insulating 4H SiC. The layer sequence was: 30 nm AlGaN doped with 2×10^{18} cm^{-3} Si atoms, 3 nm of undoped AlGaN, 50 nm of GaN doped with 5×10^{17} cm^{-3} Si atoms and a 1-μm thick GaN buffer layer. The Pt and Au Schottky gate contact was 0.7 μm long. The device isolation was achieved by ion implantation using He and As. A record total power of 2.3 W at 10 GHz was measured in a 1280-μm wide transistor at a drain voltage of 33 V. The RF power density was 2.8 mW/mm on a 320 μm wide transistor. These high values are attributed to the high thermal conductivity of the SiC substrate.

In their 1997 review, Binari and Dietrich [55] have compiled and plotted the cut-off frequencies f_t as a function of gate-length for GaAs-, GaN-, and SiC-based FETs. Their results are shown in Fig. 8.13. We have added recent results (shown by the points K and W) in the figure. The performance of GaN MODFETs compare favorably with that of SiC (scaled to the same gate length) [55].

Because of large bond strength and hardness of GaN, modifications in the methods of device processing have been made. These modifications are discussed in the next section.

8.4 Device Processing

Because of its high melting point and very high bond strength, several conventional methods of device processing do not work for GaN. Currently the main

8.4. DEVICE PROCESSING

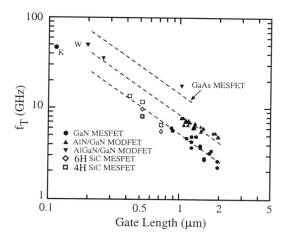

Figure 8.13: Cut-off frequencies, f_t, as a function of gate length for GaN, SiC, and GaAs FETs. The data has been compiled by Binari and Dietrich [94, and references given there in] except the two recent data points marked K for GaN HFET [95] and W for GaN MODFET [575].

issues involved in the device processing are the fabrication of low resistance ohmic contacts, stable and reproducible Schottky contacts, etching with reasonable etch-rates while maintaining surface stoichiometry, isolation and doping implantation, activation of the implanted dopants and annealing out the ion implantation damage. We have already discussed the ohmic contacts and Schottky barriers. Other topics are discussed in this section.

8.4.1 Etching

8.4.1.1 Wet etching

The etching of the III-Nitrides has been reviewed by several authors [93, 223, 587]. Because of the large bond strength it is difficult to etch the III-Nitride semiconductors using wet etching techniques. Molten salts KOH and NaOH at temperatures > 250 can be used to etch GaN at reasonable rates. However handling the molten salts is inconvenient. It is difficult to find masks which can be used with these etchants at the high temperatures involved. Recently it has been found that AlN or Al rich AlGaN alloys can be etched with KOH solution at 50-100°C [93]. It is possible to etch n-type GaN with KOH near room temperature provided 365 nm illumination of the KOH is used. For light intensity of 50 mW/cm^2 an etching rate of 300 nm min^{-1} has been demonstrated [93, and references given therein]. The etching produces undercut encroachment and does not produce very smooth surfaces. This method does not etch p-type GaN. Polycrystalline AlN has been successfully etched in several wet etchants [223]. It is more difficult to etch single crystals.

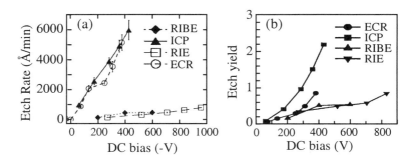

Figure 8.14: (a) Dry etch rates of GaN in Cl_2/Ar-based plasmas as a function of dc chuck bias in the case of RIE, ICP and ECR and acceleration voltage in the case of RIBE. (b) Etch yield in the same plasma and same reactors also as a function of dc chuck bias [93].

8.4.1.2 Dry etching

Conventional Reactive Ion Etching (RIE) and low pressure Reactive Beam Etching (RIBE) are not suitable as the etching rates are very small. ECR and Inductively Coupled Plasma (ICP) produce higher etching rates. The observed etching rates of GaN with the four techniques are compared in Fig. 8.14. The higher etch rates with ECR and ICP plasma are partly due to the larger ion flux and degree of dissociation and partly due to higher efficiency of reaction [93]. The etch yields with the four techniques are shown in Fig. 8.14(b). Magnetron RIE and chemically assisted ion-beam etching (CAIBE) have also been used to obtain higher etch rates [223]. A table of etch rates obtained with different plasma chemistry is given in Ref. [223, p 39]. The highest rate quoted in this table is 1300 nm/min with ICl plasma chemistry using the ECR technique.

Plasma induced damage in III-Nitrides has not been studied extensively. The damage with ECR generated plasma has been reported. The damage increases as a function of ion energy and flux. PL is adversely affected by the high density plasma. The effect depends on the experimental conditions. Further work is required in this area [223].

Selective etching of InN is often required when InN-based ohmic contacts are used. The maximum selectivity of InN over other nitrides is between 3 and 10 with Cl_2/Ar ICP etching. Side wall anisotropy is generally very good. However the side walls are rough unless the process is carefully optimized. All the etch products have not been identified. $GaCl_3$ and N_2 have been seen in the etch products.

8.4.2 Si ion implantation

Doping by ion implantation in GaN has been studied by several authors [588]-[591]. It is found necessary to anneal the implanted GaN samples at temperatures above 1300°C to obtain good electrical properties of the epilayers. At

8.4. DEVICE PROCESSING

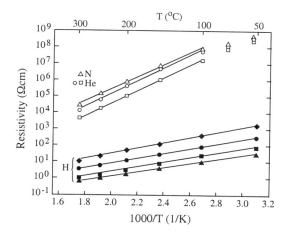

Figure 8.15: Resistivity of GaN as a function of temperature for H-, He- and N-implanted GaN samples. The solid lines are the least square fits to the data [595].

these high temperatures N_2 escapes from the GaN layers because of high pressure of nitrogen over GaN (at 1400°C equilibrium pressure of N_2 over GaN is > 1000 bar, see papers by Porowski and Grzegory [592] and Newman [593]). The equilibrium pressure of N_2 over AlN is much lower, at 1400°C it is only 10^{-8} bar. Therefore an encapsulation layer of AlN helps in preventing the escape of N_2 and decomposition of the surface. The other method of preventing the surface decomposition is to use high nitrogen pressures (15 kbar) [594, and references given therein].

In GaN some recovery of damage takes place during implantation even if the implantation is performed at liquid nitrogen temperatures. However residual damage that remains after implantation is considerable and annealing is necessary. If Si is implanted to a dose > 10^{16} cm^{-2} at 90 keV, amorphous layers of GaN are generated. Annealing up to 1100°C is not sufficient to crystallize these layers and remove the damage. On annealing the Si implanted amorphous GaN layers between 900 and 1100°C, a highly defective polycrystalline material is obtained. If the dose is lower, the GaN layers do not become amorphous but annealing up to 1100°C leaves the layers with a coarse network of extended defects [590]. From their extensive studies of Si implanted GaN layers, Tan et al. [590] concluded that it is necessary to use annealing temperature higher than 1100°C to obtain acceptable quality of implanted layers. The results of this work explain as to why low mobilities are observed in the implanted layers annealed at low temperatures.

In a recent paper Cao et al. [594] have reported high Si$^+$ implant activation efficiency in GaN using a high-temperature rapid thermal process. Undoped 2-3 μm thick GaN films were grown on sapphire at 1040°C by MOVPE. Si$^+$ ions were implanted at 25°C at an energy of 100 keV to a dose of 5×10^{15} cm^{-2}. An

AlN layer was then deposited by MOVPE or by reactive sputtering. The samples were encapsulated in a quartz ampoules under 15 psi pressure of N_2. After annealing at several temperatures in the range 1100-1500°C for 10 s, AlN layer was removed selectively from each sample by etching with KOH. The samples were characterized by scanning electron microscopy, atomic force microscopy, and Hall measurements. The quality of the surface of the encapsulated samples did not suffer degradation as a result of annealing. Hall measurements showed that activation of the implanted Si atoms and free electron density increased up to annealing temperature of 1400°C and then started decreasing. Electron mobility was also maximum (> 40 cm^{-2}/Vs) in samples annealed at 1400°C, it decreased on annealing at higher temperatures. This behavior is similar to that of GaAs and other III-V semiconductors and shows that above 1400°C, the samples start becoming compensated. Presumably this occurs because Si atoms switch sites, i.e. some of the Si_{Ga} move to Si_N sites, a process which produces self compensation. The activation efficiency of $\sim 90\%$ and electron concentration of $\sim 10^{20}$ cm^{-3} were estimated in the layers annealed at 1400°C.

8.4.3 Implantation for isolation

Implantation with light ions produces large concentration of point defects and high resistivity material. Binari et al. [595] have studied the resistivity of GaN implanted with H, He and N ions. At room temperature the resistivity of H-implanted samples becomes 1 to 10 Ωcm as shown in Fig. 8.15. The resistivity of 10 Ωcm is not high enough for isolation. Moreover annealing experiments show that the H-implanted defects anneal significantly at about 250°C. The He and N implants produce resistivity values in the range 10^8 to 10^9 Ωcm (see Fig. 8.15). The resistivity remains quite high even after 800°C anneal. Fig. 8.15 shows that the resistivity of the He- and N-implanted thermally stabilized GaN samples is $\sim 10^4$ Ωcm at room temperature. Therefore He and N implants will be useful for isolation.

Chapter 9

Summary and conclusions

9.1 Growth, defects and strain

Several advances have been made in the growth of III-V, II-VI and III-Nitride semiconductors during the recent years. With the MOVPE method III-V semiconductor epilayers are generally grown with hydrogen carrier gas. Since hydrogen is highly explosive its use involves expensive gas detecting system. As discussed in chapter 2 nitrogen has higher efficiency of energy transfer during collision with the source molecule and is more efficient in decomposing the compounds. Hardtdegen *et al.* [166] have grown high quality III-V epilayers by LPCVD using nitrogen carrier gas. They grew epilayers of GaAs/GaAs, $Al_xGa_{1-x}As$/GaAs, InP/InP and $Ga_{047}In_{0.53}As$/InP using both the hydrogen and the nitrogen as the carrier gases. For GaAs and InP the crystal quality and purity of the epilayers grown in nitrogen was comparable to those of the epilayers grown in hydrogen.

MBE has been used for the growth of most II-VI semiconductor layers for device applications. Epilayers with extremely low concentrations of extended and point defects have been grown on GaAs substrate [137]. Experimental conditions for the growth of ZnMgSSe have been optimized. ZnMgSSe is an important alloy for cladding layers because its bandgap is tunable up to ~ 4.5 eV and it forms type I heterostructure with ZnCdSe. It can be lattice matched to GaAs [409].

A breakthrough development is the two-step method for the growth of III-Nitrides on sapphire substrates perfected by Akasaki and collaborators [71]. In this method, first a thin layer of AlN or GaN is grown at a low temperature, $\sim 500°C$. The temperature is then raised to around $1025°C$ and the required Nitride epilayer is grown on the buffer layer. The large lattice mismatch between sapphire and AlN or GaN is accommodated by a large number of misfit dislocations in the buffer layer. Therefore immediately after the growth the buffer layer the strain is almost completely relaxed. However the coefficients of thermal expansion of sapphire and the Nitrides are different. The stress is therefore generated during the cool-down period. In contrast to the III-V and II-VI layers, the defect density in GaN-based epilayers is high. The typical threading dislocation density in the epilayer is 10^{10} to 10^{12} cm^{-3}. A density of 10^4 cm^{-3} affects adversely the performance of light emitters fabricated using conventional III-V semiconductors. A density of 10^7 cm^{-3} is fatal in GaAs based lasers. Several possible suggestion have been offered to account for the fact that GaN LEDs and Laser Diodes (LDs) work in spite of the presence of the large density of the defects in the devices. According to an earlier suggestion, it was assumed that the dislocations in GaN are not efficient non-radiative recombination centers. Later CL experiments showed dark-spots exactly at the locations where the dislocations are seen in the TEM. These experiments showed that dislocations do act as non-radiative recombination centers. Currently two ideas to account for the high emission efficiency of the Nitride semiconductors with high defect density are being debated. The active layer in most devices is an InGaN alloy. The distribution of In is inhomogeneous. Local regions of high In concentration act as quantum dots. According to one suggestion, the electrons and holes are confined in the dots and do not "see" the dislocations. According to the second suggestion, the average distance between the dislocations is larger than the diffusion length and therefore not many electrons or holes are able to reach the dislocations.

The defect density in the III-Nitride epilayers structures has been considerably reduced by growing the epilayers laterally on Si oxide layers [85]. The mismatch of GaN with SiC substrate is much smaller than with sapphire or GaAs substrates. Recently considerably progress has been made in the growth of c-GaN on GaAs substrates [410]. Epilayers with single phase c-GaN with good electrical and optical properties have been grown. The density of defects observed in cubic-GaN (c-GaN) epilayers grown on 3C-SiC is smaller than in the epilayers grown on GaAs by one order of magnitude. We have not come across any published work on LEDs or LDs fabricated with c-GaN at the time of this writing [410].

Number of publications on the measurements of critical thickness and strain relaxation in lattice mismatched semiconductor epilayers continues to be very large. Experimental critical thickness is larger than its values predicted by the equilibrium theories for all semiconductors including the III-Nitrides. Published work on strain and strain-relaxation in II-VI epilayers is of topical interest [125]. Strain in both small mismatch systems (ZnSe/GaAs) and large mismatch (e.g. CdTe and ZnTe epilayers grown on GaAs (100) substrates) have been studied. The behavior of the two systems is not the same. In many highly mismatched systems dislocations are nearly periodically distributed. In the low-mismatched systems the distribution is highly non-uniform. However in the highly mismatched GaN/sapphire systems the dislocations are highly clustered. Values of strain in GaN measured by different workers differ widely. Complication arises because of the large thermal mismatch between the GaN epilayers and the substrate. The values of strain in the epilayer depend on the cooling rate and other experimental conditions.

The dislocation velocity in III-V semiconductors is much larger than in Si.

9.2. BAND STRUCTURE AND ELECTRONIC PROPERTIES

The velocity of α dislocations is larger than that of the β dislocations. The dislocation velocity in the II-VI epilayers is very large. In the II-VI semiconductors the velocity is the same for the α and β dislocations. The dislocation velocity in III-Nitride layers is several orders of magnitude smaller.

9.2 Band structure and electronic properties

9.2.1 III-V semiconductors

Strain has a large effect on the band structure of zinc blende crystals. Biaxial strain is caused by lattice and/or thermal mismatch. This strain has two components: hydrostatic and shear. Hydrostatic stress changes only the band gap, degeneracy of the valence band is not lifted. The distance between the conduction band and the center of gravity of the valence bands increases under hydrostatic compression and decreases under hydrostatic tension. The shear component of the biaxial strain removes the degeneracy of the valence band multiplet. The two hole bands move in opposite directions away from the center of gravity of the two bands. Under compression the $|3/2, \pm 3/2\rangle$ HH band moves up and determines the bandgap. Under tensile strain, the $|3/2, \pm 1/2\rangle$ LH band moves up. It is closest to the conduction band and determines the bandgap. These results are consistent with observed optical properties of strained layers of zinc blende semiconductors. Recently we have reviewed [52] the optical properties of strained III-V layers. Splitting of the valence band determined using the observed PL and other optical properties agree well with the predicted values by the theory discussed in section 5.1.

Knowledge of conduction band offset ΔE_c and valence band offset ΔE_v is essential for designing heterostructure devices. Recent values of band offsets in several III-V heterostructures are given in section 5.2.2.

9.2.2 II-VI semiconductors

The band offsets of ZnSe/GaAs depend on the surface orientation of the GaAs substrate and on growth conditions. In all cases the valence band offset is a large fraction of the total bandgap difference. The band offsets of II-VI/II-VI heterojunctions can be divided in three categories:

1. The type I heterojunctions for which $\Delta E_c < \Delta E_v$. Typical examples of such heterostructures are ZnSe/ZnS$_x$Se$_{1-x}$, and ZnSe/ZnMgSSe.

2. The Type I heterojunctions for which $\Delta E_c > \Delta E_v$. Some examples are Zn$_{1-x}$Cd$_x$Se/ZnSe, CdTe/ZnTe, Zn$_{1-y}$Cd$_y$Se/Zn$_{1-x}$Mn$_x$Se ($y = 0.15$, $x = 0.16$), and Zn$_{1-x}$Mn$_x$Se/ZnSe.

3. The Type II heterostructure. A typical example is p-ZnTe/p-ZnSe heterojunction.

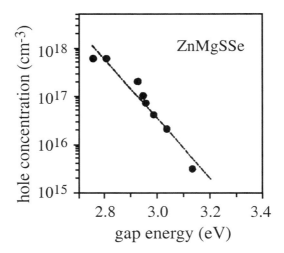

Figure 9.1: Experimental data of the hole concentration in ZnMgSSe plotted a function of gap energy. The values calculated assuming a pinning of the Fermi level at localized centers which are fixed with respect to vacuum level are shown by the solid line [410, and reference given therein].

Extensive measurements of optical properties of unstrained and strained II-VI semiconductors and alloys have been made. The properties studied include reflectivity, transmission spectra and photoluminescence. These measurements have yielded rich information about the band structure, exciton (i.e. binding energy, and transition energy of the excitons) and strain in the samples studied. The binding energy of excitons is around 25 meV in ZnSe, 12 meV in ZnTe and 10 meV in $Cd_{0.9}Mn_{0.1}Te/Cd_{0.6}Mg_{0.4}Te$ quantum wells

ZnSe, ZnS, CdSe, and CdS can be doped n-type easily. Cl is the most frequently used donor dopant. The p-type doping in wide-bandgap semiconductors is difficult. In fact the development of both ZnSe- and GaN-based semiconductors was delayed for many years because these semiconductors could not be doped p-type. The p-doping of ZnSe with nitrogen was achieved in 1990 for the first time. Nitrogen doping with plasma activated sources is now used to obtain p-type conductivity. However the maximum concentration of holes is still limited because nitrogen does not form very shallow acceptors. The energy of ionization of nitrogen in ZnSe is about 120 meV. Only a small fraction of dopants is ionized at room temperature. The large concentration of the unionized acceptor impurity reduces the hole mobility to small values. As the bandgap of ZnSe-based alloys increases, the concentration of maximum active acceptors that can be doped in the alloys decreases drastically. Results of p-doping versus bandgap in ZnMgSSe alloys are shown in Fig. 9.1.

The behavior of ZnTe, BeTe and other II-VI compounds containing Te anions is different. They can be doped p-type easily. The n-type doping of these semiconductors and their alloys is not very difficult if bandgap is not large.

9.2. BAND STRUCTURE AND ELECTRONIC PROPERTIES

CdTe and CdMnTe can be doped with chlorine or bromine to obtain an electron concentration of 10^{18} cm^{-3} provided Mn concentration in the alloy is low. If Mn concentration is more than 10%, a drastic decrease in the concentration of active donors is observed [477].

Like p doping, it is very difficult to make good ohmic contacts to p-type ZnSe. Since ZnTe can be doped p-type more easily, graded p-doped ZnSeTe layer is used for making ohmic contacts to p-type ZnSe.

Piezoelectric effect in strained II-VI layers and quantum wells modifies the optical properties. External electric field also has a large effect on the optical properties. When a perpendicular electric field is applied to the semiconductor quantum wells or superlattices, the optical absorption remains resolved but shifts to lower energies. This red shift is known as Quantum Confined Stark Effect (QCSE). QCSE and electro-optic effect have been studied in several II-VI heterostructures. As a result of these studies, modulators have been designed and fabricated. In some cases a blue shift instead of a red shift has been observed. Two possible explanations have been proposed to explain the observed blue shift. In the cases of a thin-period superlattice a miniband is formed. The effect of electric field is to break the coupling between the neighboring wells so that instead of the carriers being in the miniband, they are now confined to "single" quantum wells. This causes a blue shift. The blue shift can also be caused by the presence of large concentration of free charge carriers. The charge carriers screen the electric field and annul the existing red shift. This results in an apparent blue shift.

Extensive work has been done on the effect of magnetic field on II-VI quantum wells and superlattices in which Mn ions are present in the barrier layers. External magnetic field changes the optical and magneto-optical properties of these structures. If the thickness of the wells is not large, exciton wavefunctions penetrate the barrier layers. Magnetic polarons are formed due to exchange interactions between the excitons and magnetic spins. Optical and magneto-optical studies of these structures yield valuable information about the exciton parameters and interface structure.

9.2.3 III-Nitride semiconductors

In WZ Nitrides the crystal field splits the valence band. The combined effect of crystal field and spin-orbit coupling lifts the degeneracy of the valence band and produces three sub-valence bands, designated as $A(\Gamma_9)$, $B(\Gamma_7)$ and $C(\Gamma_7)$ bands. $A(\Gamma_9)$ is the top most band. For WZ GaN the separation between A and B bands is 6 meV and between B and C is 37 meV. The binding energy of B and C excitons is 20 meV and that of A excitons it is 18 meV. The effective mass of electrons is $0.18m_0$ and that of holes is in the range $0.6m_0$ to $1.1m_0$. Considerable research has been done on AlGaN and InGaN alloys. Absorption and bandgap measurements show that bowing parameter has large values in both alloys. In GaN the piezoelectric coefficients are larger than in the conventional III-V semiconductors. Piezoelectricity affects the band-offsets and charge-sheet density in strained quantum wells. The calculated valence band offsets do not agree well with the observed values. The valence band offsets between a pair of Nitrides have values between 0.6 and 1.7 eV.

In high quality GaN layers, strong near band edge PL due to free and bound excitons dominates. PL bands due to donor-acceptor recombination are observed on the low energy side of the near band-edge luminescence. A yellow band with strong vibronic structure is observed at ~ 2.2 eV. The centers responsible for this band have not been identified unambiguously. Photoconductivity induced in GaN by incident light persists after the incident light is switched off. It can persist up to several days. Extensive studies of PL in GaN quantum wells have been made. PL peak moves to high energies as the well width decreases. A large part of this shift is due to the quantum confinement effects. Exact determination of the quantum shift from the observed spectra is difficult because the band-filling and strain also cause shifts of the PL peak. A small molar fraction of In in InGaN active layer increases enormously the electroluminescence efficiency. The distribution of indium in InGaN alloys is non-uniform. Regions of local high concentration of In probably act as quantum dots and are responsible for the observed high efficiency.

Ge and Si are efficient shallow donors in GaN. The energy of activation of Si in GaN is 18 to 20 meV. GaN can be easily doped to more than 10^{19} cm^{-3} electrons. Room temperature electron mobility at this concentrations is ~ 100 cm^2/Vs. The mobility rises to 900 cm^2/Vs when electron concentration becomes less than 10^{17} cm^{-3}. These experimental values of mobility are smaller than the values predicted by Monte-Carlo simulations for uncompensated GaN epilayers. Monte-Carlo simulations also sow that the peak and saturation velocities in GaN are considerably larger than in Si or GaAs. The optical phonon energy is high, 90 meV. Scattering of charge-carriers by the optical phonons is not so efficient in GaN. Efficiency of n-type doping of AlGaN decreases as Al concentration in the alloy increases. Initial measurements show that the activation energy of Si atoms increases from 20 meV for GaN to 54 meV for AlGaN alloy with 8 to 20% molar fraction of Al.

The situation for p doping in GaN is not so good. Until about 10 years ago, good p-type conductivity could not be obtained in the III-Nitrides. Break through results obtained by Akasaki and co-workers in 1988/1989 [72] changed the whole scenario. Immediately after doping with Mg, GaN does not show the p-type conductivity. Low-energy electron beam irradiation (LEEBI) [72] or high temperature annealing [74, 423] is necessary to activate the Mg atoms. Hole concentration of a few times 10^{18} cm^{-3} have been achieved. The energy required to activate Mg atoms is more than 150 meV. Therefore only 1 to 5% of Mg atoms are active acceptors at room temperature. The huge concentration of Mg atoms required to produce a reasonable concentration of holes causes low hole mobility. When the hole concentration is more than 5×10^{18} cm^{-3}, hole mobility is ~ 3 cm^2/Vs. The highest measured hole mobility is in zincblende (ZB) GaN and is 40 cm^2/Vs at a hole concentration of about 8×10^{16} cm^{-3}.

A decrease in the maximum p-doping level is observed in AlGaN alloys with increase of Al content. Activation energy of Mg in GaN and AlGaN was determined by Tanaka et al. [418] by Hall effect measurements. It is about 180 meV in

$Al_{0.075}Ga_{0.925}N$, higher than the value 150 meV in GaN measured by the same authors. It was not possible to dope the $Al_xGa_{1-x}N$ layers p-type with Mg if $x > 0.13$. Akasaki and Amano have mentioned in a recent publication [81] that a hole concentration of $\sim 2 \times 10^{17}$ cm^{-3} can be obtained in MOCVD grown $Al_{0.2}Ga_{0.8}N$ layers. For the same bandgap values, the concentration of holes that can be built in GaN-based alloys is larger than in the ZnSe-based alloys. If we consider two semiconductors with same bandgaps from the ZnSe and GaN families, the hole concentration that can be built in GaN based semiconductors is about four time larger.

Because of the hardness and large bond-strength of the III-Nitrides new methods of device processing had to be developed. Wet-etching rates of GaN are not sufficiently high. Dry etching with ECR or with inductively coupled plasma produces acceptable etching rates. Doping of GaN by Si implantation requires annealing at 1400°C to activate the dopants and remove the damage due to implantation. Techniques of ion implantation for isolation have been developed. Schottky barriers with several metals deposited on GaN have been measured. Schottky barriers larger than 1 eV are obtained with Ni, Pd and Pt. Very low resistance n-type ohmic contacts have been obtained. The resistance of contacts on p-Nitrides continues to be high.

9.3 Applications and future work

9.3.1 Applications

Performance of the ZnSe-based laser diodes has become as good as that of GaAs-based lasers [53, 410]. In 1997 the differential quantum efficiency of the CW lasers had reached values of 0.6–0.7 and wall-plug conversion efficiency, up to 20% [367]. This has been possible because of high quality growth of the quantum wells, achievement of efficient optical and electronic confinement, improved p-doping, Mg containing quaternary alloy cladding layers, and low resistance p-contacts. The performance of ZnSe-based LEDs and LDs is superior as compared to that of GaN-based devices [53, 410, 411]. However the degradation of the devices and short working lifetime continue to be serious problems. Recent work leading to a significant reduction of pre-existing macroscopic and point defects has resulted in the increase of lifetime to ~ 400 hours [137]. The working lifetime of the GaN based LEDs and LDs has already reached 10,000 hours and they have been commercialized. Considerable more work will have to be done before ZnSe based devices can reach the market place.

The bond in Be-chalcogenides is not as ionic as in the other II-VI semiconductors and the bond-strength in BeTe is larger. It is predicted that II-VI alloys containing Be should be more stable against the introduction of defects. There is intense research activity on the II-VI alloys containing Be-chalcogenides. However resistance to introduction of defects or laser diodes with better performance and higher reliability have not yet been demonstrated. Unfortunately BeTe and BeS are indirect bandgap semiconductors. Band structures of other

Be-chalcogenides are not yet known.

The success in growth of high quality GaN, in p doping and in making ohmic contacts has made it possible to fabricate GaN-based LEDs in the blue and green regions and LDs in the UV region of the spectrum. Laser diodes emitting at 376 nm have been fabricated. These are the shortest wavelength semiconductor lasers. The commercial implications of these developments are enormous. These emitters will play a dominant role in optical storage, laser printing, medical diagnostics, full color display, and other areas. The bandgap of GaN is larger than that of ZnSe. This combined with the fact that doping levels of holes is better in GaN makes it possible to obtain shorter-wavelength emitting devices with GaN. The shortest wavelength at which ZnSe based devices can emit light is in the blue and green regions.

Considerable work on transistors using III-V, II-VI and III-Nitrides has been done in recent years. In the III-V area InGaAs active layer high electron mobility transistors have dominated the scene for several years. The concentration of In required for best performance is rather high: 30 to 40% mole fraction of InAs is required in the alloy. This concentration is so high that the layers grown on GaAs substrate have very small critical thickness. This makes the device structures susceptible to generation of dislocation and strain relaxation. A few years ago the emphasis shifted to InP substrates due to this problem. However InP is brittle and expensive. Large area GaAs wafers are available at a reasonable cost. Recently metamorphic HEMTs have been developed. These transistors are grown on relaxed alloy buffer layers grown on GaAs substrates and are lattice-matched to the buffer. The structures are not strained. Recent work shows that the performance of the metamorphic HEMTs is nearly as good as of those grown on InP substrates. Significant advances in high temperature and high power performance of GaN based transistors have been made. The devices fabricated on SiC substrates have shown better performance than the performance of those grown on sapphire substrates.

9.3.2 Future work

Despite these spectacular achievements, there are several areas where further research is required. In the area of III-V devices the metamorphic HEMTs have shown great promise. Further work is still required to bring this work to a stage that the devices can be used in commercial and military equipment. In the field of II-VI semiconductor devices the most urgent need is to improve their long term reliability. It is not yet clear whether use of the Be-chalcogenides to increase the hardness of the material will solve this problem. Extensive research in this area is being done. It is hoped that some conclusions will be reached in the near future. MOCVD has been used most extensively to grow Nitride epilayers for both research and fabrication of devices. Improvements in MBE growth technique are required. It is yet to be determined whether other methods of growth, e.g. laser deposition will become important. So far sapphire substrates have been used most extensively. Recent work shows that at least for transistors, SiC substrates may prove better. Further work is necessary

9.3. APPLICATIONS AND FUTURE WORK

to understand the nature of centers which give rise to the yellow band, the persistent photoconductivity and the background donors in III-Nitrides. If large GaN wafers become available, the quality of the epilayers can be improved with a reduction in the dislocation density and an over all improvement in the device performance. There is an urgent need to find acceptors with a lower activation energy to improve p doping and hole mobility. If higher p-doping becomes possible, it will also help in solving the problem of making ohmic contacts to the p-type Nitrides. If these problems are solved, many more devices will be commercialized. Scientists working in this area are looking forward to new exciting developments during the next few years.

Appendix A
Useful tables

Table A: Lattice constants a and bandgaps E_g of wide-band II-VI and other semiconductors. The values of a are at room temperature. The values of E_g are also at room temperature unless a different temperature is mentioned. In some cases more than one value of the parameter a or E_g are given. The scatter in the values shows the uncertainty in the available values of the parameter.

Semicond.	Lattice	a (Å)	E_g(eV)(T in K)	Ref.
MgS	NaCl	5.19, 5.6	−,4.5	[596, 525]
MgSe	ZB	5.45, 5.7	−,3.8	[596, 525]
ZnS	ZB	5.4102	3.56	[596]
ZnS	ZB	5.411	3.82	[361]
ZnSe	ZB	5.6686,−	−,2.830(5)	[325, 597]
ZnSe	ZB	5.6676	2.7(RT)	[596]
ZnSe	ZB	−	2.830(5)	[597]
ZnSSe[a]	ZB	−	2.957	[325]
ZnTe	ZB	6.1037,−	−,2.26, 2.35(RT)	[598, 596]
CdS	ZB	5.825	2.582(4.2), 2.501(RT)	[596]
CdSe	ZB	6.052	1.841(1.8), 1.751(RT)	[596]
CdTe	ZB	6.481	1.43(RT),1.59(1.6)	[596]
CdTe	ZB	6.482	1.60(75), 1.605(< 13)	[600]
InSb	ZB	6.4794	0.170	[347]
GaAs	ZB	5.65325	1.424	[347]
GaSb	ZB	6.096	0.725	[347]
Si		5.431	1.1 (77 K)	[601]
CaF$_2$		5.464	−	[601]
BaF$_2$		6.200	−	[601]
PbSe		6.124	0.18	[601]
PbTe		6.462	0.22	[601]
PbSnSe[a]		6.06	0	[601]

[a] Concentration of S is 0.19 and that of Sn is 0.20.
Note: 1. MgS, ZnS, ZnSe, CdS, and CdSe also exist in wurtzite form [596].
Note: 2. A more extensive tabulation of calculated and experimental values of a and E_g of II-VI semiconductors is given in [361].

Table B: Refractive indices of II-VI semiconductors.

Semiconductor	n	k	T °C	Ref.
CdTe	2.51	0.23	286	[156]
CdTe	2.49	–	175	[156]
MnTe	3.26	–	286	[156]
ZnSe	2.55	–	RT[1]	[173]
ZnS	2.35	–	286	[173]

[1]Room temperature, (not explicitly mentioned by the author [173]).

Table C: Values of Elastic constants (GPa), band deformation potentials (eV) and spin-orbit splitting parameter Λ (eV) of II-VI and other semiconductors.

Semicond.	T (K)	C_{11}	C_{12}	C_{44}	a	b	Λ	Ref.
ZnS	RT	104.0	65.0	46.2	–	–	–	[602]
ZnS	300T	106.7	66.6	–	–4.56	–0.75	0.07	[325]
ZnSe	RT	81.0	48.8	44.1	–	–	–	[602]
ZnSe	RT	82.6	49.8	–	–5.40	–1.2	0.43	[325]
ZnSe	RT	–	–	–	–4.17	–1.2	–	[274]
ZnSe	300	85.9	50.6	–	–3.0	–1.2		[326]
ZnSe	12	88.8	52.7	–	–	–		[326]
ZnTe	300	71.3	40.7	31.2	–	–	–	[602]
ZnSe	298	85.9	50.6	40.6	–	–	–	[603]
ZnTe	298	71.1	40.7	31.3	–	–	–	[603]
CdTe	RT	53.5	36.8	19.	–	-	–	[602]
CdTe		52.5	36.1	19.5	–	–	–	[279]
Si	RT	166	64	79	–	-	–	[601]
GaP	RT	140.5	62.1	70.3	–9.6	–1.65	–	[604]
InP	RT	101.1	56.1	45.6	–6.35	–2.0	–	[604]
GaAs	RT	119.0	53.8	59.5	–9.77	–1.70	–	[604]
InAs	RT	83.3	45.3	39.6	–6.0	–1.8	–	[604]
CaF_2	3	165	45	34	–	–	–	[601]
BaF_2	300	90	42	26	–	–	–	[601]
PbSe	300	124	19	17	–	–	–	[601]
PbTe	300	10.8	0.77	1.34	–	–	–	[601]

Note: Temperature variation of the elastic constants of ZnSe is discussed in [603] and pressure dependence is discussed in [605].

Appendix A

Table D: Room temperature lattice constant a and bandgaps E_g of selected semiconductors

	a(Å)	E_g(eV)	Ref.
AlP	5.4635	3.6(Γ)	[347]
AlP	5.4635	3.5(L)^a	[347]
AlAs	5.6605	2.15(X)	[347]
AlSb	6.1358	1.612(X)	[347]
GaP	5.4510	2.272(X)	[347]
GaAs	5.65325	1.424(Γ)	[347]
GaSb	6.09602	0.725(Γ)	[347]
InP	5.8689	1.350(Γ)	[347]
InAs	6.0583	0.355(Γ)	[347]
InSb	6.47943	0.170(Γ)	[347]
ZnSe	5.6676	2.70	[606]
ZnS	5.4093	–	[606]
ZnTe	6.1037	2.26	[606]
CdTe	6.4770	–	[606]
Si	5.4311	1.124(X)	[347]
Ge	5.6579	0.666(L)	[347]

^{a}Theoretical estimate

Table E: Hydrostatic deformation potentials a and shear deformation potentials b of III-V semiconductors. Values given in this table are taken from the paper of V. Swaminathan in *Indium Phosphide and Related Materials: Processing, Technology, and Devices,* Ed.: Avishay Katz, (Artech House Boston 1992), p 29. Values of a given in [607] are positive and considerably smaller.

	a(eV)	b(eV)
AlSb	−5.9	−1.35
GaP	−9.6	−1.65
GaAs	−9.77	−1.70
GaSb	−8.28	−2.0
InP	−6.35	−2.0
InAs	−6.0	−1.8
InSb	−7.7	−2.0

Table F: Values of bandgaps $E_g(X)$ of unstrained ternary alloys. The values are compiled by Tu et al. [608, p 161] from the data taken from Casey and Panish [609]. The values are for 300 K and except those values for which alternative temperatures and references are shown.

Alloy	$E_g(X)$(eV)
$Al_X In_{1-X} P$	$1.351 + 2.23X$
$Al_X Ga_{1-X} As^1$	$1.424 + 1.247X$
$Al_X Ga_{1-X} As^2$	$1.424 + 1.455X$
$Al_X In_{1-X} As$	$0.360 + 2.012X + 0.698X^2$
$Al_X Ga_{1-X} Sb$	$0.726 + 1.129X + 0.368X^2$
$Al_X In_{1-X} Sb$	$0.172 + 1.621X + 0.43X^2$
$Ga_X In_{1-X} P$	$1.351 + 0.643X + 0.786X^2$
$Ga_X In_{1-X} P^3$	$1.423 + 0.77X + 0.648X^2$
$Ga_X In_{1-X} As$	$0.36 + 1.064X$
$Ga_X In_{1-X} As^4$	$0.354 + 0.576X + 0.494X^2$
$Ga_X In_{1-X} As^5$	$0.415 + 0.6337X + 0.475X^2$
$Ga_X In_{1-X} Sb$	$0.172 + 0.139X + 0.415X^2$
$GaP_X As_{1-X}$	$1.424 + 1.150X + 0.176X^2$
$GaAs_X Sb_{1-X}$	$0.726 - 0.502X + 1.2X^2$
$InP_X As_{1-X}$	$0.360 + 0.891X + 0.101X^2$
$InAs_X Sb_{1-X}$	$0.18 - 0.41X + 0.58X^2$

[1] For $0 < X < 0.45$.
[2] For $0 < X < 0.37$, Keuch et al. [610].
[3] At low temperatures and for low Ga concentrations [611, 612].
[4] At 296 K, Kuphal et al. [613].
[5] At 2 to 4 K, Goetz et al. [614], see also [615].

Table G: Values of elastic constants, Young's modulus Y ($\times 10^{12}$ dyne/cm^2), and Poisson's ratio ν of III-V semiconductors. Values of elastic constants are taken from [608] and those of Y and ν are from [607].

	C_{11}	C_{12}	C_{44}	Y	ν
GaP	1.405	0.621	0.703	1.028	0.31
InP	1.011	0.561	0.456	0.607	0.36
GaAs	1.190	0.538	0.595	0.853	0.31
InAs	0.833	0.453	0.396	0.514	0.35

Table H: Coefficients of thermal expansion α_{th} (K^{-1}) of Si, GaAs, GaP, and InP.

S	α_{th}	Ref.
Si[1]	2.3 × 10^{-6}	[616]
GaAs	6.63 × 10^{-6}	[607]
GaP	5.91 × 10^{-6}	[607]
InP	4.56 × 10^{-6}	[607]

[1] Approximate value, see also [617].

Table I: Values of TO and LO Raman frequency ω_0 (cm^{-1}) and phonon deformation potentials $\tilde{K}_{11}$, $\tilde{K}_{12}$ and $\tilde{K}_{44}$ of semiconductors. LO values are given in parentheses [618]. The values are at 77 K for InSb, 110 K for Si and 300 K for all other semiconductors. Grüneisen parameter $\gamma = -(\tilde{K}_{11} + 2\tilde{K}_{12})/6$ can be calculated from the values of phonon deformation potentials. Numerical values of γ are given in the review of Anastassakis [618].

	ω_0	$\tilde{K}_{11}$	$\tilde{K}_{12}$	$\tilde{K}_{44}$
AlSb	317	−2.1	−2.6	−0.7
	(338)	(−1.6)	(−2.6)	(−0.3)
GaP	366	−1.35	−1.95	−0.6
	(403)	(−1.45)	(−2.5)	(−0.5)
GaAs	269	−2.4	−2.7	-0.9
	(292)	(−1.7)	(−2.4)	(−0.55)
GaSb	231	−1.9	−2.35	−1.1
	(241)			
InP	304	−2.5	−3.2	−0.5
	(351)	(−1.6)	−3.2	(−0.2)
InAs	218	−0.95	(−2.8)	−0.75
	(243)			
InSb	180	−2.45	−3.3	−0.6
	(191)			
ZnSe	205	−2.75	−4.0	−0.45
	(250)			
Diamond	1332	−2.9	−1.9	−1.2
Si	521	−1.85	−2.3	−0.7
Ge	300	−1.45	−1.95	−1.1

Table J: Room temperature (RT) values of the lattice constants a and c (Å), bandgaps E_g (eV), and high temperature thermal coefficients α_T (10^{-6}/K) (of the in-plane lattice constant a) of the III-Nitrides and related materials. Temperatures at which E_g values are valid, if different from RT, are given in the rounded brackets. H is the abbreviation for the hexagonal lattice.

Semicond.	Lattice	a	c	E_g	α_T	Ref.
GaN	WZ	3.19	5.20	3.50(1.6 K)	5.6	[619]
GaN	WZ	3.189	5.185	3.39	5.59	[99]
AlN	WZ	3.11	4.976	6.28(5 K)	–	[619]
AlN	WZ	3.112	4.982	6.2	4.2	[99]
InN	WZ	3.54	5.6994	1.89	–	[619]
InN	WZ	3.548	5.69	1.89	–	[99]
GaN	ZB	4.50	–	3.45	–	[619]
AlN	ZB	4.38	–	5.11(Theory)	–	[99]
InN	ZB	4.98	–	2.2(Theory)	–	[99]
LiGaO$_2$	WZ	3.17	5.2305	3.415	–	[306]
6H SiC	WZ	3.08	15.12	2.45	4.2	[304, 620]
3C SiC	ZB	4.36	–	Note^a	–	[304, 620]
Sapphire	H	4.758	12.991	–	7.5	[196, 99]
Sapphire	–	–	–	9.9	7.5	[599]
ZnO	W	3.250	5.213	–	8.25	[99]
ZnO	W	–	–	∼ 3.36	8.25	[623]

Note^a E_g of SiC is around 3 eV, depending on the polytype [47, p 10].

Table K: Elastic constants (in GPa) of III-Nitrides. References given are to the authors who have compiled the data. These papers should be consulted for original references.

C_{11}	C_{12}	C_{13}	C_{33}	C_{44}	C_{66}	Ref.
			AlN			
398	140	127	382	–	–	[285]
			GaN			
296	130	158	267	241	–	[621]
390	145	196	398	105	123	[622]
			InN			
271	124	94	200	–	–	[285]

Note: Values for GaN determined by other authors generally lie between the values given above [330].

Bibliography

[1] F. C. Frank, and J. Van der Merwe, One dimensional dislocations. II. Misfitting monolayers and oriented overgrowth, Proc. Roy. Soc. (London), A 198, 216-25 (1949).

[2] J. W. Matthews, Coherent interfaces and misfit dislocations, in *Epitaxial growth*, Part B, Ed.: J. W. Matthews (Academic Press, New York 1975), pp 559-609; J. W. Matthews, Defects associated with the accommodation of misfit between crystals, J. Vac. Sci. Technol. 12, 126-33 (1975).

[3] S. C. Jain, *Germanium–Silicon Strained Layers and Heterostructures*, Advances in Electronics and Electron Physics series, (Supplement 24), Editor-in-chief of the series: Peter W. Hawkes, (Academic Press, Boston 1994).

[4] W. Shockley, U. S. Patent 2569347, issued 25 September 1951.

[5] E. Kasper, H. J. Herzog, and H. Kibbel, A one dimensional SiGe superlattice grown by UHV epitaxy, App. Phys. 8, 199-205 (1975).

[6] E. Kasper, and H. J. Herzog, Elastic strain and misfit dislocation density in $Ge_{0.08}Si_{.92}$ films on silicon substrates, Thin Solid Films 44, 357-70 (1977).

[7] J. W. Matthews and A. E. Blakeslee, Defects in epitaxial multilayers, J. Crystal Growth, 27, 118-25 (1974).

[8] G. C. Osbourn, $In_xGa_{1-x}As$–$In_yGa_{1-y}As$ strained layer superlattices: A proposal for useful, new electronic materials, Phys. Rev. B 27, 5126-28 (1983)

[9] G. C. Osbourn, Strained-layer superlattices: A brief review, IEEE J. Quantum Electronics 22, 1677-81 (1986).

[10] I. J. Fritz, Role of experimental resolution in measurement of critical thickness for strained-layer epitaxy, Appl. Phys. Lett. 51, 1080-2 (1987).

[11] J. C. Bean, L. C. Feldman, A. T. Fiory, S. Nakahara, and I. K. Robinson, Ge_xSi_{1-x}/Si strained-layer superlattice grown by molecular beam epitaxy, J. Vac. Sci. Technol. A2, 436-40 (1984).

[12] H. Morkoç, B. Sverdlov, and G.-B. Gao, Strained layer heterostructures, and their applications to MODFET's, HBT's and lasers, Proceedings of the IEEE 81, pp 493-556 (1993).

[13] S. J. Pearton, J. C. Zolper, R. J. Shul, and F. Ren, GaN: Processing, defects, and devices, J. Appl. Phys. 86, 1-78 (1999).

[14] H. Kroemer, Theory of a wide gap emitter for transistors, Proc. IRE 45, 1535-7 (1957).

[15] H. Kroemer, Proc. IEEE 51, 1782 (1963).

[16] G. P. Agrawal and N. K. Dutta, *Semiconductor Lasers* (Second Edition), (Van Nostrand Reinhold, New York 1993).

[17] H. Kroemer, Heterostructure Bipolar Transistors and Integrated Circuits, Proc. IEEE 70, 13-25 (1982).

[18] H. Yamada, Future optical storage in multimedia environment and role of blue lasers, presented at *The Second International Conf. on Nitride Semiconductors*, held at Tokushima, Japan, October 27-31, 1997, Sponsored by The Japan Society of Appl. Phys. paperM1-1.

[19] H. Q. Le, G. W. Turner, and J. R. Ochoa, Turn-key, liquid-nitrogen-cooled 3.9 μm semiconductor laser package with 0.2 W CW output, Electronics Lett. 32, 2359-60 (1996).

[20] B. Lane, D. Wu, A. Rybaltowski, H. Yi, J. Diaz, and M. Razeghi, Compressively strained multiple quantum well InAsSb lasers emitting at 3.6 μm grown by metal-organic chemical vapor deposition, Appl. Phys. Lett. 70, 443-5 (1997).

[21] C. Van Hoof, J. Nemeth, S. C. Jain, G. Borghs, R. Mertens, and R. Van Overstraeten, Advances in strained layers and strained layer devices, Physics of Semiconductor Devices (Proc. of the 9th International Workshop on the Physics of Semiconductor Devices), Eds.: Vikram Kumar and S. K. Agarwal, (Narosa Publishing House, New Delhi 1998), pp 554-62.

[22] B. Grietens, S. C. Jain, C. Van Hoof, S. Nemeth, P. Van Daele and G. Borghs, Suppression of Auger recombination by strain in Sb based mid-IR lasers, Solid State Phenomena (Scitec Publication, Switzerland) 55, 20-5 (1997), see also B. Grietens, Ph. D. dissertation, K. U. 1998.

[23] Nils G. Weimann, Lester F. Eastman, Dharanipal Doppalapudi, Hock M. Ng, and Theodore D. Moustakas, Scattering of electrons at threading dislocations in GaN, J. Appl. Phys. 83, 3656-9 (1998).

[24] K. van der Zanden, Ph. D. Dissertation, K. U. (1999).

[25] S. K. O'Leary, B. E. Foutz, M. S. Shur, U. Bhaphkar, and L. F. Eastman, Electron transport in wurtzite indium nitride, J. Appl. Phys. 83, 826-9 (1998).

[26] K. J. Linthicum, T. Gehrke, D. B. Thomson, K. M. Tracy, E. P. Carlson, T. P. Smith, T. S. Zheleva, C. A. Zorman, M. Mehregany, and R. F. Davis, Process routes for low defect density GaN on various substrates employing Pendeo-epitaxial growth techniques, MRS Internet J. Nitride Semicond. Res. 4S1, G4.9(1999).

[27] Michael C. Y. Chan, Kwok-On Tsang, E. Herbert Li, and Steven P. DenBaars, Thermal annealing of InGaN/GaN strained-layer quantum well, MRS Internet J. Nitride Semicond. Res. 4S1, G6.25 (1999).

[28] M. Sumiya, T. Ohnishi, M. Tanaka, A. Ohtomo, M. Kawasaki, M. Yoshimoto, H. Koinuma, K. Ohtsuka, and S. Fuke, Control of polarity and surface morphology of GaN films deposited on c-plane sapphire, Internet J. Nitride Semicond. Res. 4S1, G6.23 (1999).

[29] Nicolas Grandjean, Jean Massies, Mathieu Leroux, Marguerite Lagt, Pierre Lefebvre, Bernard Gil, Jacques Allgre, Pierre Bigenwald, Optical and structural properties of AlGaN/GaN quantum wells grown by molecular beam epitaxy, MRS Internet J. Nitride Semicond. Res. 4S1, G11.7 (1999).

[30] Tsvetanka S. Zheleva, Scott A. Smith, Darren B. Thomson, Thomas Gehrke, Kevin J. Linthicum, Pradeep Rajagopal, Eric Carlson, Waeil M. Ashmawi, and Robert F. Davis, Pendeo-epitaxy - A new approach for lateral growth of gallium nitride structures, MRS Internet J. Nitride Semicond. Res. 4S1, G3.38 (1999).

[31] M. A. L. Johnson, Z. Yu, J. D. Brown, F. A. Koeck, N. A. El-Masry, H. S. Kong, J. A. Edmond, J. W. Cook, JR., and J. F. Schetzina, A critical comparison between MOVPE and MBE growth of III-V Nitride semoconductor maaterials for opto-electronic devices applications, MRS Internet J. Nitride Semicond. Res. 4S1, G5.10 (1999).

[32] L. L. Chao, G. S. III. Cargill, T. Marshall, E. Snoeks, J. Petruzzello, and M. Pashley, Spectral shifts associated with dark line defects in degraded II-VI laser diodes. App, Phys. Lett. 72, 1754-6 (1998).

[33] C. Giannini, E. Carlino, P. Sciacovelli, L. Tapfer, M. Sauvage-Simkin, Y. Garreau, N. Jedrecy, M. B. Veron, R. Pinchaux, Influence of the interface layer on the strain relaxation of ZnSe epitaxial layers grown by MBE on (001)GaAs, J. Physics D (Appl. Phys.) 32, A51-5 (1999).

[34] S. F. Chichibu, T. Sota, K. Wada, S. P. DenBaars, and S. Nakamura, Spectroscopic studies in InGaN quantum wells, MRS Internet J. Nitride Semicond. Res. 4S1, G2.7(1999).

[35] N. V. Edwards, A. D. Batchelor, I. A. Buyanova, L. D. Madsen, M. D. Bremser, R.F. Davis, D. E. Aspnes, and B. Monemar, Relaxation phenomena in GaN/AlN/6H-SiC heterostructures, MRS Internet J. Nitride Semicond. Res. 4S1, G3.78 (1999).

[36] H. Y. An, O. H. Cha, J. H. Kim, G. M. Yang, K. Y. Lim, E.-K. Suh, and H. J. Lee, Thermal treatment effect of the GaN buffer layer on the photoluminescence characteristics of the GaN epilayer, J. Appl. Phys. 85, 2888-93 (1999).

[37] S. R. Lee, A. F. Wright, M. H. Crawford, G. A. Petersen, J. Han, and R. M. Fiefeld, The band-gap bowing of AlGaN alloys, Appl. Phys. Lett. 74, 3344-6 (1999).

[38] Shizuo Fujita, Mitsuru Funato, Doo-Cheol Park, Yoshifumi Ikenaga, Shigeo Fujita, Electrical characterization of MOVPE-grown p-type GaN:Mg against annealing temperatuare, MRS Internet J. Nitride Semicond. Res. 4S1, G6.31 (1999).

[39] J. Wagner, A. Ramakrishnan, D. Behr, M. Maier, N. Herres, M. Kunzer, H. Obloh,and K.-H. Bachem, Composition dependence of the band gap energy of $In_xGa_{1-x}N$ layeres on GaN (≤ 0.15) grown by metal-organic chemical vapor deposition, MRS Internet J. Nitride Semicond. Res. 4S1, G2.8 (1999).

[40] J. D. Albrecht, P. P. Ruden, E. Bellotti, and K. F. Brennan, Monte Carlo simulation of Hdall Effect in n-type GaN MRS Internet J. Nitride Semicond. Res. 4S1, G6.6 (1999).

[41] *GeSi strained layers and their applications*, Eds.: A. M. Stoneham and S. C. Jain, (IOP Publishing, Bristol 1995).

[42] E. P. O'Reilly and A. R. Adams, Band-structure engineering in strained semiconductor lasers, IEEE J. Quantum Electronics 30, 366-79 (1994).

[43] R. L. Gunshor and A. V. Nurmikko, Volume Editor, II–VI Blue/Green light emitters: Device physics and epitaxial growth, Semiconductors and Semimetals, edited by R. K. Willardson, R. Weber and A. C. Beer, vol. 44, (Academic Press 1997).

[44] S. C. Jain, M. Willander, J. Narayan and R. Van Overstraeten, III-Nitrides: Growth, characterization and properties, to be published J. Appl. Phys.

[45] S. Nakamura and G. Fasol, *The Blue Laser Diodes*, (Springer-Verlag, Heidelberg 1997).

[46] *GaN*, Vol. 1, edited by J. I. Pankove and T. D. Moustakas, (Academic Press, N. Y. 1998).

BIBLIOGRAPHY 285

[47] *Group III nitride semiconductor compounds,* edited by B. Gil, (Clarendon Press, Oxford 1998).

[48] *High speed semiconductor devices,* Ed.: S. M. Sze, (Wiley Interscience 1990).

[49] S. J. Pearton and N. J. Shah, Heterostructure Field-Effect Transistors, in Ref. [48], pp 283-334.

[50] J. C. Bean, Materials and technologies, in Ref. [48], pp 13-56.

[51] E. Yablonovitch and E. O. Kane, Band structure engineering of semiconductor lasers for optical communications, J. Lightwave Technol. 6, 1292-9 (1988).

[52] S. C. Jain, M. Willander, and H. Maes, Stresses and strains in epilayers, stripes and quantum structures of III-V compound semiconductors, Semiconductor Sci. and Technol. 11, 641-71 (1996); Erratum Semiconductor Sci. and Technol. 11, 975 (1996).

[53] H. Morkoç, S. Strite, G. B. Gao, M. E. Lin, B. Sverdlov, and M. Burns, Large-band-gap SiC, III-V nitride, and II-VI ZnSe-based semiconductor device technologies, J. Appl. Phys. 76, 1363-98 (1994).

[54] I. Akasaki and H. Amano, Recent progress of the crystal growth, conductivity control and light emitters of group III Nitride semiconductors, IEDM Technical Digest 96, 231-7 (1996).

[55] S. C. Binari and H. C. Dietrich, III-V Nitride electronic devices, in *GaN and related materials*, Edited by S. J. Pearton, (Gordon and Breach Science Publishers 1997), pp 509-34.

[56] L. Pfeiffer, K. W. West, H. L. Stormer, and K. W. Baldwin, Electron mobilities exceeding 10^7 cm^2/Vs in modulation doped GaAs, Appl. Phys. Lett. 55, 1888-90 (1989).

[57] M. A. Khan, Q. Chen, J. W. Yang and C. J. Sun, Electronic devices based on GaN-AlGaN material system, paper presented at *Int'l Conf. on SiC and related materials,* published in Inst. Phys. Conf. Ser. No 142, pp 985-90 (1995).

[58] J. D. Albrecht, R. P. Wang, P. P. Ruden, M. Farahmand and K. F. Brennan, Electron transport characteristics of GaN for high temperature device modelling, J. Appl. Phys. 83, 4777-81 (1996).

[59] T. L. Tansley, E. M. Goldys, M. Goldewski, B. Zhou, and H. Y. Zou, The contribution of defects to the electrical and optical properties of GaN, in *GaN and related materials,* Edited by S. J. Pearton, (Gordon and Breach Science Publishers 1997), pp 233-94.

[60] B. Gelmont, K. Kim, and M. Shur, Monte Carlo simulation of electron transport in gallium nitride, J. Appl. Phys. 74, 1818-21 (1993).

[61] J.-Y. Duboz and M. A. Khan, Transistors and detectors based on GaN-related materials, in Ref. [47], pp 343-90.

[62] J. I. Pankove, A historical survey of research on Gallium Nitride, in *GaN and related materials*, Edited by S. J. Pearton, (Gordon and Breach Science Publishers 1997), pp 1-9.

[63] B. Monemar, Basic III-V Nitride Research - past, present and future, in Ref. [88], paper M1-2, pp 6-8.

[64] R. Juza and H. Hahn, Z. Anorg. Allgem. Chem. 234, 282 (1938), *ibid.*, 244, 133 (1940).

[65] H. Grimmeiss, and Z. H-Koelmans, Nature 14a, 264 (1959).

[66] H. P. Maruska and J. J. Tietjen, The preparation and properties of vapor-deposited single-crystalline GaN, Appl. Phys. Lett. 15, 327-9 (1969).

[67] J. I. Pankove, E. A. Miller, and J. E. Berkeyheiser, GaN blue light emitting diodes, J. Luminescence 5, 84-86 (1972); J. I. Pankove, E. A. Miller, and J. E. Berkeyheiser, GaN electroluminescent emitting diodes, RCA review 32, 383-92 (1971); J. I. Pankove, M. T. Duffy, E. A. Miller, and J. E. Berkeyheiser, Luminescence of insulating Be-doped and Li-doped GaN, J. Luminescence 8, 89-93 (1972).

[68] J. I. Pankove, and H. E. P. Schade, Photoemission from GaN, Appl. Phys. Lett. 25, 53-5 (1974).

[69] M. T. Duffy, C. C. Wang, G. D. O'Clock, S. H. McFarlane, and P. Z. Zanzucchi, Epitaxial growth and piezoelectric properties of AlN, GaN, and GaAs on sapphire or spinel, J. Elect. Mat. 2, 359-72 (1973).

[70] S. Yoshida, S. Misawa, and S. Gonda, Improvements on the electrical and luminescent properties of reactive molecular beam epitaxially grown GaN films by using AlN-coated sapphire substrates, Appl. Phys. Lett. 42, 427-9 (1983).

[71] H. Amano, I. Akasaki, T. Kozawa, K. Hiramatsu, N. Sawak, K. Ikeda, and Y. Ishi, Electron beam effects on blue luminescence of zinc-doped GaN, J. Luminescence 40-41, 121-2 (1988).

[72] H. Amano, N. Sawaki, I. Akasaki, and Y. Toyoda, Metalorganic vapor phase epitaxial growth of a high quality GaN film using an AlN buffer, Appl. Phys. Lett. 48, 353-5 (1986); H. Amano, I. Akasaki, T. Kozawa, K. Hiramatsu, N. Sawaki, K. Ikeda, and Y. Ishii, Electron Beam Effects on Blue luminescence of zinc-doped GaN, J. Luminescence 40&41, 121-2

(1988); H. Amano, M. Kito, K. Hiramatsu, and I. Akasaki, P-type conduction in Mg-doped GaN treated with low-energy electron beam irradiation (LEEBI), Jpn. J. Appl. Phys. 28, L2112-4 (1989).

[73] J. A. Vechten, J. D. Zook, and R. D. Horning, Defeating compensation in wide gap semiconductors by growing in H that is removed by low temperature de-ionizing radiation, Jpn. J. Appl. Phys. 31, 3662-3 (1992).

[74] S. Nakamura, N. Iwasa, M. Senoh, and T. Mukai, Hole compensation mechanism of P-type GaN films, Jpn. J. Appl. Phys. 31, 1258-66 (1992).

[75] M. A. Khan, R. A. Skogman, and J. M. Van Hove, Photoluminescence characteristics of AlGaN-GaN-AlGaN quantum wells, Appl. Phys. Lett. 56, 1257-9 (1990).

[76] K. Itoh, T. Kawamoto, H. Amano, K. Hiramatsu, and I. Akasaki, Metalorganic vapor phase epitaxial growth and properties of GaN/AlGaN layered structure, Jpn. J. Appl. Phys. 30 (9A), 1924-7 (1991).

[77] S. Nakamura, T. Mukai, and M. Senoh, p-GaN/n-InGaN/n-GaN double heterostructure blue-light-emitting diodes, Jpn. J. Appl. Phys. 32, L8-11 (1993).

[78] M. A. Khan, J. N. Kuznia, A. R. Bhattarai, and D. T. Olsen, Metal semiconductor field effect transistor based on single crystal GaN, Appl. Phys. Lett. 62, 1786-7 (1993).

[79] J. I. Pankove, S. S. Chang, H. C. Lee, R. Molnar, T. D. Moustakas, and B. Van Zeghbroeck, High temperature GaN/SiC heterojunction bipolar transistor with high gain, IEDM Technical Digest 94, 389-92 (1994).

[80] I. Akasaki, H. Amano, and I. Suemune, Recent progress of crystal growth, conductivity control, and light emitters of column-III nitrides, and future prospect of nitride-based laser diodes, Inst. Phys. Conf. Ser. 142, pp 7-10 (1996).

[81] I. Akasaki and H. Amano, Lasers, in *GaN*, edited by J. I. Pankove and T. D. Moustakas, (Academic Press, N. Y. 1998), Vol. 1, pp 459-72.

[82] S. Nakamura, Blue-green light-emitting diodes and violet laser diodes, MRS Bulletin 22, pp 29-35, February 1997.

[83] S. Nakamura, III-V Nitride based LEDs, in *GaN and related materials*, Edited by S. J. Pearton, (Gordon and Breach Science Publishers 1997), pp 471-507.

[84] S. Nakamura, Applications of LEDs and LDs, in *GaN*, edited by J. Pankove and T. D. Moustakas, (Academic Press, N. Y. 1998), Vol. 1, pp 431-57.

[85] S. Nakamura, The role of structural imperfections in InGaN-based blue light-emitting diodes and laser diodes, Science 281, 956–61 (1998).

[86] I. Akasaki, S. Sota, T. Tanaka, M. Koike, and H. Amano, Shortest wavelength semiconductor laser diode, Electronics Lett. 32, 1105-6 (1996).

[87] III-V Nitrides, Symposium, (Mater. Res. Soc. Pittsburgh, PA, USA, 1997), Symp. 449, Eds.: F. A. Ponce, T. D. Moustakas, I. Akasaki and B. A. Monemar (1997).

[88] The Second International Conf. on Nitride Semiconductors, held at Tokushima, Japan, October 27-31, 1997, Sponsored by The Japan Society of Appl. Phys.

[89] *Nitride semiconductors,* (Mater. Res. Soc. Pittsburgh, PA, USA, 1998), Symp. 482, Eds.: F. A. Ponce, S. P. DenBaars, B. K. Meyer, S. Nakamura and S. Strite (1998).

[90] Several interesting papers on III-Nitride materials and devices are contained in the February 1997 special issue of MRS Bulletin, Vol. 22, No. 2 (1997).

[91] *GaN and related materials,* edited by S. J. Pearton, (Gordon and Breach, N. Y. 1997).

[92] S. Nakamura, InGaN-based blue laser diodes, Selected Topics in Quantum Electronics 3, 712-8, (1997).

[93] S. J. Pearton, F. Ren, J. C. Zolper, and R. J. Shul, GaN device processing, (Mater. Res. Soc. Pittsburgh, PA, USA, 1998), Symp. 482, 961-71 (1998).

[94] S. C. Binari, GaN FETs for microwave and high temperature applications, Proc. Topical Workshop on III-V Nitrides, 1995.

[95] M. Asif Khan, Q. Chen, J. Yang, M. Z. Anwar, M. Blasingame, and M. S. Shur, Recent advances in III-V nitride electron devices, IEDM Technical Digest 96, 27-30 (1996).

[96] M. S. Shur and M. A. Khan, GaN/AlGaN heterostructure devices: Photodetectors and Field-Effect Transistors, MRS Bulletin 22, pp 44-50, February 1997.

[97] W. Kim, Ö. Aktas, A. E. Botchkarev, A. Salvador, S. N. Mohammad, and H. Morkoç, Reactive molecular beam epitaxy of wurtzite GaN: Material characteristics and growth kinetics, J. Appl. Phys. 79, 7657-66 (1996).

[98] S. N. Mohammad. W. Kim, A. Salvador, and H. Morkoç, Reactive molecular beam epitaxy for wurtzite GaN, See Ref. [90, pp 22-28].

[99] G. Popovici, H. Morkoç, and S. Noor Mohammad, Deposition and properties of group III nitrides by molecular beam epitaxy, in Ref. [47], pp 19-69.

[100] S. Strite and H. Morkoç, GaN, AlN, and InN: A review, J. Vac. Sci. Technol. B 10, 1237-65 (1992).

[101] A. M. Akhekyan, V. I. Kozlovskii, A. S. Nasibov, Yu. M. Popov, and P. V. Shapkin, Single crystals of $ZnSe_{1-x}Te_x$, $Zn_{1-x}Cd_xSe$, and $Zn_xCd_{1-x}S$ solid solutions for electron beam pumped lasers, Sov. J. Quantum Electronics, 15, 737-9 (1985).

[102] Yu. V. Korostelin, V. I. Kozlovsky, A. S. Nasibov, and P. V. Shapkin, Vapour growth of II-VI solid solution single crystals, J. Crystal Growth 159, 181-5 (1996).

[103] H. Yoshida, T. Fujii, A. Kamata, and Y. Nakata, Undoped ZnSe single crystal growth by the vertical Bridgman method, J. Crystal Growth 117, 75-79 (1992).

[104] B. J. Fitzpatrick, A review of the bulk growth of high band gap II-VI compounds, J. Crystal Growth 86, 106-10 (1988).

[105] R. Triboulet, The scope of CdTe growth for production of epitaxial substrates, Materials Forum 15, 30-4 (1991).

[106] P. Rudolph, S. Kawasaki, S. Yamashita, S. Yamamoto, Y. Usuki, Y. Konagaya, S. Matada, and T. Fukuda, Attempts to growth of undoped CdTe single crystals with high electrical resistivity, J. Crystal Growth 161, 28-33 (1996).

[107] P. Rudolph, K. Umetsu, H. J. Koh, N. Schäfer, and T. Fukuda, The state of the art of ZnSe melt growth and new steps towards twin-free bulk crystals, Materials Chem. Phys. 42, 237-41 (1995).

[108] P. Rudolph and M. Mühlberg, Basic problems of vertical Bridgman growth of CdTe, Materials Sci. and Engineering B16, 8-16 (1993).

[109] R. Triboulet, The growth of bulk ZnSe crystals, Semicond. Sci. Technol. 6, A18-A23 (1991).

[110] R. Triboulet, The scope of the low temperature growth of bulk II-VI, compounds, Proceedings of the SPIE - (The International Society for Optical Engineering, 1995), Vol. 2373, pp 24-30.

[111] M. R. Lorenz., in *Physics and chemistry of II-VI compounds*, Eds.: M. Aven and J. S. Prener, (North Holland, Amsterdam, 1967).

[112] T. Asahi, O. Oda, and Y. Koyama, Growth and characterization of 100 mm diameter CdZnTe single crystals by the vertical gradient method, J. Crystal Growth 161, 20-7 (1996).

[113] S. V. Ivanov, S. V. Sorokin, P. S. Kop'ev, J. R. Kim, H. D. Jung, and H. S. Park, Composition, stoichiometry and growth rate control in molecular beam epitaxy of ZnSe based ternary and quaternary alloys, J. Crystal Growth 159, 16-20 (1996).

[114] H. Okada, T. Kawanaka, and S. Ohmoto, Study of ZnSe phase diagram by differential thermal analysis, J. Crystal Growth 165, 31-6 (1996).

[115] L. Shcherbak, P. Feichouk, and O. Panchouk, Effect of CdTe "postmelting", J. Crystal Growth 161, 16-9 (1996).

[116] G Landwehr, A. Waag, F. Fischer, H.-J. Lugauer, and K. Schüll, Blue emitting heterostructure laser diodes, Physica E 3, 158-68 (1998).

[117] H. A. Mar and R. M. Park, Observation of strain effects and evidence of gallium autodoping in molecular-beam-epitaxial ZnSe on (100)GaAs, J. Appl. Phys. 60, 1229-32 (1986).

[118] G. Cantwell, W. C. Harsch, H. L. Cotal, B. G. Markey, S. W. S. McKeever, and J. E. Thomas, Growth and characterization of substrate-quality ZnSe single crystals using seeded physical vapor transport, J. Appl. Phys. 71, 2931-6 (1992).

[119] L. H. Kuo, K. Kimura, S. Miwa, T. Yasuda, and T. Yao, Dependence of generation and structures of stacking faults on growing surface stoichiometries in ZnSe/GaAs, J. Appl. Phys. 80, 1408-10 (1996).

[120] G. Kalpana, G. Pari, A. Mookerjee, and A. K. Bhattacharya, *Ab initio* Electronic band structure calculations for Beryllium chalcogenides, International J. Modern Phys. 12, 1975-84 (1998).

[121] D. J. Olego, Effects of ZnSe epitaxial growth on the surface properties of GaAs, Appl. Phys. Lett. 51, 1422-4 (1987).

[122] J. Petruzzello, B. L. Greenberg, and R. Dalby, Structural properties of the ZnSe/GaAs system grown by molecular-beam epitaxy, J. Appl. Phys. 63, 2299-303 (1988).

[123] T. Yokogawa, H. Sato, and M. Ogura, Dependence of elastic strain on thickness for ZnSe films grown on lattice-mismatched materials, Appl. Phys. Lett. 52, 1678-80 (1988).

[124] R. L. Gunshor, L. A. Kolodziejski, A. V. Nurmikko, and N. Otsuka, Molecular-Beam epitaxy of II-VI semiconductors, in *Strained-layer superlattices: Materials science and technology*, Volume Editor: T. M. Pearsall, Semiconductors and semimetals, Eds.: R. K. Willardson and A. C. Beer, (Academic Press, Boston 1991), pp 337-409.

BIBLIOGRAPHY

[125] K. Pinardi, Uma Jain, S. C. Jain, H. E. Maes, R. Van Overstraeten, and M. Willander, Critical thickness and strain relaxation in lattice mismatched II-VI semiconductor layers, J. Appl. Phys. 83, 4724-33 (1998).

[126] Jui-ichi Nishizawa, K. Itoh, Y. Okuno, And F. Sakurai, Blue light emission from ZnSe p-n junctions, J. Appl. Phys. 57, 2210-6 (1985).

[127] R. L. Gunshor, N. Otsuka, and A. V. Nurmikko, Blue lasers on the horizon, IEEE Spectrum, pp 28-33 (1993).

[128] A. M. Glass, K. Tsai, R. B. Sylsma, R. D. Feldman, D. H. Olsen, and R. F. Austin, Room-temperature optically pumped $Cd_{0.25}Zn_{0.75}Te/ZnTe$ quantum well lasers grown on GaAs substrate, Appl. Phys. Lett. 53, 834-6 (1988).

[129] J. Han and R. L. Gunshor, MBE growth and electrical properties of wide bandgap ZnSe-based II-VI semiconductors, in Ref. [43], pp 1-58.

[130] R. M. Park, M. B. Troffer, C. M. Rouleau, J. M. DePuydt, and M. A. Haase, p-type ZnSe by nitrogen atom beam doping during molecular beam epitaxial growth, Appl. Phys. Lett. 57, 2127-9 (1990).

[131] K. Ohkawa, T. Karasawa, and T. Mitsuyu, Characteristics of p-type ZnSe layers grown by molecular beam epitaxy with radical doping, Jpn. J. Appl. Phys. 30, Part 2, L152-5 (1991).

[132] K. Ohkawa and T. Mitsuyu, p-type ZnSe layers grown by molecular beam epitaxy with nitrogen radical doping, J. Appl. Phys. 70, 439-41 (1991).

[133] M. A. Hasse, J. Qui, J. M. DePuydt, and H. Cheng, Blue-green laser diodes, Appl. Phys. Lett. 59, 1272-4 (1991).

[134] H. Jeon, J. Ding, W. Patterson, A. V. Nurmikko, W. Xie, D. C. Grillo, M. Kobayashi and R. L. Gunshor, Blue-green injection laser diodes in (Zn,Cd)Se/ZnSe quantum wells, Appl. Phys. Lett. 59, 3619-21 (1991).

[135] S. Heun, J. J. Paggel, L. Sorba, S. Rubini, and A. Franciosi, Interface composition and stacking fault density in II-VI/III-V heterostructures, Appl. Phys. Lett. 70, 237-9 (1997).

[136] N. V. Sochinskii, V. Munoz, V. Bellani, L. Vina, E. Dieguez, E. Alves, M. F. da Silva, J. C. Soares, and S. Bernardi, Appl. Phys. Lett. 70, 1314-6 (1997).

[137] E. Kato, H. Noguchi, M. Nagai, H. Okuyama, S. Kijima, and A. Ishibashi, Significant progress in II-VI blue-green laser diode lifetime, Electronics Letters 34, 282-4 (1998).

[138] A. Waag, F. Fischer, K. Schüell, T. Baron, H. J. Lugauer, Th. Litz, U. Zehnder, W. Ossau, T. Gerhard, M. Keim, G. Reuscher, and G. Landwehr, Laser diodes based on beryllium-chalcogenides, Appl. Phys. Lett. 70, 280-2 (1997).

[139] W. S. Kuhn, B. Qu'Hen, and O. Gorochov, The metal-organic vapour phase of ZnTe: I. Properties and decomposition kinetics of organometallic, Prog. Crystal Growth and Charact. 31, 1-44 (1995).

[140] H. P. Wagner, S. Festner, H. Stanzl, G. Kudlek, and N. Presser, High-density and optical nonlinearity effects in epitaxially grown ZnTe layers, Adv. Mat. for Optics and Electronics 3, 33-9 (1994).

[141] S. C. Jain, H. Maes, K. Pinardi, and I. De Wolf, Stresses and strains in lattice-mismatched stripes, quantum wires, quantum dots, and substrates in Si technology, J. Appl. Phys. 79, 8145-65 (1996).

[142] S. C. Jain, A review of our recent work on Raman scattering and a tribute to K. S. Krishnan, Current Science 75, 1222-32 (1998).

[143] M. A. Hayat, *Basic techniques for transmission electron microscopy*, (Academic press, Orlando Fla. 1986); Peter R. Buseck, High-resolution transmission electron microscopy and associated techniques, (Oxford University Press, New York N.Y. 1988).

[144] J. Vanhellemont, K. G. F. Janssens, S. Frabbonu, R. Balboni, and A. Armigliato, Electron Microscopy Techniques for the assessment of localised stress distributions in semiconductors, Electrochemical Soc. Proc. 95-30, 174-93 (1995).

[145] R. Bierwolf, H. Hohenstein, F. Phillipp, O. Brandt, G. E. Crook and K. Ploog, Direct measurement of local lattice distortions in strained layer structures by HREM, Ultramicroscopy 49, 273-85 (1993).

[146] M. O. Möller, V. Beyersdorfer, D. Hommel, T. Behr, H. Heinke, T. Lippmann, and G. Landwehr, Structural characterization and interfacial studies of ZnSe based heterostructures on GaAs, J. Crystal Growth 143, 162-71 (1994).

[147] W. Faschinger, P. Juza, A. Pesek, and H. Sitter, Strain and relaxation in ZnSe/CdSe superlattices, Materials Sci. Forum, Vol. 143-147, 1617-22 (1994).

[148] P. F. Fewster, X-ray diffraction from low-dimensional structures, Semicond. Sci. Technol. 8, 1915-34 (1993).

[149] S. Bauer, M. Huber, C. Rüth, and W. Gebhardt, Strain relaxation in thin ZnTe epilayers on GaAs and ZnSe/GaAs, Materials Science Forum 143-147, 531-5 (1994).

[150] S. Fuji, Y. Kawakami and S. Fujita, MO(GS)MBE and photo-MO(GS)MBE of II-VI semiconductors, J. Crystal Growth 164, 196-201 (1996).

[151] S. Németh, B. Grietens, H. Bender, and G. Borghs, Molecular beam epitaxial growth of bulk $AlAs_{0.16}Sb_{0.84}$ and $AlAs_{0.16}Sb_{0.84}/InAs$ superlattices on lattice-matched InAs substrates, Jpn. J. Appl. Phys. 36, 3426-28 (1997).

[152] N. Grandjean and J. Massies, Real time control of $In_xGa_{1-x}N$ molecular beam epitaxy growth, Appl. Phys. Lett. 72, 1078-80 (1998).

[153] S. Nakamura, *In situ* monitoring of GaN growth using interference effects, Jpn. J. Appl. Phys. 30, 1620-7 (1991).

[154] M. Mesrine, N. Grandjean and J. Massies, Efficiency of NH_3 as nitrogen source for GaN molecular beam epitaxy, Appl. Phys. Lett. 72, 350-2 (1998).

[155] D. E. Aspnes, Real-time optical diagnostics for epitaxial growth, Surface Sci. 307-309, 1017-27 (1994).

[156] G. J. Glanner, H. Sitter, and W. Faschinger, Evaluation of growth temperature, refractive index, and layer thickness of thin ZnTe, MnTe, and CdTe films by *in situ* visible laser interferometry, Appl. Phys. Lett. 65, 998-1000 (1994).

[157] G. B. Stringfellow, Organometallic vapour phase epitaxy reaction kinetics, in *Handbook of Crystal Growth*, Vol. 3, Ed. D. T. J. Hurle, (Elsevier Science B. V. 1994), pp 491-540.

[158] W. Grieshaber, C. Bodin, J. Cibert, J. Gaj, Y. M. d'Aubigyné, A. Wasiela, and G. Feuillet, Rough versus dilute interfaces in semiconductor heterostructures: The role of growth conditions, Appl. Phys. Lett. 65, 1287-9 (1994).

[159] J. K. Furdyna, M. Qazzaz, G. Yang, L. Montes, S. H. Xin, and H. Luo, Investigation of strain in II-VI semiconductor superlattices using electron paramagnetic resonance of Mn^{++}, Acta Physica Polonica A 88, 607-18 (1995).

[160] M. A. Herman and H. Sitter, MBE growth physics: application to device technology, Microelectronics Journal 27, 257-96 (1996).

[161] R. L. Gunshor and A. V. Nurmikko, Blue-laser CD technology, Scientific American, July 1996, pp. 34-37.

[162] N. Grandjean and J. Massies, GaN and $Al_xGa_{1-x}N$ molecular beam epitaxy monitored by reflection high-energy electron diffraction, Appl. Phys. Lett. 71, 1816-8 (1997).

[163] W. S. Kuhn, B. Qu'Hen, O. Gorochov, R. Triboulet and W. Gebhardt, The metal organic vapour phase of ZnTe: II. Analysis of growth conditions, Prog. Crystal Growth and Charact. 31, 45-117 (1995).

[164] W. S. Kuhn, A. Lusson, B. Qu'Hen, C. Grattepain, H. Dumont, O. Gorochov, S. Bauer, K. Wolf, M. Wörz, T. Reisinger, A. Rosenauer, H. P. Wagner, H. Stanzl, and W. Gebhardt, The metal organic vapour phase of ZnTe: III. Correlation of growth and layer properties, Prog. Crystal Growth and Charact. 31, 119-117 (1995).

[165] M. Behet, R. Hövel, A. Kohl, A. Mesquida Küsters, B. Opitz and K. Heime, MOVPE growth of III-V compounds for optoelectronic and electronic applications, Microelectronics Journal 27, 297-334 (1996).

[166] H. Hardtdegen, Chr. Ungermanns, M. Hollfelder, T. Raafat, R. Carius, St. Hasenöhrl and H. Lüth, A new approach towards low-pressure metallorgnic vapour phase epitaxy of (AlGa)As using triethylgalllium and dimethylethylaminealane, J. Crystal Growth 145, 478-84 (1994); H. Hardtdegen and P. Giannoules, III-Vs, Vol. 8, March 1995, (see also references to earlier work of Hardtdegen and co-workers given in these papers).

[167] E. H. C. Parker, Ed.: *The technology and Physics of Molecular Beam Epitaxy*, (Plenum New York, 1985).

[168] L. L. Chang and K. Ploog, Eds.: *Molecular Beam Epitaxy and Heterostructures*, (Nijhoff, Dordrecht, 1985).

[169] W. Tong, B. K. Wagner, T. K. Tran, W. Ogle, W. Park, and C. J. Summers, Kinetics of chemical beam epitaxy for high quality ZnS film growth, J. Crystal Growth 164, 202-7 (1996).

[170] J. S. Foord, C. L. Levouer, and G. J. Davies, An investigation of ZnSe growth by chemical beam epitaxy using modulated beam scattering and related techniques, J. Crystal Growth 164, 190-5 (1996).

[171] G. Borghs, unpublished.

[172] V. M. Donnely and A. Robertson, Kinetic modelling of GaAs chemical beam epitaxy, Surface Science 293, 93-106 (1993).

[173] J. G. Nelson, Summary abstract: Epitaxial growth of ZnS and ZnSe on the low index faces of GaAs using atomic layer epitaxy, J. Vac. Sci. Technol. A 5, 2140-1 (1987).

[174] T. Yao, and Y. Okada, Phosphorus acceptor levels in ZnSe grown by molecular beam epitaxy, Jpn. J. Appl. Phys. 25, 821-7 (1986).

[175] F. Briones, I. González, and A. Ruiz, Atomic Layer Molecular Beam Epitaxy (Almbe) of III-V compounds: Growth modes and applications, Appl. Phys. A 49, 729-37 (1989).

[176] J. M. Gerard, J. Y. Marzin, B. Jusserand, F. Glas, and J. Primot, Structural and optical properties of high quality InAs/GaAs short-period superlattices grown by migration-enhanced epitaxy, Appl. Phys. Lett. 54, 30-32 (1989).

[177] Y. C. Chen, P. K. Bhattacharya, and J. Singh, Reflection high-energy electron diffraction studies of the growth of strained InGaAs on GaAs substrate by migration-enhanced epitaxy, Appl. Phys. Lett. 57, 692-4 (1990).

[178] M. Seta, H. Asahi, S. G. Kim, and S. Gonda, Gas source molecular beam epitaxy/migration enhanced epitaxy growth of InAs/AlSb superlattices, J. Appl. Phys. 74, 5033-7 (1993).

[179] C. D. Lee, S. I. Min, and S. K. Chang, High purity ZnSe epilayers grown by atmospheric double zone metalorganic atomic layer epitaxy, J. Crystal Growth 159, 108-11 (1996).

[180] M. Ishii, S. Iwai, T. Ueki, and Y. Aoyagi, Observation and control of surface morphology of AlP grown by atomic layer epitaxy, Appl. Phys. Lett. 71, 1044-6 (1997).

[181] D. B. Eason, Z. Yu, W. C. Hughes, W. H. Roland, C. Boney, J. W. Cook, Jr., and J. F. Schetzina, High-brightness blue and green light-emitting diodes, Appl. Phys. Lett. 66, 115-7 (1995).

[182] H. Wenisch, K. Schüll, T. Behr, D. Hommel, G. Landwehr, D. Siche, P. Rudolph, and H. Hartmann, (Cd,Zn)Se multi-quantum-well LEDs: homoepitaxy on ZnSe substrates and heteroepitaxy on (In,Ga)As/GaAs buffer layers, J. Crystal Growth 159, 26-31 (1996).

[183] H. H. Farrell, J. L. de Miguel, and M. C. Tamargo, Reflection high-energy electron diffraction electron-stimulated desorption from ZnSe(100)(2×1)-Se surfaces J. Appl. Phys. 65, 4084-6 (1989).

[184] R. Venkatasubramanian, N. Otsuka, J. Qui, L. A. Klolodziejski, and R. L. Gunshor, Incorporation processes in MBE growth of ZnSe, Molecular Beam Epitaxy 1988, Workbook of the Fifth International Conference 1988 MBE-V, Sapporo. Tokyo Inst. Technol, Tokyo, Japan; 1988, pp 294-7.

[185] D. L. Mathine, S. M. Durbin, R. L. Gunshor, M. Kobayashi. D. R. Menke, Z. Pei, J. Gonsalves, N. Otsuka, Q. Fu, M. Hagerott, A. V. Nurmikko, ZnTe/AlSb/GaSb heterostructures by molecular beam epitaxy, Appl. Phys. Lett. 55, 268-70 (1989).

[186] M. Yoneta, M. Ohishi, H. Saito, and T. Hamasaki, Low temperature molecular beam epitaxial growth of ZnS/GaAs(001) by using elemental sulphur source, J. Crystal Growth 127, 314-317 (1993).

[187] Correlation of room temperature photoluminescence to structural properties of ZnSSe/ZnSe superlattices grown by metalorganic vapour phase epitaxy, J. Appl. Phys. 74, 5880-2 (1993).

[188] A. Abounadi, M. Di Blasio, D. Bouchara, J. Calas, M. Averous, O. Briot, T. Cloitre, R. L. Aulombard, and B. Gill, Reflectivity and photoluminescence measurements in ZnS epilayers grown by metal-organic chemical-vapor deposition, Phys. Rev. B 50, 11677-83 (1994).

[189] Sz. Fujita and Sg. Fujita, Growth and characterization of ZnSe-based II–VI semiconductors by MOVPE, in Ref. [43], pp 59-81 (1997).

[190] M. Ohishi, K. Ohmori, Y. Fujii, H. Saito, and S. Tiong, Homo-epitaxial growth of ZnSe by MBE, J. Crystal Growth 86, 324-8 (1988).

[191] M. R. Gokhale, K. X. Bao, P. D. Healey, F. C. Jain, and J. E. Ayers, Factors influencing low-temperature photo-assisted OMVPE growth of ZnSe, J. Crystal Growth 165, 25-30 (1996).

[192] O. Pagès, M. A. Renucci, O. Briot, and R. L. Aulombard, Coupled LO–plasmon modes in semi-insulating GaAs of ZnSe/GaAs heterojunctions, J. Appl. Phys. 80, 1128-35 (1996).

[193] R. D. Vispute, J. Narayan, H. Wu, and K. Jagannadham, Epitaxial growth of AlN thin films on silicon (111) substrates by pulsed laser deposition, J. Appl. Phys. 77, 4724-7 (1995).

[194] C. R. Abernathy, Growth of Group III Nitrides from Molecular Beams, in *GaN and related materials*, Edited by S. J. Pearton, (Gordon and Breach Science Publishers 1997), pp 11-51

[195] F. A. Ponce, Defects and interfaces in GaN epitaxy, MRS Bulletin 22, pp 51-57, February 1997.

[196] S. D. Hersee, J. C. Ramer, and K. J. Malloy, The microstructure of metalorganic-chemical-vapor-deposition GaN on sapphire, MRS Bulletin 22, pp 45-51, July 1997.

[197] K. Dovidenko, S. Oktyabrsky, J. Narayan, and M. Razeghi, Aluminum nitride films on different orientations of sapphire and silicon, J. Appl. Phys. 79, 2439-45 (1996).

[198] T. C. Fu, N. Newman, E. Jones, J. S. Chan, X. Liu, M. D. Rubin, N. W. Cheung, and E. R. Weber, The influence of nitrogen ion energy on the quality of GaN films grown with molecular beam epitaxy, J. Elect. Mat. 24, 249-55 (1995).

[199] A. V. Blant, T. S, Cheng, C. T. Foxon, J. C. Bussey, S. V. Novikov, V. V. Tret'yakov, Studies of group III-nitride growth on silicon, III-V Nitrides (Mater. Res. Soc. Pittsburgh, PA, USA, 1997), Symposium 449, 465-9 (1997).

[200] R. D. Vispute, J. Narayan, H. Wu, and K. Jagannadham, Epitaxial growth of AlN thin films on silicon (111) substrates by pulsed laser deposition, J. Appl. Phys. 77, 4724-7 (1995).

[201] S. C. Binari, J. M. Redwing, and W. Kruppa, AlGaN/GaN HEMTs grown on SiC substrates, Electronics Lett. 33, 242-243 (1997).

[202] K. A. Brown, S. A. Ustin, L. Lauhon, and W. Ho, Supersonic jet epitaxy of aluminum nitride on silicon (100), J. Appl. Phys. 79, 7667-71 (1996).

[203] A. Nakadaira, and H. Tanaka, Metalorganic vapor-phase epitaxy of cubic $Al_xGa_{1-x}N$ alloy on a GaAs (100) substrate, Appl. Phys. Lett. 70, 2720-2 (1997).

[204] C.-K. Sun, F. Vallée, S. Keller, J. E. Bowers, and S. P. DenBaars, Femtosecond studies of carrier dynamics in InGaN, Appl. Phys. Lett. 70, 2004-6 (1997).

[205] T. Beierlein, S. Strite, A. Dommann and D. D. J. Smith, Properties of InGaN deposited on glass at low temperature, MRS Internet-J. Nitride-Semiconductor Research, vol. 2, 1997, Universal Resource Locator http://nsr.mij.mrs.org/2/29.

[206] K. Iwata, H. Asahi, K. Asami, R. Kuroiwa, and S. Gonda, Gas Source molecular beam epitaxy growth of GaN on C-, A-, R- and M-plane sapphire and silica glass substrates, Jpn. J. Appl. Phys. 36, Part 2, L661-4 (1997).

[207] R. F. Davis, M. J. Paisley, Z. Sitar, D. J. Kester, K. S. Ailey, K. Linthicum, L. B. Rowland, S. Tanaka, R. S. Kern, Gas-Source molecular beam epitaxy of III-V nitrides, J. Crystal Growth 178, 87-101 (1997).

[208] S. Nakamura, M. Senoh, S. Nagahama, N. Iwasa, T. Yamada, T. Matsushita, H. Kiyoku, and Y. Sugimoto, T. Kozaki, H. Umemoto, M. Sano, and K. Chocho, InGaN/GaN/AlGaN-based laser diodes with modulation-doped strained-layer superlattices grown on an epitaxially laterally overgrown GaN substrate, Appl. Phys. Lett. 72, 211-3 (1998).

[209] T. S. Zheleva, O.-H. Nam, M. D. Bremser, and R. F. Davis, Dislocation density reduction via lateral epitaxy in selectively grown GaN structures, Appl. Phys. Lett. 71, 2472-4 (1997).

[210] J. A. Freitas Jr., O.-H. Nam, and R. F. Davis, Optical characterization of lateral epitaxial overgrown GaN layers, Appl. Phys. Lett. 72, 2990-2 (1998).

[211] A. Sakai, H. Sunakawa, and A. Usui, Defect structure in selectively grown GaN films with low threading dislocation density, Appl. Phys. Lett. 71, 2259-61 (1997).

[212] T. S. Zheleva, S. A. Smith, D. B. Thomson, K. J. Linthicum, and R. F. Davis, unpublished.

[213] C. Sasaoka, H. Sunakawa, A. Kimura, M. Nido, A. Usui, and A. Sakai, High-quality MQW on low-dislocation-density GaN substrate by hydride vapor-phase epitaxy, J. Crystal Growth 189/190, 61-66 (1998).

[214] S. Nakamura, M. Senoh, S. Nagahama, N. Iwasa, T. Yamada, T. Matsushita, H. Kiyoku, and Y. Sugimoto, InGaN/GaN/AlGaN-based laser diodes with modulation-doped strained-layer superlattice, Jpn. J. Appl. Phys. 36 (12A), L1568-71 (1997).

[215] O. H. Nam, T. S. Zheleva, D. B. Thomson, and R. F. Davis, Organometallic vapor phase lateral epitaxy of low defect density GaN layers, (Mater. Res. Soc. Pittsburgh, PA, USA, 1998), Symp. 482, 301-6 (1998).

[216] D. Walker, X. Zhang, A. Saxler, P. Kung, and M. Razeghi, $Al_xGa_{1-x}N$ ($0 \leq x \leq 1$) ultraviolet photodetectors grown on sapphire by metal-organic chemical-vapor deposition, Appl. Phys. Lett. 70, 949-51 (1997).

[217] C. J. Sun, J. W. Yang, B. W. Lim, Q. Chen, M. Zubair Anwar, and M. Asif Khan, Mg-doped green light emitting diodes over cubic (111) $MgAl_2O_4$ substrate, Appl. Phys. Lett. 70, 1444-6 (1997).

[218] S. Nakamura, M. Senoh, S. Nagahama, N. Iwasa, T. Yamada, T. Matsushita, H. Kiyoku, and Y. Sugimoto, InGaN-based multi-quantum-well-structure laser diodes, Jpn. J. Appl. Phys. 35, L74-76 (1996)

[219] S. Nakamura, GaN growth using GaN buffer layer, Jpn. J. Appl. Phys. 30 (No. 10A), L1705-7 (1991).

[220] T. Kachi, K. Tomita, K. Itoh, and H. Tadano, A new buffer layer for high quality GaN grown by metalorganic vapor phase epitaxy, Appl. Phys. Lett. 72, 704-6, (1998).

[221] S. Yoshida, Photoluminescence measurement of InGaN and GaN grown by gas-source molecular beam epitaxy method, J. Appl. Phys. 81, 7966-9 (1997).

[222] T. D. Moustakas, and R. J. Molnar, Growth and doping by ECR-assisted MBE, (Mater. Res. Soc. Pittsburgh, PA, USA, 1993), Symp. 281, 253-63 (1993).

[223] J. C. Zolper, and R. J. Shul, Implantation and dry etching of Group-III-Nitride semiconductors, see pp 36-43 in Ref. [90].

[224] R. C. Powell, N.-E. Lee, Y.-W. Kim, and J. E. Greene, Heteroepitaxial wurtzite and zinc-blende structure GaN grown by reactive-ion molecular-beam epitaxy: Growth kinetics, microstructure and properties, J. Appl. Phys. 73, 189-204 (1993).

[225] N. Grandjean and J. Massies, Ultraviolet GaN light-emitting diodes grown by molecular beam epitaxy using NH_3, Appl. Phys. Lett. 72, 82-4 (1998).

[226] R. C. Powell, N.-E. Lee. and J. E. Greene, Growth of GaN(0001)1 × 1 on Al_2O_3(0001) by gas-source molecular beam epitaxy, Appl. Phys. Lett. 60, 2505-7 (1992).

[227] A. Rizzi, D. Freundt, D. Holz, and H. Lüth, Valence band offsets at the AlN/SiC interface: Evidence of internal piezoelectric fields, in Ref. [88], paper Th1-1, pp 416-7.

[228] A. Rizzi, 6H-SiC(0001)/AlN/GaN epitaxial heterojunctions and their valence band offsets, Nato ASI series, Heterostructure Epitaxy and Devices, Ed.: Peter Korols, to be published.

[229] W. J. Meng, J. A. Sell, T. A. Parry, L. E. Rehn, and P. M. Baldo, Growth of Al nitride thin films on Si(111) and Si(001): Structural characteristics and development of intrinsic stresses, J. Appl. Phys. 75, 3446-55 (1994).

[230] B. A. Ferguson, and C. B. Mullins, Supersonic jet epitaxy of III-nitride semiconductors, J. Crystal Growth 178, 134-46 (1997).

[231] D. Korakakis, H. M. Ng, K. F. Ludwig, Jr., and T. D. Moustakas, Doping studies of n- and p-$Al_xGa_{1-x}N$ grown by ECR assisted MBE, in Proc. (Mater. Res. Soc. Pittsburgh, PA, USA, 1997), Symp. 449, 233-8 (1997).

[232] V. Mamutin, V. Jmerik, T. Shubina, V. Ratnikov, S. Ivanov, P. Kop'v, M. Karlsteen, U. Södervall, and M. Willander, Plasma-assisted MBE growth of GaN and InGaN on different substrates, IX MBE International Conference, Cannes, 1998, to be published in J. Crystal Growth.

[233] R. Singh, and T. D. Moustakas, Growth of InGaN films by MBE at the growth temperature of GaN, (Mater. Res. Soc. Pittsburgh, PA, USA, 1996), Gallium Nitride and Related Materials, First International Symposium, 163-8 (1996).

[234] M. S. Minsky, S. B. Fleischer, A. C. Abare, J. E. Bowers, and E. L. Hu, Characterization of high-quality InGaN/GaN multiquantum wells with time-resolved photoluminescence, Appl. Phys. Lett. 72, 1066-8 (1998).

[235] S. C. Jain, A. H. Harker and R. A. Cowley, Misfit-strain and misfit-dislocations in lattice mismatched epitaxial layers, Philos. Mag. A 75 (6), 1461-515 (1997).

[236] J. P. Hirth, and J. Lothe, *Theory of Dislocations* (McGraw-Hill, New York 1968).

[237] U. Jain, S. C. Jain and A. H. Harker, Nucleation of dislocation loops in strained epitaxial layers, J. Appl. Phys. 77, 103-5 (1995).

[238] S. C. Jain, T. J. Gosling, J. R. Willis, D. H. J. Totterdell, and R. Bullough, A new study of critical layer thickness, stability and strain relaxation in pseudomorphic Ge_xSi_{1-x} strained epilayers, Phil. Mag. A 65, 1151-67 (1992).

[239] S. C. Jain, T. J. Gosling, J. R. Willis, R. Bullough and P. Balk, A theoretical comparison of the stability characteristics of capped and uncapped GeSi strained epilayers, Solid-State Electron. 35, 1073-9 (1992).

[240] A. Atkinson and S. C. Jain, The energy of finite system of misfit dislocations in epitaxial strained layers, J. Appl. Phys. 72, 2242-8 (1992); A. Atkinson and S. C. Jain, The energy of systems of misfit dislocations in epitaxial strained layers, Thin Solid Films 222, 161-5 (1992); A. Atkinson and S. C. Jain, A new approach to calculating the energy of systems of misfit dislocations in strained epitaxial layers, J. Phys. Condens. Matter 5, 4595-600 (1993).

[241] U. Jain, S. C. Jain, A. Atkinson, J. Nijs, R. Mertens and R. Van Overstraeten, Energy of arrays on nonperiodic interacting dislocations in semiconductor strained epilayers: Implications for train relaxation, J. Appl. Phys. 73, 1773-80 (1993).

[242] U. Jain, S. C. Jain, J. Nijs, J. R. Willis, R. Bullough, R. Mertens and R. Van Overstraeten, Effect of dislocation-dislocation interactions on critical-layer-thickness and strain relaxation in strained Ge_xSi_{1-x} layers with 60° and 90° dislocations, Solid-State Electron. 36, 331-7 (1993).

[243] B. W. Dodson, Metastability in Si/Ge strained layer structures, in *Heteroepitaxy on Silicon: Fundamentals, Structure and Devices,* Edited by H. K. Choi, R. Hull, H. Ishiwara, and R. H. Nemanich (Mater. Res. Soc. Pittsburgh, PA, USA, 1988), Symp. 116, pp. 491-503.

[244] C. G. Tuppen, and C. J. Gibbings, A quantitative analysis of strain relaxation by misfit dislocation glide in $Si_{1-x}Ge_x$/Si heterostructures, J. Appl. Phys. 68, 1526-34 (1990).

[245] R. Hull, J. C. Bean, D. Bahnck, L. J. Peticolas, Jr., K. T. Short, and F. C. Unterwald, Interpretation of dislocation propagation velocities in strained Ge_xSi_{1-x}/Si(100) heterostructures by the diffusive kink pair model, J. Appl. Phys. 70, 2052-65 (1991).

[246] D. C. Houghton, Misfit dislocation dynamics in $Si_{1-x}Ge_x$/(100) Si: Uncapped alloy layers, buried strained layers, and multiple quantum wells, Appl. Phys. Lett. 57, 1434-6 (1990); D. C. Houghton, Nucleation rate and glide velocity of misfit dislocations in SiGe/(100) Si heterostructures, Appl. Phys. Lett. 57, 2124-6 (1990).

[247] D. C. Houghton, Strain relaxation kinetics in $Si_{1-x}Ge_x$/Si heterostructures, J. Appl. Phys. 70, 2136-51 (1991).

BIBLIOGRAPHY

[248] F. Louchet, D. Cochet Muchy, and Y. Brechet, Investigation of dislocation mobilities in germanium in the low-temperature range by *in situ* straining experiments, Philos. Mag. A 57, 327-35 (1988); F. Louchet, Evidence of a transition in glide mechanism of dislocations in silicon by weak-beam *in situ* straining experiments, Inst. Phys. Conf. Ser. No. 60, 35-8 (1981).

[249] J. M. Bonar, R. Hill, J. F. Walker, and R. Malik, Observations of new misfit dislocation configurations and slip systems at ultrahigh stresses in the (Al)GaAs/In$_x$Ga$_{1-x}$As/GaAs(100) system, Appl. Phys. Lett. 60, 1327-9 (1992).

[250] D. C. Paine, D. J. Howard, D. Luo, R. N. Sacks, and T. C. Eschrich, The relaxation of In$_x$Ga$_{1-x}$As/GaAs strained multilayers, (Mater. Res. Soc. Pittsburgh, PA, USA, 1990), Symp. 160, 123-8 (1990).

[251] K. Sumino, and I. Yonenaga, Mechanical behaviour of semiconductors in terms of dislocations dynamics,, Solid State Phenomena 19-20, 295-310, (1991).

[252] Yu. A. Osip'yan, V. F. Petrenko, A. V. Zaretskiĭ, Properties of II-VI semiconductors associated with moving dislocations, Advances in Phys. 35, 115-88 (1986).

[253] C. Levade, A. Faress, and G. Vanderschaeve, A TEM *in situ* investigation of dislocation mobility in II-VI semiconductor compound ZnS, A quantitative study of the cathodoplastic effect, Philos. Mag. A 69, 855-70 (1994).

[254] L. Sugiura, J. Nishio, M. Onomura, and S. Nunoue, Nitride laser diodes with InGaN based MQW structures, (Mater. Res. Soc. Pittsburgh, PA, USA, 1998), Symp. 482, 1157-67 (1998).

[255] L. Sugiura, Dislocation motion in GaN light-emitting devices and its effect on device lifetime, J. Appl. Phys. 81, 1633-8 (1998).

[256] T. J. Gosling, S. C Jain and A. H. Harker, The Kinetics of strain relaxation in lattice-mismatched semiconductor layers, Physica Status Solidi (a) 146, 713-34 (1994).

[257] R. Hull, and J. C. Bean, Nucleation of misfit dislocations in strained-layer epitaxy in the Ge$_x$Si$_{1-x}$/Si system, J. Vac. Sci. Technol. A 7, 2580-5 (1989).

[258] D. J. Eaglesham, E. P. Kvam, D. M. Maher, C. J. Humphreys, and J. C. Bean, Dislocation nucleation near the critical thickness in GeSi/Si strained layers, Phil. Mag. A 59, 1059-73 (1989).

[259] C. G. Tuppen, and C. J. Gibbings, Misfit dislocation multiplication processes in SiGe alloys for $x < 0.15$, Appl. Phys. Lett. 56, 54-6 (1990).

[260] V. Higgs, P. Kightley, P. J. Goodhew, and P. D. Augustus, Metal-induced dislocation nucleation for metastable SiGe/Si, Appl. Phys. Lett. 59, 829-31 (1991).

[261] R. Hull, J. C. Bean, D. Bahnck, J. M. Bonar, and L. J. Peticolas, Dynamic observation of misfit dislocations in strained layer heterostructures, Proc. Microsc. Semicond. Mater. Conf., Inst. Phys. Conf. Ser. No. 117, 497-508 (1991).

[262] R. Hull, J.C. Bean and C. Buescher, A phenomenological description of strain relaxation in Ge_xSi_{1-x}/Si heterostructures, J. Appl. Phys. 66, 5837-43 (1989).

[263] G. Cohen-Solal, F. Bailly, and M. Barbé. Critical thickness in heteroepitaxial growth of zinc-blende semiconductor compounds, J. Crystal Growth 138, 68-74 (1994).

[264] D. J. Olego, K. Shahzad, J. Petruzzello, and D. A. Cammack, Depth profiling of elastic strains in lattice-mismatched semiconductor heterostructures and strained-layer superlattices, Phys. Rev. B 36, 7674-7 (1987).

[265] K. Shahzad, D. J. Olego, and D. A. Cammack, Thickness dependence of strains in strained-layer superlattices, Appl. Phys. Lett. 52, 1416-8 (1988).

[266] G. I. Hobson, B. Khamsehpour, K. E. Singer, and W. S. Truscott, MBE growth of pseudomorphic $GaAs_ySb_{1-y}$ in a GaAs host lattice, J. Crystal Growth 95, 220-3 (1989).

[267] S-L. Weng, Experimental studies of misfit dependence of critical layer thickness in pseudomorphic InGaAs single-strained quantum-well structures, J. Appl. Phys. 66, 2217-9 (1989).

[268] P. J. Orders and B. F. Usher, Determination of critical layer thickness in $In_xGa_{1-x}As$/GaAs heterostructures by x-ray diffraction, Appl. Phys. Lett. 50, 980-2 (1987).

[269] P. J. Gourley, I. J. Fritz, and L. R. Dawson, Controversy of critical layer thickness for InGaAs/GaAs strained-layer epitaxy, Appl. Phys. Lett. 52, 377-9 (1988).

[270] T. Taguchi, Y. Takeuchi, K. Matugatani, Y. Ueno, T. Hattori, Y. Sugiyama, and M. Tacano, Critical layer thickness of $In_{0.80}Ga_{0.20}As$/$In_{0.52}Al_{0.48}As$ heterostructures, J. Crystal Growth, 134, 147 (1993).

[271] D. C. Houghton, C. J. Gibbings, C. G. Tuppen, M. H. Lyons, and M. A. G. Halliwell, Equilibrium critical thickness for $Si_{1-x}Ge_x$ strained layers on (100) Si, Appl. Phys. Lett. 56, 460-2 (1990).

[272] T. G. Andersson, Z. G. Chen, V. D. Kulakovskii, A. Uddin, and J. T. Vallin, Variation of critical layer thickness with In content in strained quantum wells grown by molecular beam epitaxy, Appl. Phys. Lett. 51, 752-4 (1987).

[273] D. C. Houghton, C. J. Gibbings, C. G. Tuppen, M. H. Lyons and M. A. G. Halliwell, The structural and stability of uncapped versus buried Ge_xSi_{1-x} strained layers through high temperature processing, Thin Solid Films 183, 171-82 (1989).

[274] T. Yao, Y. Okada, S. Matsui, and K. Ishida, The effect of lattice deformation on optical properties and lattice parameters of ZnSe grown on (100)GaAs, J. Crystal growth 81, 518-23 (1987).

[275] G. Horsburgh, K. A. Prior, W. Meredith, I. Galbraith B. C. Cavenett, C. R. Whitehouse, G. Lacey, A. G. Cullis, P. J. Parbrook, P. Möck, and K. Mizuno, Topography measurements of the critical thickness of ZnSe grown on GaAs, Appl. Phys. Lett. 72, 3148-50 (1998).

[276] J. Neugebauer, and C. G. Van de Walle, Native defects and impurities in cubic and wurtzite GaN, (Mater. Res. Soc. Pittsburgh, PA, USA, 1994), Symp. 339, 687-92 (1994).

[277] A. C. Chami, E. Ligeon, R. Danielou, J. Fontenille, G. Lentz, N. Magnea, and H. Mariette, Strain relaxation in low lattice mismatch epitaxy of $CdTe/Cd_{0.97}Zn_{0.03}Te$(001) by channeling, Appl. Phys. Lett. 52, 1874-6 (1988).

[278] J. Cibert, Y. Gobil, Le Si Dang, S. Tatarenko, G. Feuillet, P. H. Jouneau, and K. Saminadayar, Critical thickness in epitaxial CdTe/ZnTe, Appl. Phys. Lett. 56, 292-4 (1990).

[279] C. Fontaine, J. P. Gaillard, S. Magli, A. Million, and J. Piaguet, Relaxation of stresses in CdTe layers grown by molecular beam epitaxy, Appl. Phys. Lett. 50, 903-5 (1987).

[280] S. Kret, G. Karczewski, A. Zakrzewski, P. Dluzewski, A. Dubon, T. Wojtowicz, J. Kossut, C. Delamarre, and J. Y. Laval, Lattice parameter relaxation during MBE of $ZnTe/Cd_{1-x}Zn_xTe/Cd_{0.5}Mn_{0.5}Te$ buffer layers by RHEED and HRTEM, Acta Physica Polonica A 88, 795-8 (1995).

[281] J. H. Li, G. Bauer, J. Stangl, L. Vanzetti, L. Sorba, and A. Franciosi, Strain and structural characterization of $Zn_{1-x}Cd_xSe$ structures grown on GaAs and InGaAs (001) substrates, J. Appl. Phys. 80, 81-8 (1996).

[282] K. Pinardi, S. C. Jain, M. Willander, A. Atkinson, H. E. Maes, and R. Van Overstraeten, A method to interpret micro-Raman experiments made to measure nonuniform stresses: Application to local oxidation of silicon, J. Appl. Phys. 84, 2507-12 (1998).

[283] H. J. Osten, H. P. Zeindl, and E. Bugiel, Consideration about the critical thickness for pseudomorphic $Si_{1-x}Ge_x$ growth on Si(100), J. Crystal Growth 143, 194-9 (1994).

[284] J. R. Waldrop and R. W. Grant, Mesuremnent of AlN/GaN (0001) heterojunction band offsets by X-ray photoemission spectroscopy, Appl. Phys. Lett. 68, 2879-81 (1996).

[285] G. Martin, A. Botchkarev, A. Rockett, and H. Morkoç, Valence band discontinuitiews of wurtzite GaN, AlN, and InN heterojunctions measured by x-ray photoemission spectroscopy, Appl. Phys. Lett. 68, 2541-3 (1996).

[286] S. Zemon, S. K. Shastry, P. Norris, C. Jagannath, and G. Lambert, Photoluminescence and excitation spectra of GaAs grown on Si, Solid State Commun. 58, 457-60 (1986).

[287] M. Sugo, N. Uchida, A. Yamamoto, T Nishioka, and M. Yamaguchi, Residual strains in heteroepitaxial III-V semiconductor films on Si(100) substrates, J. Appl. Phys. 65, 591-5 (1989).

[288] D. S. Wuu, R. H. Horng, and M. K. Lee, Strain variations in heteroepitaxial InP-on-Si grown by low-pressure metalorganic chemical vapor deposition, J. Appl. Phys. 54, 2244-6 (1989).

[289] H.-H. Wehmann, G.-P. Tang and A. Schlachetzki, Strain relaxation and threading dislocation density in lattice-mismatched semiconductor systems, Solid State Phenomena 32-33, 445-50 (1993).

[290] M. Grundmann, A. Krost, and D. Bimberg, Crystallographic and optical properties of InP/Si(100) grown by low temperature MOCVD process, Surface Science 267, 47-9 (1992).

[291] T. D. Harris, M. G. Lamont, R. Sauer, R. M. Lum, and J. K. Klingert, Near-gap photoluminescence of GaAs grown directly on silicon, J. Appl. Phys. 64, 5110-6 (1988).

[292] A. Sacedón, F. Calle, A. L. Alvarez, E. Calleja, and E. Munõz, R. Beanland, and P. Goodhew, Relaxation of InGaAs layers grown on (111)B GaAs, Appl. Phys. Lett. 65, 3212-4 (1994).

[293] K. Lu, P. A. Fisher, J. L. House, E. Ho, C. A. Coronado, G. S. Petrich, G.-C. Hua, and N. Otsuka, Gas source molecular beam epitaxy growth of ZnSe on novel buffer layers, J. Vac. Sci. Technol. B 12, 1153-5 (1994).

[294] S. Tatarenko, P. H. Jouneau, K. Saminadayar, and J. Eymery, Investigation of the epitaxial growth mechanism of ZnTe on (001) CdTe, J. Appl. Phys. 77, 3104-10 (1995).

[295] J. Massies and N. Grandjean, Oscillation of the lattice relaxation in layer-by-layer epitaxial growth of highly strained materials, Phys. Rev. Lett. 71, 1411-4 (1993).

BIBLIOGRAPHY 305

[296] S. C. Jain, H. Maes, and K. Pinardi, Stresses in strained GeSi stripes and quantum structures: calculation using the finite element method and determination using micro-Raman and other measurements, Thin Solid Films 292, 218-26 (1997).

[297] C. W. Snyder, B. G. Orr, D. Kessler, and L. M. Sander, Effect of strain on surface morphology of highly strained InGaAs films, Phys. Rev. Lett. 66, 3032-5 (1991).

[298] O. Brandt, K. Ploog, R. Bierwolf, and M. Hohenstein, Breakdown of continuum elasticity theory in the limit of monatomic films, Phys. Rev. Lett. 68, 1339-42 (1992).

[299] A. F. Schwartzman and R. Sinclair, Metastable and equilibrium defect structure of II-VI/GaAs interfaces, J. Electronic Materials 20, 805-14 (1991).

[300] U. Jain, S. C. Jain, A. H. Harker and R. Bullough, Nucleation of dislocation loops in epitaxial layers, J. Appl. Phys. 77, 103-8 (1995).

[301] T. W. Kim, H. L. Park, and J. Y. Lee, Interfacial stages of the ZnTe/GaAs strained heterostructures grown by temperature-gradient vapor transport deposition at low temperature, Appl. Phys. Lett. 64, 2526-8 (1994).

[302] T. Kozawa, T. Kachi, H. Kano, H. Nagase, N. Koide, and K. Manabe, Thermal stress in GaN epitaxial layers grown on sapphire substrates, J. Appl. Phys. 77, 4389-92 (1995).

[303] S. F, Yoon, X. B. Li, and M. Y. Kong, Some properties of gallium nitride films grown on (0001) oriented sapphire substrates by gas source molecular beam epitaxy, J. Crystal Growth 180, 27-33 (1997).

[304] J. W. Ager III, T. Suski, S. Ruvimov, J. Krueger, G. Conti, E. R. Weber, M. D. Bremser, R. Davis, and C. P. Kuo, Intrinsic and thermal stress in Gallium Nitride epitaxial films, (Mater. Res. Soc. Pittsburgh, PA, USA, 1997), Symp. 449, 775-9 (1997).

[305] H. Siegle, A. Hoffmann, L. Eckey, C. Thomsen, T. Detchprohm, K. Hiramatsu, T. Davis, and J. W. Steeds, Depth-profile of the excitonic luminescence in Gallium-Nitride layers, (Mater. Res. Soc. Pittsburgh, PA, USA, 1997), Symp. 449, 677-81 (1997).

[306] S. Limpijumnong, W. R. L. Lambrecht, B. Segall, and K. Kim, Band structure and cation ordering in $LiGaO_2$, (Mater. Res. Soc. Pittsburgh, PA, USA, 1997), Symp. 449, 905-10 (1997).

[307] K. Hiramatsu, Detchprohm, and I. Akasaki, Relaxation mechanism of thermal stresses in the heterostructure of GaN grown on sapphire by vapor phase epitaxy Jpn. J. Appl. Phys. 32T, 1528-32 (1993).

[308] S. C. Jain, K. Pinardi, M. Willander, A. Atkinson, H. E. Maes, and R. Van Overstraeten, Raman spectra of GeSi strained quantum wires, Semicond. Sci. Technol. 12, 1507-9 (1997).

[309] N. Lovergine, L. Liaci, J. D. Ganiere, G. Leo, A. V. Drigo, F. Romanato, A. M. Mancini, and L. Vasanelli, Inhomogeneous strain relaxation and defect distribution of ZnTe layers deposited on (100)GaAs by metalorganic vapor phase epitaxy, J. Appl. Phys. 78, 229-35 (1995).

[310] S. C. Jain, K. Pinardi, H. E. Maes, R. Van Overstraeten. M. Willander, and A. Atkinson, Dislocations in GaN/sapphire: their distribution and effect on stress and optical properties, (Mater. Res. Soc. Pittsburgh, PA, USA, 1998), Symp. 482, 875-80 (1998).

[311] K. Dovidenko, S. Oktyabrsky, and J. Narayan, Characteristics of stacking faults in AlN thin films, J. Appl. Phys. 82, 4296-9 (1997).

[312] K. Dovidenko, S. Oktyabrsky, and J. Narayan, J. Mater. Res. to be published.

[313] T. Sugahara, H. Sato, M. Hao, Y. Nagi, S. Kurai, S. Tottori, K. Yamashita, K. Nishino, L. T. Romano, and S. Sakai, Direct evidence that dislocations are non-radiative recombination centers in GaN, Jpn. J. Appl. Phys. 37, L398-400 (1998).

[314] S. Ruvimov Z. W. Liliental, C. Dieker, J. Washburn, M. Koike, and H. Amano, TEM/HREM analysis of defects in GaN epitaxial layers grown by MOVPE on SiC and sapphire, Gallium Nitride and Related Materials II, (Mater. Res. Soc. Pittsburgh, PA, USA, 1997), Symposium 468, 287-92 (1997).

[315] P. Vennegues, B. Beaumont, M. Vaille, and P. Gibbart, Microstructure of GaN epitaxial films at different stages of the growth process on sapphire (0001), J. Crystal Growth 173, 249-59 (1997).

[316] J. L. Rouviere, M. Arlery, R, Niebuhr, K. H. Bachem, and O. Briot, Correlation between surface morphologies and crystallographic structures of GaN layers grown by MOCVD on sapphire, J. Nitride Semiconductor Research 1, Article 33 (1996).

[317] J. L. Rouviere, M. Arlery, R. Niebuhr, K. H. Bachem, and O. Briot, Transmission electron microscopy characterization of GaN layers grown by MOCVD on sapphire, Materials Science and Engineering B 43, 161-6 (1997).

[318] S. S. Lester, F. A. Ponce, M. G. Craford, and D. A. Steigerwald, High dislocation densities in high-efficiency GaN-based light emitting diodes, Appl. Phys. Lett. 66, 1249-51 (1995).

[319] B. K. Meyer, A. Hoffmann, and P. Thurian, Defect spectroscopy in the nitrides, in Ref. [47], pp 242-306.

[320] S. J. Rosner, E. C. Carr, M. J. Ludowise, G. Girolami, and H. Erikson, Correlation of cathodoluminescence inhomogeneity with microstructural defects in epitaxial GaN grown by metalorganic chemical-vapor deposition, Appl. Phys. Lett. 70, 420-2 (1997).

[321] S. Chichibu, T. Azuhata, T. Sota, and S. Nakamura, Luminescence from localized states in InGaN epilayers, Appl. Phys. Lett. 70, 2822-4 (1997).

[322] P. De Pauw, R. Mertens, R. Van Overstraeten, and S. C. Jain, Solid State Electronics 25, 67 (1984).

[323] R. Singh, W. D. Herzog, D. Doppalapudi, M. S. Unlu, B. B. Goldberg, T. D. Moustakas, MBE growth and optical characterization of InGaN/AlGaN multi-quantum wells, III-V Nitrides, (Mater. Res. Soc. Pittsburgh, PA, USA, 1997), Symposium 449, 185-90 (1997).

[324] W. Knap et al., Cyclotron resonance and quantum Hall effect studies of the two-dimensional electron gas confined at the GaN/AlGaN interface, Appl. Phys. Lett. 70, 2123-5 (1997).

[325] K. Shahzad, D. J. Olego, and C. G. Van de Walle, Optical characterization and band offsets in $ZnSe-ZnS_xSe_{1-x}$ strained-layer superlattices, Phys. Rev. B 38, 1417-26 (1988).

[326] K. Ohkawa, T. Mitsuyu, and O. Yamazaki, Effect of biaxial strain on exciton luminescence of heteroepitaxial ZnSe layers, Phys. Rev. B 38, 12465-9 (1989).

[327] M. Suzuki and T. Uenoyama, Electronic and optical properties of GaN-based quantum wells, in Ref. [47], pp 307-42.

[328] G. D. Chen, M. Smith, J. Y. Lin, H. X. Jiang, S.-H. Wei, M. A. Khan and C. J. Sun, Fundamental optical transitions in GaN, Appl. Phys. Lett. 68, 2784-6 (1996).

[329] S.-H. Wei, and A. Zunger, Valence band splitting and band offsets of AlN, GaN and InN, Appl. Phys. Lett. 69, 2710-21 (1996).

[330] J. J. Song, and W. Shan, Optical properties and lasing in GaN, in Ref. [47], pp 182-241.

[331] W. Götz, N. M. Johnson, C. Chen, H. Liu, C. Kuo, and W. Imler, Activation energies of Si donors in GaN, Appl. Phys. Lett. 68, 3144-6 (1996).

[332] W. Shan, T. J. Schmidt, X. H. Yang, S. J. Hwang, J. J. Song, and B. Goldenberg, Temperature dependence of interband transitions in GaN grown by metalorganic chemical vapor deposition, Appl. Phys. Lett. 66, 985-7 (1995).

[333] B. Monemar, J. P. Bergman, and I. A. Buyanova, Optical characterization of GaN and related materials, in *GaN and related materials*, Edited by S. J. Pearton, (Gordon and Breach Science Publishers 1997), pp 85-139.

[334] C. G. Van de Walle, Doping of wide bandgap II-VI compounds–theory, in Ref. [43], pp 122-162.

[335] S. Rubini, E. Milocco, L. Sorba, and A. Franciosi, Transitivity of the band offsets in II-VI/III-V heterojunctions, J. Crystal Growth 184-185, 178-82 (1998).

[336] P. Rodriguez-Hernandez, M. Gonzalez-Diaz, and A. Munoz, First principles calculations of the band offset of the ZnSe/BeTe heterojunctions, Applied Surface Science 123-124, 445-8 (1998).

[337] E. T. Yu, M. C. Phillips, J. O. McCaldin, and T. C. McGill, Measurement of the CdSe/ZnTe valence band offset by x-ray photoelectron spectroscopy, J. Vac. Sci. Technol. B 9, 2233-7 (1991).

[338] J. Tersoff, Band lineups at II-VI heterojunctions: Failure of the Common-anion rule, Phys. Rev. Lett. 56, 2755-8 (1986).

[339] J. Tersoff, Theory of semiconductor heterojunctions: the role of quantum dipoles, Phys. Rev. B 30, 4874-7 (1984); see also J. Tersoff, in *Heterojunction Band Discontinuities*, edited by F. Capasso and G. Margaritondo, (North-Holland, New York, 1987), chapter 1.

[340] W. A. Harrison, Elementary theory of heterojunctions, J. Vac. Sci. Technol. 14, 1016-21 (1977).

[341] A. D. Katnani, and G. Margaritondo, Empirical rule to predict heterojunction band discontinuities, J. Appl. Phys. 54, 2522-5 (1983).

[342] F. L. Schuermeyer, P. Cook, E. Martinez, and J. Tantillo, Band-edge alignment in heterostructures, Appl. Phys. Lett. 55, 1877-8 (1989).

[343] A. Ichii, Y. Tsou, and E. Garmire, An empirical rule for band offsets between III-V alloy compounds, J. Appl. Phys. 74, 2112-3 (1993).

[344] J.-L. Shieh, J.-I. Chyi, R.-J. Lin, R.-M. Lin and J.-W. Pan, Band Offsets of $In_{0.3}Ga_{0.7}As/In_{0.29}Al_{0.71}As$ heterojunction grown on GaAs substrate, Electronics Lett. 30, 2172-3 (1994).

[345] Y. B. Li, P. J. Tang, C. C. Phillips, R. A. Stradling, W. T. Yuen, D. J. Bain, C. M. Ciesla, M. Livingstone, I. Galbraith, C. R. Pidgeon, and L. Hart, Band alignments and offsets in In(As,Sb)/InAs superlattices, Proc. 23rd International Conference on the Physics of Semiconductors, (World Scientific, Singapore 1996), pp 1755-8, vol. 3.

[346] S. H. Wei and A. Zunger, InAsSb/InAs: A Type I or a type-II band alignment, Phys. Rev. B 55, 12039-44 (1995).

BIBLIOGRAPHY

[347] A. Tiwari, and D. J. Frank, Empirical fit to band discontinuities and barrier heights in III-V alloy systems, Appl. Phys. Lett. 60, 630-2 (1992).

[348] A. V. Platonov, D. R. Yakovlev, U. Zehnder, V. P. Kochereshko, W. Ossau, F. Fischer, T. Litz, A. Waag, and G. Landwehr, Homogeneous line-width of the direct exciton in a type-II ZnSe/BeTe quantum well, J. Crystal Growth 184-185, 801-5 (1998).

[349] W. J. Walecki, A. V. Nurmikko, N. Samarth, H. Luo, J. K. Furdyna, and N. Otsuka, Band offsets and exciton confinement in $Zn_{1-y}Cd_ySe/Zn_{1-x}Mn_xSe$ quantum wells, Appl. Phys. Lett. 57, 466-8 (1990).

[350] Y. Yang, Y. Yang, W. Li, W. Li, L. Yu, X. Xiong, S. Wang, and H. Huang, Study of the band offset for ZnSe(100)/GaAs(100) heterojunctions grown by hot wall epitaxy, J. Crystal Growth 158, 455-8 (1996).

[351] T. Yokogawa, T. Ishikawa, and J. L. Merz, and T. Taguchi, Spectroscopic characterization of band discontinuity in free-standing CdZnS/ZnS strained layer superlattices, J. Appl. Phys. 75, 2189-93 (1994).

[352] M. Ukita, F. Hiei, K. Nakano, and A. Ishibashi, Band offset at p-ZnTe/p-ZnSe heterointerface, Appl. Phys. Lett. 66, 209-11 (1995).

[353] T. Miyajima, F. P. Logue, J. F. Donegan, J. Hagarty, H. Okuyama, A. Ishibashi, and Y. Mori, Quasi-two-dimensional exciton in ZnSe/ZnMgSSe single quantum well, Appl. Phys. Lett. 66, 180-2 (1995).

[354] K. Iwata, H. Asahi, T. Ogura, J. Sumino, S. Gonda, Y. Kawaguchi, and T. Matsuoka, Low temperature grown Be-doped InAlP band offset reduction layer to p-type ZnSe, Seventh International Conf. Indium Phosphide and related materials, May 9–13, 1995, Sapporo, Japan, pp 183-6.

[355] K. Nishi, H. Ohyama, T. Suzuki, T. Mitsuyu, and T. Tomonasu, Determining the band discontinuities of ZnSe/GaAs and ZnMgSSe/GaAs heterojunctions using free electron laser, Appl. Phys. Lett. 70, 2171-3 (1997).

[356] K. Shum, L. Zheng, N. Dai, and M. C. Tamargo, Observation of quantum carrier confinement near $Zn_{0.61}Cd_{0.39}Se$/InP heterointerface, Appl. Phys. Lett. 69, 4200-2 (1996).

[357] P. J. Klar, D. Wolverson, J. J. Davies, W. Heimbrodt, and M. Happ, Determination of the chemical valence-band offset for $Zn_{1-x}Mn_xSe$/ZnSe multiple-quantum-well structures of high x, Phys. Rev. B 57, 7103-13 (1998).

[358] M. Nagelstrasser, H. Droge, F. Fischer, T. Litz, A. Waag, G. Landwehr, and H. P. Steinruck, Band discontinuities and local interface composition in BeTe/ZnSe heterostructures, J. Appl. Phys. 83, 4253-7 (1998).

[359] C. Guenaud, E. Deleporte, A. Filoramo, Ph. Lelong, C. Delalande, C. Morhain, E. Tournie, and J. P. Faurie, Band offset determination of the $Zn_{1-x}Cd_xSe/ZnSe$ interface, J. Crystal Growth 184-185, 839-43 (1998).

[360] T. Surkova, W. Giriat, M. Godlewski, P. Kaczor, A. J. Zakrzewski, S. Permogorov, and L. Tenishev, Energy level position of Ni and band offsets in $Zn_{1-x}Cd_xSe$:Ni and ZnS_xSe_{1-x}:Ni, Acta Physica Polonica A 88, 925-8 (1995).

[361] X. Chen, X. Hua, J. Hu, J.-M. Langlois, and W. A. Goddard III, Band structures of II-VI semiconductors using Gaussian basis function with separable *ab initio* pseudopotentials: Application to prediction of band offsets, Phys. Rev. B 53, 1377-87 (1996).

[362] R. M. Park, and A. Mar, Molecular beam epitaxial growth of high quality ZnSe on (100) Si, Appl. Phys. Lett. 48, 529-31 (1986).

[363] Ch. Maierhofer, S. Kulkarni, M. Alonso, T. Reich, and K. Horn, Valence band offset in ZnS layers on Si(111) grown by molecular beam epitaxy, J. Vac. Sci. Technol. B 9, 2238-43 (1991).

[364] R. Cingolani, Optical properties of excitons in ZnSe-based quantum well heterostructures, in Ref. [43], pp 163-226.

[365] I. Gorczyca and N. E. Christensen, ZnS/ZnSe strained layer superlattices under pressure, Phys. Rev. B 48, 17202-8 (1993).

[366] J. Menendéz, A. Pinczuk, J. P. Valladares, R. D. Feldman, and R. F. Austin, Resonance Raman scattering in CdTe-ZnTe superlattices, Appl. Phys. Lett. 50, 1101-3 (1987).

[367] A. V. Nurmikko and A. Ishibashi, II-VI diode lasers: A current view of device performance and issues, in Ref. [43], pp 227-270.

[368] Q. X. Zhao and M. Willander, Theoretical study of shallow acceptor states under the influence of both a confinement potential and deformation potential, Phys. Rev. B 57, 13033-8 (1998).

[369] Y. Fu, M. Willander, W. Lu, X. Q. Liu, S. C. Shen, C. Jagdish, M. Gal, J. Zou, and D. J. H. Cockayne, Strain effects in a GaAs-$In_{0.25}Ga_{0.75}As$-$Al_{0.5}Ga_{0.5}As$ asymmetric quantum wire, to be published in Phys. Rev. B.

[370] Q. X. Zhao, N. Magnea, and M. Willander, The influence of strained ZnTe islands on excitonic recombination and the self-organizing effects of the ZnTe islands in a wide CdTe-based quantum well, J. Phys. Condens. Matter 10, 1839-54 (1998).

[371] A. Sitnikova, S. Sorokin, I. Sedova, T. Shubina, A. Toropov, S. Ivanov, L. Falk, and M. Willander, TEM studies of self-organizing phenomena in CdSe fractional monolayers in a ZnSe matrix, Thin Solid Films 336, 76-9 (1998).

[372] A. Mycielski, L. Kowalczyk, and B. Witkowska, Growth and characterization of the ternary ZnSe-based compounds obtained by low temperature vapour transport, J. Crystal Growth 159, 191-4 (1996).

[373] K. Strössner, S. Ves, C. K. Kim, and M. Cardona, Pressure dependence of the lowest direct absorption edge of ZnTe, Solid St. Commun. 61, 275-8 (1987).

[374] Yu. V. Korostelin, V. I. Kozlovsky, A. S. Nasibov, and P. V. Shapkin, Vapour growth and characterization of bulk ZnSe single crystals, J. Crystal Growth 161, 51-9 (1996).

[375] E. Tournié, C. Morhain, G. Neu, J.-P. Faurie, R. Triboulet, and J. O. Ndap, Photoluminescence study of ZnSe single crystals by solid-phase re-crystallization, Appl. Phys. Lett. 68, 1356-8 (1996).

[376] E. Tournié, C. Morhain, G. Neu, J.-P. Faurie, R. Triboulet, and J. O. Ndap, Structural and optical characterization of study of ZnSe single crystals by solid-phase re-crystallization, J. Appl. Phys. 80, 2983-9 (1996).

[377] K. Pinardi and S. C. Jain, unpublished.

[378] A. Krost, W. Richter, D. R. T. Zahn, K. Hingerl, and H. Sitter, Chemical reaction at the ZnSe/GaAs interface detected by Raman spectroscopy, Appl. Phys. Lett. 57, 1981-2 (1990).

[379] H. Babucke, V. Egorov, P. Thiele, F. Henneberger, M. Rabe, J. Grieshe, N. Hoffmann, and K. Jacobs, Optical properties of $Zn_{1-x}Cd_xSe/ZnSe$ quantum well excitons in an electric field: Experiment and model calculations, Physica Status Solidi (a) 152, 161-70 (1995).

[380] A. A. Toropov, T. Shubina, S. V. Sorokin, A. V. Lebedev, B. N. Kyutt, M. Willander, S. V. Ivanov, M. Karlsteen, G. P. Pozina, J. P. Bergman, and B. Monemar, Optical selection of localized states excitonic states in CdSe based nanostructures, Phys. Rev. B R2510-13 (1999).

[381] R. H. Miles, T. C. McGill, S. Sivanathan, X. Chu, and J. P. Faurie, Structure of CdTe/ZnTe superlattices, J. Vac. Sci. Technol. B 5, 1263-7 (1987).

[382] G. Monfroy, S. Sivanathan, X. Chu, and J. P. Faurie, Molecular beam epitaxial growth of a novel strained-layer superlattice system: CdTe-ZnTe, Appl. Phys. Lett. 49, 152-4 (1986).

[383] W. S. Li, Y. B. Chi, Y. M. Li, X. W. Fan, B. J. Yang, D. Z. Shen, and Y. M. Lu, Photoluminescence of $ZnSe/Zn_{1-x}Cd_xSe$ strained superlattices under hydrostatic pressure, Thin Solid Films 266, 307-10 (1995).

[384] S. J. Hwang, W. Shan, J. J. Song, Z. O. Zhu, and T. Yao, Effect of hydrostatic pressure on strained CdSe/ZnSe single quantum wells, Appl. Phys. Lett. 64, 2267-9 (1994).

[385] E. Griebl, A. Stier, M. Krenzer, M. Kastner, T. Reisinger, H. Preis, and W. Gebhardt, Investigation of the dependence of subband transitions in ZnSe/Zn$_{1-x}$Mg$_x$Se quantum wells by PLE, J. Crystal Growth 184-185, 853-6 (1998).

[386] J. Feldmann, G. Peter, E. O. Göbel. P. Dawson, K. Moore, C. Foxon and R. J. Elliott, Linewidth dependence of radiative exciton lifetimes in quantum wells, Phys. Rev. Lett. 59, 2337-40 (1987).

[387] J. Martinez-Pastor, A. Vinattieri, L. Carraresi, M. Colocci, Ph. Roussignol, and G. Weimann, Temperaature dependence of exciton lifetime in GaAs/Al$_x$Ga$_{1-x}$As single quantum wells, Phys. Rev. B 47, 10456-60 (1993).

[388] C. V. Reddy, K. Balakrishnan, H. Okumura, and S. Yoshida, The origin of persistent photoconductivity and its relationship with yellow luminescence in molecular beam epitaxy grown undoped GaN, Appl. Phys. Lett. 73, 244-6 (1998).

[389] B. Gil, O. Briot, and R. L. Aulombard, Valence-band physics and the optical properties of GaN epilayers grown onto sapphire with wurtzite symmetry, Phys. Rev. B 52, R17028-31 (1995).

[390] X. B. Li, D. Z. Sun, J. P. Zhang, and M. Y. Kong, Influence of rapid thermal annealing on the optical properties of gallium nitride grown by gas-source molecular-beam epitaxy, Appl. Phys. Lett. 72, 936-8, (1998).

[391] D. M. Hansen, R. Zang, N. R. Perkins, S. Safvi, L. Zang, K. L. Bray, and T. F. Kuech, Photoluminescence of erbium-implanted GaN and *in situ*-doped GaN:Er, Appl. Phys. Lett. 72, 1244-6, (1998).

[392] W. Shan, B. D. Little, J. J. Song, Z. C. Feng, M. Schurman, and R. A. Stall, Optical transitions in In$_x$Ga$_{1-x}$N alloys by metalorganic chemical vapor deposition, Appl. Phys. Lett. 69, 3315-7 (1996).

[393] S. Nakamura and T. Mukai, High quality InGaN films grown on GaN films, Jpn. J. Appl. Phys. 31, L457-59 (1992).

[394] S. Nakamura, T. Mukai, and M. Senoh, Si-doped InGaN films grown on GaN films, Jpn. J. Appl. Phys. 32, L16-9 (1993).

[395] T. Matsuoka, H. Tanaka, T. Sasaki, and A. Katsui, Wide-gap semiconductor (InGa)N, Inst. Phys. Conf. Ser. 106, 141-6 (1990), (paper presented at the GaAs and Related Compounds Conf. in 1989 in Japan).

BIBLIOGRAPHY

[396] M. D. McCluskey, C. G. Van de Walle, C. P. Master, L. T. Romano, and N. M. Johnson, Large band gap bowing of $In_xGa_{1-x}N$ alloys, Appl. Phys. Lett. 72, 2725-6 (1998).

[397] Y. Narukawa, Y. Kawakami, M. Funato, S. Fujita, S. Fujita, and S. Nakamura, Role of self formed InGaN quantum dots for exciton localization in the purple laser diode emitting at 420 nm, Appl. Phys. Lett. 70, 981-3 (1997).

[398] D. Behr, J. Wagner, A. Ramakrishnan, H. Obloh, and K.-H. Bachem, Evidence for compositional inhomogeneity in low In content (InGa)N obtained by resonant Raman scattering, Appl. Phys. Lett. 73, 241-3 (1998).

[399] P. A. Crowell, D. K. Young, S. Keller, E. L. Hu, and D. D. Awschalom, Near field scanning optical spectroscopy of an InGaN quantum well, Appl. Phys. Lett. 72, 927-9 (1998).

[400] S. Keller, U. K. Mishra, S. P. DenBaars and W. Seifert, Spiral growth of InGaN nanoscale islands on GaN, Jpn. J. Appl. Phys. 37, L431-4 (1998).

[401] Y. Koide, H. Itoh, M. R. H. Khan, K. Hiramatsu, N. Sawaki, and I. Akasaki, Energy band-gap bowing parameter in an $Al_xGa_{1-x}N$ alloy, J. Appl. Phys. 61, 4540-3 (1987).

[402] H. Angerer, D. Brunner, F. Freudenberg, O. Ambacher, M. Stutzmann, R. Höpler, T. Metzger, E. Born, G. Dollinger, A. Bergmaier, S. Karsch, and H.-J. Körmer, Appl. Phys. Lett. 71, 1504-6 (1997).

[403] S. C. Jain, and D. J. Roulston, A Simple Expression for Band Gap Narrowing (BGN) in heavily doped Si, Ge, GaAs and Ge_xSi_{1-x} strained Layers, Solid State Electronics 34, 453-465 (1991); The bandgap narrowing and its effects on the properties of moderately and heavily doped germanium and silicon, S. C. Jain, R. P. Mertens and R. J. Van Overstraeten, Advances in Electronics and Electron Physics, Ed: P. W. Hawkes, (Academic Press 1991), Vol. 82, pp 197-275.

[404] S. P. DenBaars, U. K. Mishra, Y.-F. Wu, D. Kapolnek, P. Kozodoy, M. Mack, J. S. Speck, and S. Keller, Growth of high quality (In,Ga,Al)N/GaN heterostructure materials and devices by atmospheric pressure MOCVD in *III-Nitride, SiC and Diamond Materials for Electronic Devices* (Mater. Res. Soc. Pittsburgh, PA, USA, 1996), Symp. 423, 23-31 (1996).

[405] E. Ho and L. A. Kolodziejski, Gaseous source UHV epitaxy technologies for wide bandgap II–VI semiconductors, in Ref. [43], pp 83-119 (1997).

[406] A. Waag, T. Litz, F. Fischer, H. J. Lugauer, T. Baron, K. Schuell, U. Zehnder, T. Gerhard, U. Lunz, M. Keim, G. Reuscher, and G. Landwehr, Novel beryllium containing II-VI compounds: basic properties and potential applications, J. Crystal Growth 184-185, 1-10 (1998).

[407] V. Wagner, J. Geurts, T. Gerhard, Th. Litz, H. J. Lugauer, F. Fischer, A. Waag, G. Landwehr, Th. Walter, and D. Gerthsen, Determination of BeTe phonon dispersion by Raman spectroscopy on BeTe/ZnSe-superlattices, Appl. Surface Science 123-124, 580-4 (1998).

[408] U. Zehnder, D. R. Yakovlev, W. Ossau, T. Gerhard, F. Fischer, H. J. Lugauer, M. Keim, G. Reuscher, T. Litz, A. Waag, K. Herz, G. Bacher, A. Forchel, G. Landwehr, Optical properties of laser diodes and heterostructures based on beryllium chalcogenides, J. Crystal Growth 184-185, 541-4 (1998).

[409] A. Ishibashi, II–VI Blue-green laser diodes, IEEE J. Selected Top. in Quantum Electronics 1, 741-8 (1995).

[410] K. Lischka, Epitaxial ZnSe and cubic GaN: wide-band-gap semiconductors with similar properties? Physica Status Solidi B 202, 673-81 (1997).

[411] T. I. Kim, M. C. Yoo, E. S. Oh, M. H. Jeon, Y. J. Park, T. Kim, and J. W. Lee, Wide bandgap semiconductor laser - challenge for the future, SPIE - (The International Society for Optical Engineering) vol. 3001, 88-100 (1997).

[412] H.-J. Lugauer, A. Waag, L. Worschech, W. Ossau, and G. Landwehr, Generation of atomic group V materials for the p-type doping of wide gap II-VI semiconductors using a novel plasma cracker, J. Crystal Growth 161, 86-9 (1996).

[413] ISOLDE Collaboration, Transmutation doping of wide-bandgap II-VI compounds, J. Crystal Growth 161, 82-5 (1996).

[414] H. L. Hwang, K. Y. J. Hsu, and H. Y. Ueng, Fundamental studies of p-type doping of CdTe, J. Crystal Growth 161, 73-81 (1996).

[415] V. Valdna, F. Buschmann, and E. Mellikov, Conductivity conversion in CdTe layers, J. Crystal Growth 161, 164-7 (1996).

[416] P. Perlin, T. Suski, H. Teisseyre, M. Leszczynski, I. Grezgory, J. Jun, S. Prowski, P. Boguslavski, J. Bernholc, J. C. Chervin, A. Polian, and T. D. Moustakas, Towards the identification of the dominant donor in GaN, Phys. Rev. Lett. 75, 296-9 (1995).

[417] S. Nakamura, T. Mukai, and M. Senoh, Si- and Ge-doped GaN films grown with GaN buffer layers, Jpn. J. Appl. Phys. 31, 2883-8 (1992).

[418] T. Tanaka, A. Watanabe, A. Amano, H. Kobayashi, I. Akasaki, S. Yamazaki, and M. Koike, P-type conduction in Mg-doped GaN and $Al_{0.08}Ga_{0.92}N$ grown by metalorganic vapor phase epitaxy, Appl. Phys. Lett. 65, 593-4 (1994).

[419] D. C. Look and R. J. Molnar, Degenerate layer at GaN/sapphire interface: Influence on Hall-effect measurements, Appl. Phys. Lett. 70, 3377-9, (1997).

[420] W. Götz, L. T. Romano, J. walker, N. M. Johnson, and R. J. Molnar, Hall-effect analysis of GaN films grown by hydride vapor phase epitaxy, Appl. Phys. Lett. 72, 1214-6, (1998).

[421] S. Nakamura, T. Mukai, and M. Senoh, *In situ* monitoring and Hall measurements of GaN grown with GaN buffer layers, J. Appl. Phys. 71, 5543-9 (1992).

[422] S. Nakamura, M. Senoh, and T. Mukai, Highly P-type Mg-doped GaN films grown with GaN buffers, Jpn. J. Appl. Phys. 30 (No. 10A), L1708-11 (1991).

[423] S. Nakamura, T. Mukai, M. Senoh, and N. Iwasa, Thermal annealing effects on p-type Mg-doped GaN films, Jpn. J. Appl. Phys. 31, L139-42 (1992).

[424] W. Götz, N. M. Johnson, J. Walker, D. P. Bour, and R. A. Street, Activation of acceptors in Mg-doped GaN by metalorganic chemical vapor deposition, Appl. Phys. Lett. 68, 667-9 (1996).

[425] N. Akutsu, H. Tokunaga, I. Waki, A. Yamaguchi, and K. Matsumoto, Epitaxial growth and properties of Mg-doped GaN film produced by atmospheric MOCVD system with three layered laminar flow gas injection, (Mater. Res. Soc. Pittsburgh, PA, USA, 1998), Symp. 482, 113-8 (1998).

[426] M. E. Lin, G. Xue, G. L. Zhou, J. E. Greene, and H. Morkoç, *p*-type zinc-blende GaN on GaAs substrates, Appl. Phys. Lett. 63, 932-3 (1993).

[427] M. Miyachi, T. Tanaka, Y. Kimura, and H. Ota, The activation of Mg in GaN by annealing with minority-carrier injection, Appl. Phys. Lett. 72, 1101-3 (1998).

[428] O. Brandt, H. Yang, H. Kostial, and K. H. Ploog, High *p*-type conductivity in cubic GaN/GaAs(113)A by using Be as the acceptor and O as the codopant, Appl. Phys. Lett. 69, 2707-9 (1996).

[429] I. Akasaki, H. Amano, M. Kito, and K. Hiramatsu, Photoluminescence of Mg-doped GaN and electroluminescence of GaN p-n junction LED, J. Lumn. 48/49, 666-70 (1991).

[430] M. A. L. Johnson, Z. Yu, C. Boney, W. C. Hughes, J. W. Cook, Jr., J. F. Schetzina, H. Zhao, B. J. Skromme, and J. A. Edmond, Reactive MBE growth of GaN and GaN:H on GaN/SiC substrates, (Mater. Res. Soc. Pittsburgh, PA, USA, 1997), 449, 215-20 (1997).

[431] A. E. Hughes and S. C. Jain, Metal Colloids in Ionic Crystals, Advances in Physics 28, 717-828 (1979).

[432] C. Stampfl, and C. G. Van de Walle, Doping of $Al_xGa_{1-x}N$, Appl. Phys. Lett. 72, 459-61 (1995).

[433] F. Ren, Contacts in III-Nitrides, in *GaN and related materials*, Edited by S. J. Pearton, (Gordon and Breach Science Publishers 1997), pp 433-69

[434] R. W. Miles, B. Ghosh, S. Duke, J. R. Bates, M. J. Carter, P. K. Dutta and R. Hill, Formation of low resistance contacts to p-CdTe by annealing autocatalytically deposited Ni–P alloy coatings, J. Crystal Growth 161, 148-52 (1996).

[435] A. Waag, F. Fischer, H. J. Lugauer, T. Litz, J. Laubender, and G. Landwehr, Light emitting BeMgZnSe-ZnSe diodes on p-GaAs using BeTe-ZnSe pseudogradings, *Blue Laser and Light Emitting Diodes*, Ohmsha, Tokyo, Japan; pp 131-4 (1996).

[436] M. E. Lin, F. Y. Huang, Z. F. Fan, L. H. Allen, and H. Morkoç, Low resistance ohmic contacts on wide band-gap GaN, Appl. Phys. Lett. 64, 1003-5 (1994).

[437] H. Ishikawa, K. Nakamura, T. Egawa, T. Jimbo, and M. Umeno, Pd/GaN Schottky diode with barrier height of 1.5 eV and a reasonably effective Richardson coefficient, Jpn. J. Appl. Phys. 37 (1A/B), L7-9, (1998).

[438] J. T. Trexler, S. J. Pearton, P. H. Holloway, M. G. Mier, K. R. Evans, and R. F. Karlicek, Comparison of Ni/Au, Pd/Au. and Cr/Au metallizations for ohmic contacts to p-GaN, Mat. Res. Soc. Symp. Proc. 449, 1091-6 (1997).

[439] M. Mamor, F. D. Auret, S. A. Goodman, and G. Myburg, Electrical characterization of defects introduced in p-$Si_{1-x}Ge_x$ during electron-beam deposition of Sc Schottky barrier diodes, Appl. Phys. Lett. 72, 1069-71 (1998).

[440] L. S. Yu, D. J. Qiao, Q. J. King, S. S. Lau, K. S. Boutros, and J. M. Redwing, Ni and Ti Schottky barriers on n-AlGaN grown on SiC substrates, Appl. Phys. Lett. 73, 238-40 (1998).

[441] M. Murakami, Y. Koido, T. Oku, H. Mori, and C. J. Uchibori, Ohmic contacts to wide gap compound semiconductors, Electrochemical Soc. Proc. 97-21, 286-97 (1997).

[442] B. P. Luther, S. E. Mohney, T. N. Jackson, M. Asif Khan, Q. Chen, and J. W. Yang, Investigation of aluminum and titanium/aluminum contacts to n-type Gallium Nitride, (Mater. Res. Soc. Pittsburgh, PA, USA, 1997), Symp. 449, 1097-102 (1997).

[443] Z.-F. Fan, S. N. Mohammad, W. Kim, O. Aktas, A. E. Botchkarev and H. Morkoç, Very low resistance multilayer ohmic contact to n-GaN, Appl. Phys. Lett. 68, 1672-4 (1996).

[444] A. T. Ping, M. Asif Khan, and I. Adesida, Ohmic contacts to n-type GaN using Pd/Al metallization, J. Electron. Mat. 25, 819-24 (1996).

[445] M. W. Cole, D. W. Eckart, W. Y. Han, R. L. Pfeffer, T. Monahan, F. Ren, C. Yuan, R. A. Stall, S. J. Pearton, Y. Li, and Y. Lu, Thermal stability of W ohmic contacts to n-type GaN, J. Appl. Phys. 80, 278-81 (1996).

[446] M. W. Cole, D. W. Eckart, W. Y. Han, T. Monahan, F. Ren, and S. J. Pearton, A structural and chemical investigation of WSi ohmic contacts to n$^+$-GaN, (SOTAPOCS XXV) Electrochemical Soc. Pennington, NJ, USA, 1996, pp 271-7.

[447] M. W. Cole, F. Ren, and S. J. Pearton, Materials characterization of WSi contacts to n$^+$-GaN as a function of rapid thermal annealing temperature, J. Electrochem. Soc. 144, L275-7 (1997).

[448] J. D. Dow and D. Redfield, Electroabsorption in semiconductors: The excitonic absorption edge, Phys. Rev. B 1, 3358-71 (1970).

[449] F. Binet, J. Y. Duboz, E. Rosencher, F. Scholz, and V. Harle, Electric field effects on excitons in gallium nitride Phys. Rev. B 54, 8116-21 (1996).

[450] J. Y. Duboz, F. Binet, E. Rosencher, F. Scholz, and V. Haerle, Electric field effects on excitons in gallium nitride, Materials Sci. Engineering B43, 269-73 (1997).

[451] S. Gupta, R. K. Jain, and M. A. Khan. First measurement of the electro-optic effect in GaN, CLEO '95. Summaries of Papers Presented at the Conference on Lasers and Electro-Optics (IEEE Cat. No. 95CH35800). Opt. Soc. America, Washington, DC, USA 1995, pp p.142-3.

[452] S. J. Pearton, C. R. Abernathy, J. D. MacKenzie, J. R. Mileham, R. J. Shul, S. P. Kilcoyne, M. Hagerott-Crawford, F. Ren, W. S. Hobson, and J. M. Zavada, Fabrication of novel III-N and III-V modulator structures by ECR plasma etching, Surface/Interface and Stress effects in electronic material nanostructures symposium, (Mater. Res. Soc. Pittsburgh, PA, USA, 1996), pp 115-20

[453] D. A. B. Miller, Joseph S. Weiner, and D. S. Chemla, Electric field dependence of linear optical properties in quantum well structures: Waveguide electroabsorption and sum rules, IEEE J. Quantum Electronics QE22, 1816-30 (1986).

[454] I. Bar-Joseph, C. Klingshirn, D. A. B. Miller, D. S. Chemla, U. Koren, and B. I. Miller, Quantum-confined Stark effect in InGaAs/GaAs quantum wells grown by organometallic phase epitaxy, Appl. Phys. Lett. 50, 1010-2 (1987).

[455] M. J. Snelling, D. R. Harken, A. L. Smirl, and E. Towe, Electronic polarization modulation in [110]-oriented GaAs-InGaAs multiple quantum wells, IEEE J. Quantum Electronics 33, 1114-22 (1997).

[456] A. Hartwit, C. Hsu, F. Agulló-Rueda, and L. L. Chang, Observation of miniband formation in the $CdTe/Cd_{1-x}Mn_xTe$ quantum well system, Appl. Phys. Lett. 57, 1769-71 (1990).

[457] P. J. Thompson, S. Y. Wang, G. Horsburgh, T. A. Steele, and K. A. Prior, II-VI quantum confined Stark effect waveguide modulators, Appl. Phys. Lett. 68, 946-8 (1996).

[458] H. Babucke, P. Thiele, T. Prasse, M. Rabe, and F. Henneberger, ZnSe-based electro-optic waveguide modulators for the blue-green spectral range, Semicond. Sci. Technol. 13, 200-6 (1998).

[459] K.-K. Law, K. Smekalin, G. M. Haugen, G. D. U'Ren, and M. A. Haase, Stark effect on absorption of CdZnSSe-ZnSSe quantum wells, IEEE Photonics Technol. Lett. 8, 263-5 (1996).

[460] I. V. Bradley, J. P. Creasey, and K. P. O'Donnell, Space-charge effects in type-II strained layer superlattices, J. Crystal Growth 184-185, 728-31 (1998).

[461] I. V. Bradley, J. P. Creasey, K. P. O'Donnell, B. Neubauer, and D. Gerthsen, CdS-ZnSe intrinsic Stark superlattices, J. Crystal Growth 184-185, 718-22 (1998).

[462] D. L. Smith and C. Mailhiot, Optical properties of strained-layer superlattices with growth axis along [111], Phys. Rev. Lett. 58, 1264-7 (1987).

[463] R. D. Underwood, P. Kozodoy, S. Keller, S. P. DenBaars, and U. K. Mishra, Piezoelectric surface barrier lowering applied to InGaN/GaN field emitter arrays, Appl. Phys. Lett. 73, 405-7 (1998).

[464] J. Cibert, R. André, and Le Si Dang, Piezoelectric effect in strained CdTe-based heterostructure, Acta Physica Polonica A 88, 591-600 (1995).

[465] X. Chen, P. J. Parbrook, C. Trager-Cowan, B. Henderson, K. P. O'Donnell, M. P. Halsall, J. J. Davies, J. E. Nicholls, P. J. Wright, and B. Cockayne, Time-resolved optical studies of piezoelectric effects in wurtzite strained-layer superlattices, Semicond. Sci. Technol. 5, 997-1000 (1990).

[466] F. Bernardini, V. Fiorentini, and D. Vanderbilt, Spontaneous polarization and piezoelectric constants of III-V nitrides, Phys. Rev. B 56, 10024-7 (1997).

[467] F. Bernardini, V. Fiorentini, and D. Vanderbilt, Offsets and polarization at strained AlN/GaN polar interfaces, III-V Nitrides, (Mater. Res. Soc. Pittsburgh, PA, USA, 1997), Symposium 449, 923-8 (1997).

[468] P. M. Asbeck, E. T. Yu, S. S. Lau, G. J. Sullivan, J. Van Hove, and J. Redwing, Piezoelectric charge densities in AlGaN/GaN HFETs, Electronics Letters 33, 1230-1 (1997).

[469] J. Wang, J.-B. Jeon, K. W. Kim, and M. A. Littlejohn, Self-consistent study of strained wurtzite GaN quantum-well lasers, Proc. IEEE/Cornell Conf. *Advanced Concepts in High Speed Semiconductor Devices and Circuits* (IEEE, New York, NY, 1997), pp 306-13; see also J. Wang, J.-B. Jeon, K. W. Kim, and M. A. Littlejohn, Self-consistent calculation of mode spectra and modulation response in wurtzite GaN quantum-well lasers, IEEE Photonics Technol. Lett. 10, 51-3 (1998).

[470] T. Deguchi, A. Shikanai, K. Torii, T. Sota, S. Chichibu, and S. Nakamura, Luminescence spectra from InGaN multiquantum wells heavily doped with Si, Appl. Phys. Lett. 72, 3329-31 (1998).

[471] S.-H. Park, and S.-L. Chuang, Piezoelectric effects on electrical and optical properties of wurtzite GaN/AlGaN quantum well lasers, Appl. Phys. Lett. 72, 3103-5 (1998).

[472] S. F. Chichibu, A. C. Abare, M. S. Minsky, S. Keller, S. B. Fleischer, J. E. Bowers, E. Hu, U. K. Mishra, L. A. Coldren, S. P. DenBaars, and T. Sota, Effective band gap inhomogeneity and piezoelectric field in InGaN/GaN multiquantum well structures, Appl. Phys. Lett. 73, 2006-8 (1998).

[473] C. J. Sun, M. Zubair Anwar, Q. Chen, J. W. Yang, M. Asif Khan, M. S. Shur, A. D. Bykhovski, M. L.-Weber, C. Kisielowski, M. Smith, J. Y. Lin, and H. X. Xiang, Quantum shift of band-edge stimulated emission in InGaN-GaN multiple quantum well light-emitting diodes, Appl. Phys. Lett. 71, 2978-80 (1997).

[474] S. Muensit, I. L. Guy, The piezoelectric coefficient of gallium nitride thin films, Appl. Phys. Lett. 72, 1896-8 (1998).

[475] G. Borghs, and J. De Boeck, Growth and characterization of GaMnAs, a III-V diluted magnetic semiconductor, (Mater. Res. Soc. Pittsburgh, PA, USA, 1999), Spring meeting, San Francisco April 1999, (invited paper, unpublished).

[476] *Diluted magnetic semiconductors,* Vol. Editors: J. K. Furdyna and J. Kossut, in the series *Semimetals and Semiconductors,* Eds.: R. K. Willardson and A. C. Beer, (Academic Press, N. Y. 1988); *Semimagnetic Semiconductors and Diluted Magnetic Semiconductors,* Eds.: M. Averons and M. Balkanski, (Plenum London 1991).

[477] S. Scholl, H. Schäfer, A. Waag, D. Hommel, K. von Schierstedt, B. Kuhn-Heinrich, and G. Landwehr, Two-dimensional Shubnikov-de Haas oscillations in modulation-doped CdTe/CdMnTe quantum-well structures, Appl. Phys. Lett. 62, 3010-2 (1993).

[478] S. Takeyama, G. Grabecki, S. Adachi, Y. Takagi, T. Wojtowicz, G. Karczewski, and J. Kossut, Exciton magnetic-optical study of single quantum wells, $Cd_{1-x}Zn_xTe/Cd_{1-x'-y}Zn_{x'}Mn_yTe$, Acta Physica Polonica A 88, 945-8 (1995).

[479] G. Mackh, W. Ossau, D. R. Yakovlev, G. Landwehr, T. Wojtowicz, G. Karczewski, and J. Kossut, Exciton magnetic polaron features in photoluminescence excitation spectra of CdTe-(CdMn)Te quantum wells with high Mn contents, Acta Physica Polonica A 88, 849-52 (1995).

[480] J. X. Shen, W. Ossau, F. Fischer, A. Waag, and G. Landwehr, Optically detected SdH oscillations in CdTe/(CdMg)Te and CdTe/(CdMnMg)Te modulation doped quantum wells, Acta Physica Polonica A 88, 1033-7 (1995).

[481] H. Mariette, I. Lawrence, S. Haacke, and W. W. Rühle, Transfer of excitons by a magnetic field in II-VI asymmetric double quantum wells, Annals de Physique, Colloque C2, supplément au n°3, Vol. 20, 143-148 (1995).

[482] S. Lee, F. Michl, U. Roessler, M. Dobrowolska, and J. K. Furdyna, Magnetoexcitons and Landau levels in strained ZnTe and ZnSe layers, J. Crystal Growth 184-185, 1105-9 (1998).

[483] Y. Oka, K. Yanata, S. Takano, K. Egawa, K. Matsui, M. Takahashi, and H. Okamoto, Exciton dynamics in diluted magnetic II-VI semiconductor nanostructures, J. Crystal Growth, 184-185, 926-30 (1998).

[484] A. K. Bhattacharjee, and C. Benoit-a-la-Guillaume, Magnetic polaron associated with a hole in zinc-blende semimagnetic semiconductors, J. Crystal Growth 184-185, 912-16 (1998).

[485] S. Takeyama, Y. G. Semenov, T. Karasawa, G. Karczewski, T. Wojtowicz, and J. Kossut, An approach to the exciton magnetic polaron bifurcation problem studied in $Cd_{1-x}Mn_xTe$-CdTe-$Cd_{1-y}Mg_yTe$ asymmetric single quantum wells, J. Crystal Growth 184-185, 907-11 (1998).

[486] M. Kutrowski, G. Karczewski, G. Cywinski, M. Surma, K. Grasza, E. Lusakowska, J. Kossut, T. Wojtowicz, R. Fiederling, D. R. Yakovlev, G. Mackh, U. Zehnder, and W. Ossau, Growth by molecular beam epitaxy and magneto-optical studies of (100)- and (120)-oriented digital magnetic quantum well structures, Thin Solid Films 306, 283-90 (1997).

[487] J. Stankiewicz, F. Palacio, and F. Villuendas, Direct measurements of bound magnetic polaron magnetization in Ga-doped $Cd_{1-x}Mn_xTe$, Acta Physica Polonica A 92, 976-80 (1997).

[488] I. A. Merkulov, K. V. Kavokin, G. Mackh, B. Kuhn-Heinrich, W. Ossau, G. Landwehr, and D. R. Yakovlev, Luminescence polarization and spontaneous lowering of symmetry caused by magnetic-polaron formation in

semimagnetic-semiconductor quantum wells, Phys. Solid-State 39, 1859-63 (1997), (Translated from: Fizika-Tverdogo-Tela, 39, p.2079-84 Nov. 1997).

[489] L. Bryja, M. Ciorga, J. Misiewicz, and P. Becla, Photoluminescence of $Cd_{0.9}Mn_{0.1}Se_{0.3}Te/_{0.7}$ in magnetic field, Proc. SPIE - (The International Society for Optical Engineering) vol. 3178, 266-9 (1997).

[490] L. Bryja, M. Ciorga, A. Bohdziewicz, J. Misiewicz, and P. Becla, Photoluminescence of quaternary manganese based semimagnetic semiconductors, Spectrochimica Acta Part A (Molecular-and Bimolecular-Spectroscopy) 54A, 1671-4 (1998).

[491] R. Fiederling, D. R. Yakovlev, W. Ossau, G. Landwehr, A. Merkulov, K. V. Kavokin, T. Wojtowicz, M. Kutrowski, K. Grasza, G. Karczewski, and J. Kossut, Exciton magnetic polarons in (100)- and (120)-oriented semimagnetic digital alloys (Cd,Mn)Te, Phys. Rev. B 58, 4785-92 (1998).

[492] T. Dietl, From magnetic polarons to ferromagnetism, Acta Physica Polonica A 94, 111-23 (1998).

[493] Y. Imanaka, and N. Miura, Cyclotron resonance in ZnCdSe/ZnSe multi-quantum wells at ultra-high magnetic fields up to 150 T, Physica-B 249-251, 932-6 (1998).

[494] T. Yasuda, I. Mitsuishi, and H. Kukimoto, Metal organic vapor phase epitaxy of low-resistivity p-type ZnSe, Appl. Phys. Lett. 52, 57-9 (1988).

[495] M. W. Prairie, III-V semiconductors for mid-infrared lasers: progress and future trends, SPIE - (The International Society for Optical Engineering) vol. 2397, 322-32 (1995).

[496] C. H. Grein, P. M. Young, and H. Ehrenreich, Theoretical performance of $InAs/In_xGa_{1-x}Sb$ superlattice based midwave infrared lasers, J. Appl. Phys. 76, 1940-2 (1994).

[497] C. M. Ciesla, B. N. Murdin, C. R. Pidgeon, R. A. Stradling, C. C. Phillips, M. Livingstone, I. Galbraith, D. A. Jaroszynski, C. J. G. M. Langerak, P. J. P. Tang, and M. J. Pullin, Suppression of Auger recombination in arsenic-rich $InAs_{1-x}Sb_x$ strained layer superlattices, J. Appl. Phys. 80, 2994-7 (1996).

[498] E. R. Youngdale, J. R. Meyer, C. A. Hoffman, F. J. Bartoli, C. H. Grein, P. M. Young, H. Ehrenreich, R. H. Miles, and D. H. Chow, Auger lifetime enhancement in $InAs-Ga_{1-x}In_xSb$ superlattices, Appl. Phys. Lett. 64, 3160-2 (1994).

[499] Y. Zou, J. S. Osinsky, P. Grodzinski, P. D. Dapkus, W. C. Rideout, W. F. Sharfin, J. Schlafer, and F. D. Crawford, Experimental study of Auger recombination, gain, and temperature sensitivity of 1.5 μm compressively

strained semiconductor lasers, IEEE J. Quantum Electronics 29, 1565-75 (1993).

[500] S. R. Kurtz and R. M. Biefeld, Heterostructures and infrared emitters with compressed InAsSb layers (invited), 7th Int. Conf. on Narrow Gap Semicond., Santa Fe 9-12 January 1995, Inst. Phys. Conf. Series No. 144 (IOP Publishing, Bristol 1995), pp 18-23.

[501] Y.-H. Zhang, R. H. Miles, and D. H. Chow, InAs-InAs$_x$Sb$_{1-x}$ Type II superlattice midwave infrared lasers grown on InAs substrates, IEEE J. Selected Topics Quantum Electronics 1, 74956 (1995).

[502] W. Xie, D. C. Grillo, R. L. Gunshor, M. Kobayashi, H. Jeon, J. Ding, A. V. Nurmikko, G. C. Hua, and N. Otsuka, Room temperature blue light emitting p-n diodes from Zn(S,Se)-based multiple quantum well structures, Appl. Phys. Lett. 60, 1999-2001 (1992).

[503] W. Faschinger, R. Krump, G. Brunthaier, S. Ferreira and H. Sitter, ZnMgSeTe light emitting diodes, Appl. Phys. Lett. 65, 3215-7 (1994).

[504] S. Nakamura, T. Mukai, and M. Senoh, Candela-class high-brightness InGaN/AlGaN double-heterostructure blue-light-emitting diodes, Appl. Phys. Lett. 64, 1687-9 (1994).

[505] K. Nakanishi, I. Suemune, Y. Fujii, Y. Kuroda, M. Yamanishi, Extremely-low-threshold and high-temperature operation in a photopumped ZnSe/ZnSSe blue laser, Appl. Phys. Lett. 59, 1401-3 (1991).

[506] H. Okuyama, F. Hiei, and K. Akimoto, Optically pumped blue lasing in ZnSe-ZnMgSSe double heterostructures at room temperature, Jap. J. Appl. Phys. 31, L340-2, (1992).

[507] H. Okuyama, Y. Morinaga, and K. Akimoto, High-temperature blue lasing in photopumped ZnSSe-ZnMgSSe double heterostructures, J. Crystal Growth 127, 335-8 (1993).

[508] A. Toda, T, Asano, K. Funato, F. Nakamura, and Y. Mori, Epitaxial growth of ZnMgSe on GaAs substrate by metalorganic chemical vapor deposition, J. Crystal Growth 145, 537-40 (1994).

[509] A. Toda, T. Margalith, D. Imanishi, K. Yanashima, and A. Ishibashi, MOCVD-grown blue-green laser diode, Electronics Lett. 31, 1921- 2 (1995).

[510] S. V. Ivanov, A. A. Toropov, S. V. Sorokin, T. V. Shubina, I. V. Sedova, A. A. Sitnikova, P. S. Kop'ev, Zh. I. Alferov, H.-J. Lugauer, G. Reuscher, M. Keim, F. Fischer, A. Waag, and G. Landwehr, CdSe fractional-monolayer active region of molecular beam epitaxy grown green ZnSe-based lasers Appl. Phys. Lett. 74, 498-500 (1999).

[511] A. L. Gurskii, I. P. Marko, E. V. Lutsenko, G. P. Yablonskii, H. Kalisch, H. Hamadeh, and M. Heuken, High-temperature optically pumped lasing in ZnMgSSe/ZnSe heterostructures grown by metalorganic vapor phase epitaxy, Appl. Phys. Lett. 73, 1496-8 (1998).

[512] L. Zeng, B. X. Yang, A. Cavus, W. Lin, Y. Y. Luo, M. C. Tamargo, Y. Guo, and Y. C. Chen, Red-green-blue photopumped lasing from ZnCdMgSe/ZnCdSe quantum well laser structures grown on InP, Appl. Phys. Lett. 72, 3136-8 (1998).

[513] Y. Guo, G. Aizin, Y. C. Chen, L. Zheng, Abdullah Cavus, and M. C. Tamargo, Photo-pumped ZnCdSe/ZnCdMgSe blue-green quantum well lasers grown on InP substrates, Appl. Phys. Lett. 70, 1351-3 (1997).

[514] H. Jeon, V. Kozlov, P. Kelkar, A. V. Nurmikko, D. C. Grillo, J. Han, M. Ringle, and R. L. Gunshor, Optically pumped blue-green vertical-cavity surface-emitting laser, CLEO '95, Summaries of papers presented at the conference on lasers and electro-optics, (Opt. Soc. America 1995), pp 33-4.

[515] A. S. Nasibov, V. I. Kozlovsky, P. V. Reznikov, Ya. K. Skasyrsky, and Yu. M. Popov, Full colour TV projector based on A_2B_6 electron-beam pumped semiconductor lasers, J. Crystal Growth 117, 1040-5 (1992).

[516] R. R. Rice, J. F. Shanley, and N. F. Ruggieri, Dual-use applications for e-beam-pumped semiconductor lasers Proc. SPIE - (The International Society for Optical Engineering), vol. 2374, 225-31 (1995).

[517] V. I. Kozlovsky, A. S. Nasibov, Yu. M. Popov, and Ya. K. Skasyrsky, Full colour L-CRT projector, Proc. SPIE - (The International Society for Optical Engineering), vol. 2407, 312-20 (1995).

[518] V. I. Kozlovsky, E. A. Shcherbakov, E. M. Dianov, A. B. Krysa, A. S. Nasibov, and P. A. Trubenko, Electron-beam pumped laser structures based on MBE grown ZnCdSe/ZnSe superlattices, J. Crystal Growth 159, 609-12 (1996).

[519] J-M. Bonard, J-D. Ganiere, L. Vanzetti, J-J. Paggel, L. Sorba, A. Franciosi, D. Herve, and E. Molva, Combined transmission electron microscopy and cathodoluminescence studies of degradation in electron-beam-pumped $Zn_{1-x}Cd_xSe$/ZnSe blue-green lasers, Appl. Phys. Lett. 84, 1263-73 (1998).

[520] M. A. Haase, P. F. Baude, M. S. Hagedorn, J. Qiu, J. M. DePuydt, and H. Cheng, Low-threshold buried-ridge II-VI laser diodes, Appl. Phys. Lett. 63, 2315-7 (1993).

[521] J. M. DePuydt, M. A. Haase, S. Guha, J. Qui, H. Cheng, B. J. Wu, G. E. Höfler, G. M-Haugen, M. S. Hagedorn, and P. F. Baude, Room temperature II-VI lasers with 2.5 mA threshold, J. Crystal growth 138, 667-76 (1994).

[522] D. C. Grillo, Y. Fan, J. Han, L. He, R. L. Gunshor, A. Salokatve, M. Hagerott, H. Jeon, A. V. Nurmikko, G. C. Hua, and N. Otsuka, Pseudomorphic separate confinement heterostructure blue-green diode lasers, Appl. Phys. Lett. 63, 2723-5 (1993).

[523] J. M. Gaines, R. R. Drenten, K. W. Haberern, T. Marshall, P. Mensz, and J. Petruzzello, Blue-green injection lasers containing pseudomorphic $Zn_{1-x}Mg_xS_ySe_{1-y}$ cladding layers and operating up to 394 K, Appl. Phys. Lett. 62, 2462-4 (1993).

[524] N. Nakayama, S. Itoh, H. Okuyama, M. Ozawa, T. Ohata, K. Nakano, M. Ikeda, A. Ishibashi, and Y. Mon, Continuous-wave operation of 489.9 nm blue laser diode at room temperature, Electronics Lett. 29, 2194-5 (1993).

[525] Y. Morinaga, H. Okuyama, and K. Akimoto, Photopumped blue lasers with ZnSSe-ZnMgSSe double heterostructure and attempt at doping in ZnMgSSe, Jpn. J. Appl. Phys. 32, 678-80 (1993).

[526] H. Okuyama, F. Hiei, M. Ozawa, and K. Akimoto, ZnSe/ZnMgSSe blue laser diode, Electronics Lett. 28, 1798-9 (1992).

[527] H. Okuyama, E. Kato, S. Itoh, N. Nakayama, T. Ohata, and A. Ishibashi, Operation and dynamics of ZnSe/ZnMgSSe double heterostructure blue laser diode at room temperature, Appl. Phys. Lett. 66, 656-8 (1995).

[528] S. Itoh, S. Taniguchi, T. Hino, R. Imoto, K. Nakano, N. Nakayama, M. Ikeda, A. Ishibashi, Room temperature laser operation of wide band-gap II-VI laser diodes, Materials Sci. Eng. B 43, 55-9 (1997).

[529] R. L. Gunshor, J. Han, G. C. Hua, A. V. Nurmikko, and H. Jeon, Growth issues for blue–green laser diodes, J. Crystal growth 159, 1-10 (1996).

[530] S. Guha and J. Petruzzello, Defects and degradation in wide-band II-VI-based structures and light-emitting devices, in Ref. [43], pp 271-318.

[531] S. Itoh, S. Tomiya, R. Imoto, and A. Ishibashi, Heterointerface control of ZnSe based II–VI laser diodes, Appl. Surface Sci. 117/118, 719-24 (1997).

[532] S. Taniguchi, T. Hino, S. Itoh, K. Nakano, N. Nakayama, A. Ishibashi, and M. Ikeda, 100 h II-VI blue-green laser diode, Electronics Lett. 32, 552-3 (1996).

[533] Y. Sanaka, H. Okuyama, S. Kijima, E. Kato, H. Noguchi, and A. Ishibashi, II-VI laser diode with low operating voltage and long device lifetime, Electronics Letters 34, 1891-2 (1998).

[534] F. Nakanishi, H. Doi, N. Okuda, T. Matsuoka, K. Katayama, A. Saegusa, H. Matsubara, T. Yamada, T. Uemura, M. Irikura, and S. Nishine, Low-threshold room-temperature CW operation of ZnSe-based blue/green laser diodes grown on conductive ZnSe substrates, Electronics Letters 34, 496-7 (1998).

[535] H. Wenisch, K. Ohkawa, A. Isemann, M. Fehrer, and D. Hommel, Planar homo-epitaxial laser diodes grown on aluminium-doped ZnSe substrates, Electronics Letters 34, 891-3 (1998).

[536] D. Albert, B. Olszowi, W. Spahn, J. Nuernberger, K. Schuell, M. Korn, V. Hock, M. Ehinger, W. Faschinger, and G. Landwehr, Optical properties and defect characterization of ZnSe laser diodes grown on tellurium-terminated GaAs, J. Crystal Growth 184-185, 571-4 (1998).

[537] W. Faschinger, W. Spahn, J. Nuernberger, A. Gerhard, M. Korn, K. Schuell, D. Albert, H. Ress, R. Ebel, R. Schmitt, B. Olszowi, M. Ehinger, and G. Landwehr, Application of wide gap II-VI compounds as emitters and detectors, Physica Status Solidi B 202, 695-706 (1997).

[538] M. Legge, S. Bader, G. Bacher, H.-J. Lugauer, A. Waag, A. Forchel, and G. Landwehr, High temperature operation of II-VI ridge-waveguide laser diodes, Electronics Lett. 34, 2032-4 (1998).

[539] S. Nakamura, M. Senoh, and T. Mukai, High power GaN p-n junction blue-light-emitting diodes, Jpn. J. Appl. Phys. 30, L1998-2001 (1991).

[540] S. Nakamura, T. Mukai, and M. Senoh, High brightness InGaN/AlGaN double heterostructure blue-green light-emitting diodes, J. Appl. Phys. 76, 8189-91 (1994).

[541] S. Nakamura, M. Senoh, N. Iwasa, S. Nagahama, T. Yamada, and T. Mukai, Super bright green InGaN single-quantum-well-structure light emitting diodes, Jpn. J. Appl. Phys. 34, L1332-5 (1995).

[542] S. Nakamura, M. Senoh, N. Iwasa, and S. Nagahama, High-brightness InGaN blue, green and yellow light-emitting diodes with quantum well structures, Jpn. J. Appl. Phys. 34, L797-9 (1995).

[543] S. Nakamura, M. Senoh, N. Iwasa, S. Nagahama, High power InGaN single-quantum-well-structure blue and violet light emitting diodes, Appl. Phys. Lett. 67, 1868-70 (1995).

[544] S. Nakamura, Zn doped InGaN growth and InGaN/AlGaN double heterostructure blue light-emitting diodes, J. Crystal Growth 145, 911-17 (1994).

[545] K. Koga and T. Yamaguchi, Single crystals of SiC and their applications to blue LEDs, Progress Crystal Growth and Character. 23, 127-51 (1991).

[546] M. G. Craford, LEDs challenge the incandescents, Circuits and Devices, 24-9 (1992).

[547] H. Sugawara, K. Itaya, and G. Hatakoshi, Emission properties of InGaAlP visible light emitting diodes employing a multiple quantum well active layer, Jpn. J. Appl. Phys. 33, 5784-7 (1994).

[548] T. J. Schmidt, X. H. Yang, W. Shan, J. J. Song, A. Salvador, W. Kim, O. Aktas, A. Botchkarev, and H. Morkoç, Room-temperature stimulated emission in GaN/AlGaN separate confinement heterostructures grown by molecular beam epitaxy, Appl. Phys. Lett. 68, 1820-2 (1996).

[549] M. A. Khan, D. T. Olsen, J. M. Van Hove, and J. N. Kuznia, Vertical-cavity, room-temperature stimulated emission from photopumped GaN films deposited over sapphire substrates using low-pressure metalorganic chemical vapor deposition, Appl. Phys. Lett. 58, 1515-17 (1991).

[550] S. T. Kim, H. Amano, I. Akasaki, and N. Koide, Optical gain of optically pumped $Al_{0.1}Ga_{0.9}N$/GaN double heterostructure at room temperature, Appl. Phys. Lett. 64, 1535-6 (1994).

[551] I. Akasaki, H. Amano, S. Sota, S. Sakai, T. Tanaka, and M. Koike, Stimulated emission by current injection from an AlGaN/GaN/GaInN quantum well device, Jpn. J. Appl. Phys. 34, L1517-19 (1995).

[552] M. A. Haase, J. Qui, J. M. DePuydt, and H. Cheng, Blue-green laser diodes, Appl. Phys. Lett. 59, 1272-4 (1991).

[553] S. Nakamura, M. Senoh, S. Nagahama, N. Iwasa, T. Yamada, T. Matsushita, H. Kiyoku, and Y. Sugimoto, InGaN-based-Multi-Quantum-Well-Structure laser diodes, Jpn. J. Appl. Phys. 35, L74-76 (1996).

[554] S. Nakamura, M. Senoh, S. Nagahama, N. Iwasa, T. Yamada, T. Matsushita, H. Kiyoku, and Y. Sugimoto, Characteristics of InGaN multi-quantum-well-structure laser diodes, Appl. Phys. Lett. 68, 3269-71 (1996).

[555] S. Nakamura, M. Senoh, S. Nagahama, N. Iwasa, T. Yamada, T. Matsushita, H. Kiyoku, and Y. Sugimoto, InGaN multi-quantum-well structure laser diodes grown on $MgAl_2O_4$ substrate, Appl. Phys. Lett. 68, 2105-7 (1996).

[556] S. Nakamura, M. Senoh, S. Nagahama, N. Iwasa, T. Yamada, T. Matsushita, H. Kiyoku, and Y. Sugimoto, Continuous-wave operation of InGaN multi-quantum-well-structure laser diodes at 233 K, Appl. Phys. Lett. 69, 3034-6 (1996).

[557] S. Nakamura, M. Senoh, S. Nagahama, N. Iwasa, T. Yamada, T. Matsushita, Y. Sugimoto, and H. Kiyoku, Room-temperature continuous-wave operation of InGaN multi-quantum-well-structure laser diodes with a long lifetime, Appl. Phys. Lett. 70, 868-70 (1997).

[558] S. Nakamura, M. Senoh, S. Nagahama, N. Iwasa, T. Yamada, T. Matsushita, Y. Sugimoto, and H. Kiyoku, Room-temperature continuous-wave operation of InGaN multi-quantum-well-structure laser diodes with a lifetime of 27 hours, Appl. Phys. Lett. 70, 1417-1419 (1997).

[559] S. Kijima, H. Okuyama, Y. Sanaka, T. Kobayashi, S. Tomiya, and A. Ishibashi, Optimized ZnSe:N/ZnTe:N contact structure of ZnSe-based II–VI laser diodes, Appl. Phys. Lett. 73, 235-7 (1998).

[560] N. Nakayama, A. Ishibashi, and N. Eguchi, ZnSe-based room-temperature CW lasers, Review of Laser Engineering 25, 504-9 (1997).

[561] Y. Cordier, S. Bollaert, M. Zaknoune, J. Dipersio, D. Ferre, InAlAs/InGaAs metamorphic high electron mobility transistors on GaAs substrate: Influence of indium content on material properties and device performance, Japn. J. Appl. Phys. (Part 1) 38, 1164-8 (1999).

[562] M. Zaknoune, B. Bonte, C. Gaquiere, Y. Cordier, Y. Druelle, D. Theron, and Y. Crosnier, InAlAs/InGaAs metamorphic HEMT with high current density and high breakdown voltage, IEEE Electron Dev. Lett. 19, 345-7 (1998).

[563] K. van der Zanden, M. Behet, and G. Borghs, GaAs MANTECH (1999), pp 201-4; K. van der Zanden, D. Schreurs, B. Nauwelaers, and G. Borghs, to be published; K. van der Zanden, D. Schreurs, B. Nauwelaers, and G. Borghs, to be published.

[564] H. Happy, S. Bollaert, H. Foure, and A. Cappy, Numerical analysis of device performance of metamorphic $In_yAl_{1-y}As/In_xGa_{1-x}As$ ($0.3 \leq x \leq 0.6$) HEMT's on GaAs substrate, IEEE Trans. Electron Dev. 45, 2089-95 (1998).

[565] G. D. Studtmann, R. L. Gunshor, L. A. Kolodziejski, M. R. Melloch, J. A. Cooper, Jr., R. F. Pierret, D. P. Munich, C. Choi and N. O. Otsuka, Pseudomorphic ZnSe/n-GaAs doped-channel field-effect transistors by interrupted molecular beam epitaxy, Appl. Phys. Lett. 52, 1249-51 (1988).

[566] M. K. Lee, H. C. Liao, C. C. Hu, and M. Y. Yeh, Heterojunction ZnSe/InP field-effect transistor by metalorganic chemical vapor deposition, Materials Chem. Phys. 38, 398-401 (1994).

[567] A. Z. H. Wang, W. A. Anderson, B. J. Wu, M. Haase, and T. J. Mountziaris, A ZnSSe metal-semiconductor field effect transistor, Proc. IEEE/Cornell Conf. on Advanced Concepts in High Speed Semiconductor Devices and Circuits, pp 517-21 (1995).

[568] H. M. Dauplaise, K. Vaccaro, A. Davis, G. O. Ramseyer, and J. P. Lorenzo, Analysis of thin CdS layers on InP for improved metal–insulator–semiconductor devices, J. Appl. Phys. 80, 2873-82 (1996).

[569] T. C. Tseng, L. S. Ji, and C. H. Ru, DC and AC characteristics of ZnSe/Ge/GaAs and ZnSe/InGaAs/GaAs lattice-matched heterojunction bipolar transistors grown by metal-organic chemical vapor deposition, Jap. J. Appl. Phys. Part 2 vol 33, L1759-61 (1994).

[570] S. L. Wen, ZnSe/Si visible-sensitivity emitter bipolar transistors, Solid St. Electronics 41, 7-10 (1997).

[571] M. A. Khan, J. N. Kuznia, A. R. Bhattarai, and D. T. Olson, High mobility transistor based on GaN-Al$_x$Ga$_{1-x}$N heterojunctions, Appl. Phys. Lett. 63, 1214-5 (1993).

[572] M. Asif Khan, M. S. Shur, B. T. Dermott, and A. Higgins, Microwave operation of GaN/AlGaN doped-channel heterostsructure field effects transistors, IEEE Electron Dev. Lett. 17, 325-8 (1996).

[573] M. Asif Khan, M. S. Shur, B. T. Dermott, A. Higgins, J. Burm, W. Schaff, and L. F. Eastman, Short channel GaN/AlGaN doped channel heterostsructure field effect transistors with 36.1 GHz cutoff frequency, Electronics Lett. 32, 357-8 (1996).

[574] Y.-F. Wu, S. Keller, P. Kozodoy, B. P. Keller, P. Parikh, D. Kapolnek, S. P. DenBaars, and U. K. Mishra, Bias dependent Microwave Performance of AlGaN/GaN MODFET's up to 100 V, IEEE Electron Dev. Lett. 18, 290-2 (1997).

[575] Y.-F. Wu, B. P. Keller, S. Keller, N. X. Nguyen, M. Le, C. Nguyen, T. J. Jenkins, I. T. Kehias, S. P. DenBaars, and U. K. Mishra, Short channel AlGaN/GaN MODFET's with 50-GHz f_T and 1.7-W/mm output-power at 10 GHz, IEEE Electron Dev. Lett. 18, 438-40 (1997).

[576] Y.-F. Wu, B. P. Keller, O. Fini, S. Keller, T. J. Jenkins, I. T. Kehias, S. P. DenBaars, and U. K. Mishra, High Al-content AlGaN/GaN MODFET's for ultrahigh performance, IEEE Electron Dev. Lett. 19, 50-3 (1998).

[577] Z. Fan, S. N. Mohammad, O. Aktas, A. E. Botchkarev, A. Salvador, and H. Morkoç, Suppression of leakage currents and their effects on the electrical performance of AlGaN/GaN modulation doped field-effect transistors, Appl. Phys. Lett. 69, 1229-31 (1996).

[578] Y.-F. Wu, B. P. Keller, S. Keller, D. Kapolnek, P. Kozodoy, S. P. DenBaars, and U. K. Mishra, Very high breakdown voltage and large transconductance realized on GaN heterojunction field effect transistors, Appl. Phys. Lett. 69, 1438-40 (1996).

[579] S. C. Binari, GaN FETs for high temperature and microwave applications, State-of-the-Art Program on Compound Semiconductors (SOTAPOCS XXIII), Electrochem. Soc. 95-21, 136-43 (1995).

[580] S. M. Sze, Physics of semiconductor devices, (John Wiley N. Y. 1981).

[581] A. T. Ping, Q. Chen, J. W. Yang, M. A. Khan, and I. Adesida, High performance 0.25 μm gate-length doped-channel AlGaN/GaN heterostructure field effect transistors grown on p-type SiC substrates, IEDM Technical Digest 97, 561-4 (1997).

BIBLIOGRAPHY

[582] C. Weitzel, L. Pond, K. Moore, and M. Bhatnagar, Effect of device temperature on RF FET power density, Materials Science Forum 264-268, pt.2, 969-72 (1998).

[583] R. Gaska, J. Yang, A. Osinsky, M. Asif Khan, A. O. Orlov, G. L. Snider, and M. S. Shur, Electron transport in AlGaN-GaN heterostructures grown on 6H-SiC substrates, Appl. Phys. Lett. 72, 707-9 (1998).

[584] R. Gaska, J. Yang, A. Osinsky, M. Asif Khan, and M. S. Shur, Novel high power AlGaN/GaN HFETs on SiC substrates, IEDM Technical Digest 97, 565-8 (1997).

[585] A. T. Ping, Q. Chen, J. W. Yang, M. A. Khan, I. Adesida, DC and microwave performance of high-current AlGaN/GaN heterostructure Field Effect Transistors grown on p-type SiC substrates, IEEE Electron Dev. Lett. 19, 54-6 (1998).

[586] G. J. Sullivan, M. Y. Chen, J. A. Higgins, J. W. Yang, Q. Chen, R. L. Pierson, and B. T. McDermott, High-power 10-GHz operation of AlGaN HFT's on insulating SiC, IEEE Electron Dev. Lett. 19, 198-200 (1998).

[587] S. J. Pearton, and R. J. Shul, Etching of III Nitrides, in *GaN*, edited by J. I. Pankove and T. D. Moustakas, (Academic Press, N. Y. 1998), Vol. 1, pp 103-26.

[588] S. J. Pearton, C. B. Vartuli, J. C. Zolper, C. Yuan, and R. A. Stall, Ion implantation doping and isolation of GaN, Appl. Phys. Lett. 67, 1435-1437 (1995).

[589] J. C. Zolper, J. Han, R. M. Biefeld, S. B. Van Deusen, W. R. Wampler, S. J. Pearton, J. S. Williams, H. H. Tan, R. J. Karlicek, Jr., and R. A. Stall, Implantation activation annealing of Si-implanted gallium nitride at temperatures > 1100 °C, (Mater. Res. Soc. Pittsburgh, PA, USA, 1997), Symp. 468, 401-6 (1997).

[590] J. C. Tan, J. S. Williams, J. Zou, D. J. H. Cockayne, S. J. Pearton, and J. C. Zolper, Annealing of ion implanted gallium nitride, Appl. Phys. Lett. 72, 1190-2 (1998).

[591] J. C. Zolper, R. M. Biefeld, S. B. Van Deusen, W. R. Wampler, D. J. Reiger, S. J. Pearton, J. S. Williams, H. H. Tan, R. F. Karlicek Jr., and R. A. Stall, Si implantation activation annealing of GaN up to 1400 °C, J. Electron. Mater. 27, 179-84 (1998).

[592] S. Porowski and I. Grzegory, Growth of GaN single crystals under high nitrogen pressure, in *GaN and related materials*, edited by S. J. Pearton, (Gordon and Breach, N. Y. 1997), pp 295-313.

[593] N. Newman, Thermochemistry of III-N semiconductors, in *GaN*, edited by J. I. Pankove and T. D. Moustakas, (Academic Press, N. Y. 1998), Vol. 1, pp 55-101.

[594] X. A. Cao, C. R. Abernathy, R. K. Singh, S. J. Pearton, M. Fu, V. Sarvepalli, J. A. Sekhar, J. C. Zolper, D. J. Rigger, J. Ham, T. J. Drummond, R. J. Shul, and R. G. Wilson, Ultrahigh Si^+ implant efficiency in GaN using a high-temperature rapid thermal process system, Appl. Phys. Lett. 73, 229-31 (1998).

[595] S. C. Binari, H. C. Dietrich, G. Kelner, I. B. Rowland, K. Doverspike, and D. K. Wickenden, H, He and N implant isolation of n-type GaN, J. Appl. Phys. 78, 3008-11 (1995).

[596] Landolt-Börnstein, Semiconductors, Vol. 17 Sub-Vol. b: Physics of II-VI and I-VII compounds, semimagnetic semiconductors, (Springer-Verlag Berlin-Heidelberg 1982).

[597] K. Mohammed, D. J. Olego, P. Newbury, D. A. Cammack, and H. Cornelissen, Quantum confinement and strain effects in $ZnSe-ZnS_XSe_{1-X}$ strained-layer superlattices, Appl. Phys. Lett. 50, 1820-2 (1987).

[598] Z. P. Guan, Y. M. Lu, L. C. Chen, B. J. Yang, and X. W. Fan, Optical and structural properties of ZnTe-ZnS strained-layer superlattices, Physica Status Solidi (b) 185, 137-49 (1994).

[599] C. Kittel, Introduction to Solid State Physics, John Wiley & Sons, London (1968).

[600] J. Krustok, V. Valdna, K. Hjelt, and H. Collan, Deep center luminescence in p-type CdTe, J. Appl. Phys. 80, 1757-62 (1996).

[601] H. Zogg, S. Blunier, A. Fach, C. Maissen, P. Müller, S. Todoropol, V. Meyer, G. Kostorz, A. Dommann, and T. Richmond. Thermal-mismatch-strain relaxation in epitaxial CaF_2, BaF_2/CaF_2 and $PbSe/BaF_2/CaF_2$ layers on Si(111) after many temperature cycles, Phys. Rev. B 50, 10801-10 (1994).

[602] R. M. Martin, Elastic properties of ZnS structure semiconductor, Phys. Rev. B 1, 4005-11 (1970).

[603] B. H. Lee, Elastic constants of ZnTe and ZnSe between 77°-300°K, J. Appl. Phys. 41, 2984-7 (1970).

[604] V. Swaminathan, Properties of InP and related materials, in *Indium Phosphide and Related Materials: Processing, Technology, and Devices*, Ed.: Avishay Katz, (Artech House Boston 1992), p 29.

[605] B. H. Lee, Pressure dependence of second-order elastic constants of ZnTe and ZnSe, J. Appl. Phys. 41, 2988-90 (1970).

BIBLIOGRAPHY

331

[606] M. Shur, *Physics of Semiconductor Devices*, (Prentice Hall, Englewood Cliffs 1990), pp 632-3.

[607] A. Adachi, Material parameters of $In_{1-x}Ga_xAs_yP_{1-y}$ and related binaries, J. Appl. Phys. 53, 8775-92 (1982).

[608] K.-N. Tu, J. W. Mayer and L. C. Feldman, *Electronic thin Film Science for Electrical Engineers and Material Scientists*, (Maxwell Publishing Co., New York 1992).

[609] H. C. Casey, Jr. and M. B. Panish, *Heterostructure Lasers*, Part B (Academic Press, New York, 1978).

[610] T. F. Kuech, D. J. Wolford, R. Potemski, J. A. Bradley, and K. H. Kelleher, Dependence of the $Al_xGa_{1-x}As$ band edge on alloy composition based on the absolute measurement of x, Appl. Phys. Lett. 51, 505-7 (1987).

[611] P. Merle, D. Auvergne, and H. Mathieu, Conduction band structure of GaInP, Phys. Rev. B 15, 2032-47 (1977).

[612] A. Bensaada, A. Chennouf, R. W. Cochrane, J. T. Graham, and R. Leonelli, Misfit strain, relaxation, and band-gap shift in $Ga_xIn_{1-x}P/InP$ epitaxial layers, J. Appl. Phys. 75, 3024-9 (1994).

[613] E. Kuphal, A. Pöcker, and A. Eisenbach, Relation between photoluminescence wavelength and lattice mismatch in metalorganic vapor-phase epitaxy InGaAs/InP, J. Appl. Phys. 73, 4599604 (1993).

[614] K. H. Goetz, D. Bimberg, H. Jürgensen, J. Solders, A. V. Solomonov, G. F. Glinski, and M. Razeghi, Optical and crystallographic properties and impurity incorporation of $Ga_xIn_{1-x}As$ ($0.44 < x < 0.49$) grown by liquid phase epitaxy, vapor phase epitaxy, and metal organic chemical vapor deposition, J. Appl. Phys. 54, 4543-52 (1983).

[615] T. Y. Wang and G. B. Stringfellow, Strain effects on $Ga_xIn_{1-x}As/InP$ single quantum wells grown by organometallic vapour-phase epitaxy with $0 \leq x \leq 1$, J. Appl. Phys. 67, 344-52 (1990).

[616] S. M. Hu, Stress related problems in silicon technology, J. Appl. Phys. 70, R53-R79 (1991).

[617] *Thermophysical Properties of Matter,* Vol. 13 of *TPRC Data Series,* edited by Y. S. Touloukian, R. K. Kirby, R. E. Taylor, and T. Y. R. Lee, (Plenum, New York, 1977).

[618] E. Anastassakis, Strain characterization in semiconductor structures and superlattices, in *Light Scattering in Semiconductor Structures and Superlattices*, Ed.: D. J. Lockwood and J. F. Young, NATO ASI Series B, pp 173-96, New York 1991.

[619] M. B. Nardelli, K. Rapcewicz, E. L. Briggs, C. Bungaro, and J. Bernholc, Theory of interfaces in wide-gap nitrides, Mat. Res. Soc. Symp. Proc. 449, 893-8 (1997).

[620] J. A. Majewski, M. Städele, and P. Vogl, Stability and band offsets of SiC/GaN, SiC/AlN, and AlN/GaN heterostructures, Mat. Res. Soc. Symp. Proc. 449, 917-22 (1997).

[621] V. A. Savastenko and A. U. Sheleg, Study of the elastic properties of Gallium Nitride, Physica Status Solidi (a) 48, K135-9 (1978).

[622] A. Polian, M. Grimsditch, and I. Grzegony, Elastic constants of gallium nitride, J. Appl. Phys. 79, 3343-4 (1996).

[623] J. F. Muth, R. M. Kolbas, A. K. Sharma, S. Oktyabrsky, and J. Narayan, Excitonic Structure and Absorption Coefficient Measurements of ZnO Epitaxial Films, J. Appl. Phys. 85, 7884-87 (1999).

Index

December 2, 1999
2AAA
activation energy
    of Mg in AlGaN 175
    of Mg in GaN 172-174
    of Si in AlGaN 174
    of Si in GaN 164
ALE see growth, epitaxial
AlGaN 84, 85-86
    critical thickness 84
    islanding 85
    heterostructure mobility 5
AlN 7, 275, 50-51, 85-86
    critical thickness 85
    RHEED oscillations 85
AlAs 2, 277
AlP 2, 277
AlSb 2, 277
Auger recombination
    suppression by strain 207-210
    measurement of 210-211

2BBB
band alignments 15, 17
band offsets
    III-V heterostructures 121-122
    II-VI heterostructures 122-125, 267
    III-Nitride heterostructures 125-126
bandgaps
    AlGaN 154
    II-VI semiconductors 13
    III-V semiconductors 2
    III-Nitrides 7
    InGaN 151-152
bandstructure
    Wurtzite III-Nitrides 118-121
    Zinc blende semiconductors 115-118, 267
bowing parameter
    AlGaN 154
    InGaN 152-153
breakdown, gate and drain 251
breakdown field
    GaAs 8
    GaN 8
    Si 8
    SiC 8

2CCC
carrier concentration
    electrons in GaN 162-165
    holes in GaN 170-172
    in AlGaN 174
    in InGaN 176
    holes in ZnMgSSe 268
    holes in II-VI semiconductors 161
CBED 19
CdS 10
    crystal structure 10
    ionicity 10
CdTe 10, 40
    coefficient of thermal expansion 10
    crystal structure 10
    etch-pit density 10
    hardness 10
    melting point 10
    thermal conductivity 10
    vapor pressure 10
    x-ray rocking curve half-width 10

characterization 19-24
    electron microscopy 19
    magnetic methods 22-24
    optical methods 22-24
    RHEED
    RHEED oscillations 23
    X-ray diffraction 19-20
cohesive energy 12
covalency 12
critical thickness
    experimental values 79-84
    of capped layers 82
    determined by islanding 84-85
    of GeSi layers 81
    of III-Vs layers 80
    of II-Vi layers 83-84
    of superlattices 78
    theory 77
cyclotron resonance 206

2DDD
defect clusters 109
detectors, optical 229
dislocation s
    activation energy in GaN 70-71
    blocking 75-77
    dipoles 63
    energy 60-63
    kink model 67
    multiplication 74
    non-periodic arrays
    nucleation 72
    propagation 65-72
    periodic arrays 60, 100
    spacing 91
    velocity 69-70

2EEE
electrical properties
    alloys AlGaN and InGaN 174-176
    doping, n-type 159, 163-165
    doping, p-type 160-162, 170-174
    contacts, ohmic 176-179, 182-185
    Schottky barriers 176, 179-182

electric field, effect of
    Franz-Keldysh effect 185-187
    experimental results
        CdTe/CdMnTe quantum wells 187-188
        CdZnSe/ZnSe quantum wells 188-190
        CdZnSSe multiple quantum wells 188-190
    on GaN layers 192
    Piezoelectric effect 192-199
    Quantum Confined Stark Effect (QCSE) 185-187
transmission electron microscopy (TEM), 19, 266, see also characterization
EMP bifurcation 206
epr 102-103
etching
    dry 262
    wet 261

2FFF

2ggg
GaAs 2, 8, 275, 276, 277
    bandgap 2.275
    coefficient of thermal expansion 10
    etch-pit density 10
    hardness 10
    lattice constant 2, 275
    melting point 10
    thermal conductivity 10
    vapor pressure 10
    mobility 5
    x-ray rocking curve half-width 10
GaN
    bandgap 7, 275
    critical thickness 84, 85, 86
    cubic 7
    lattice constant 7, 275
GaP 2, 277, 278, 279
growth, epitaxial
    ALE 32

INDEX

II-VI semiconductors 38-44
  CdTe 40
  photo-assisted 43-44
  ZnS 41-43
  ZnSe 38-40
  ZnTe 40
III-V semiconductors
  in nitrogen carrier gas 35-38
  highly mismatched layers 35
  MBE 25-27, 33
  MEMBE 32-33
  MMBE 32-33
  MOMBE 30-32
  MOVPE 27-30, 33
  substrates 24-25

2HHH

HBT 252, 253-254
heterostructures
  unstrained III-V-based 4
  strained III-V based 5
  band offsets 121- 126
highly mismatched layers 35, 99-102
historical developments
  III-Nitrides 8-10
  II-VI semiconductors 13-14
HREM 19
TEM 19

2III

III-V heterostructures 121-122
II-VI heterostructures 122-125
III-Nitride heterostructures 125-12126
III-Nitride 7
InAs 2, 277, 278, 279
InGaN 85
  critical thickness 85
  RHEED oscillations 85
InP 2, 277, 278, 279
  coefficient of thermal expansion 10
  etch-pit density 10
  hardness 10
  melting point 10
  thermal conductivity 10
  vapor pressure 10
  x-ray rocking curve half-width 10
InSb 2, 277, 278, 279
ion implantation
  for Si-doping 262-264
  for isolation 263, 264
interfaces 110-114, 202-203

2JJJ

2KKK

2LLL

lattice constants
  III-V semiconductors 2
  II-VI semiconductors 13
  III-Nitride semiconductors 7
lasers
  applications 243-244
  GaN based
    comparison with II-VI lasers 242
    emission mechanism 241-242
    Fabri-Perot cavity mode 236
    laser diodes 238-241
    on $MgAl_2O_4$ substrate 236
    on sapphire 236, 238-241
    optically pumped 237-238
  II-VI e-beam-pumped 219-220
  II-VI laser diodes
    life-time 227
    aging 226
    degradation 223-228
    structure and performance 220-223
    II-VI photo-pumped 216-219
  mid-IR 3, 212-213
LEDs
  applications 242-243
  GaN-based
    comparison with other LEDs 235-236
    blue and green 232-233
    fabrication 231
    performance 231-235
  III-V-based 236

short-wavelength 3
SiC based 236
ZnSe based 214-216
LiGaO$_2$ 7

2MMM
magnetic methods, see characterization
Magnetic field, effect of
    magnetic polarons 199-200
    on diluted magnetic semiconductors 199-201, 202-205
    on optical properties 201-202
    on quantum dots
    on transport properties 200-201
    ZnTe epilayers 202
    ZnSe epilayers 202
material parameters 14-17
MBE 25-27, 33, 265
MgS 13, 275
minority carriers 109-110, 241
misfit strain 59-60
mobility
    low temperature
        electron and hole mobility in Si 5
        electron and hole mobility in Ge 5
        electron mobility in GaAs 5
        electron mobility in AlGaAs/GaAs 5
    room temperature mobility in III-Vs 15-16
MODFETs 255
modulators II-VI-bases 228-229

2NNN
nitrogen carrier gas 35-38

2OOO
optical properties
    AlGaN 154-155
    III-V semiconductors 126-127
    II-VI semiconductors 127-129
        effect of strain 129-133
    III-Nitrides

        photoluminescence 145-147
        exciton binding energy 147
        yellow band 147148
        persistent photoconductivity 147-148
        effect of strain 148-150
        effect of temperature 150-151
        InGaN alloy 151-152

2PPP
periodic dislocation arrays 60, 100
photo-assisted growth, see growth, epitaxial
photoluminescence (PL)
    III-V Sb containing superlattices
pinch-off 255
processing, device 260-262

2QQQ
quantum wells and superlattices
    of diluted magnetic alloys 203-205
    II-VI superlattices 134-144
    effect of hydrostatic pressure 144-145
    III-Nitrides quantum wells 156-157
    quantum confinement 133

2RRR
RHEED, see characterization
RHEED oscillations, see characterization

2SSS
sapphire 7
Schottky barriers 180
Schottky contact, see electrical properties
series resistance 251
Si 7, 8, 277
    coefficient of thermal expansion 10
    etch-pit density 10
    hardness 10
    melting point 10

INDEX

thermal conductivity 10
vapor pressure 10
x-ray rocking curve half-width 10
SiC 7, 8
stokes shift 241
strain
   in GeSi layers
   in III-Nitride layers 107-109
   in III-V layers 79-80
   in II-VI layers 82-84
   in magnetic superlattices 102-103
   oscillations 104-106
strain relaxation
   by irradiation 97-99
   due to patches 106
   in initial stages 99-100
   with increase of thickness 97-98
   with low mismatch 94
   with high mismatch 99-100
strained layer devices
   III-Nitride devices 3, 231-242, 253-260
   III-V devices 4-6, 207-213, 245-251
   II-VI devices 214-231, 251-253
strained layers, evolution 1-4
stress, excess 66
substrates, see Growth, epitaxial
switches, II-VI-based 228-229

2TTT

transconductance 251, 255
transistors
   blue light sensitive 252
   depletion mode 252
   II-VI transistors 251-253
   InGaAs HEMTs
      evolution 246
      lattice matched (LM) 245-246
      metamorphic 246-251
      modulation doped 245
      pseudomorphic (PM)
   GaN based
      f_t and f_{max} values 256

      pinch-off 255
      on SiC 258260
      on sapphire 254-258
      Pt/Au Schottky gate 255
      transconductance 255

2UUU

VV

2WWW

wide bandgap semiconductors
   II-VI semiconductors
      crystal structure 10
      phase diagram 11
      bond energy 12
   III-Nitride semiconductors
      bond energy 12
workfunction 180

2XXX

x-ray diffraction, see characterization

2YYY

yellow band 147148

2ZZZ

ZnS 41-43
   crystal structure 10
   ionicity 10
ZnSe 38-40
   coefficient of thermal expansion 10
   crystal structure 10
   etch-pit density 10
   hardness 10
   ionicity 10
   melting point 10
   phase diagram 11
   thermal conductivity 10
   vapor pressure 10
   x-ray rocking curve half-width 10
ZnTe 40
   crystal structure 10
   ionicity 10